VoIP
HANDBOOK
Applications, Technologies,
Reliability, and Security

VoIP
HANDBOOK

Applications, Technologies, Reliability, and Security

Edited by
Syed A. Ahson
Mohammad Ilyas

CRC Press
Taylor & Francis Group
Boca Raton London New York

CRC Press is an imprint of the
Taylor & Francis Group, an **informa** business

CRC Press
Taylor & Francis Group
6000 Broken Sound Parkway NW, Suite 300
Boca Raton, FL 33487-2742

© 2009 by Taylor & Francis Group, LLC
CRC Press is an imprint of Taylor & Francis Group, an Informa business

No claim to original U.S. Government works
Printed in the United States of America on acid-free paper
10 9 8 7 6 5 4 3 2 1

International Standard Book Number-13: 978-1-4200-7020-0 (Hardcover)

Visit the Taylor & Francis Web site at
http://www.taylorandfrancis.com

and the CRC Press Web site at
http://www.crcpress.com

Contents

Preface

Voice over Internet Protocol (VoIP) is emerging as an alternative to regular public telephones. IP telephone service providers are moving quickly from low-scale toll bypassing deployments to large-scale competitive carrier deployments. This is giving enterprise networks the opportunity and choice of supporting a less expensive single network solution rather than multiple separate networks. Voice deployment over packet networks has experienced tremendous growth over the last four years. The number of worldwide VoIP customers reached 38 million at the end of 2006 and it is projected that there will be approximately 250 million by the end of 2011. VoIP is a reality nowadays and each day, more and more individuals use this system to phone around the world. There are many common programs that make it easy to use VoIP such as Skype, MSN Messenger, VoIPcheap, and so on.

The evolution of the voice service to VoIP from the circuit-switched voice is due to the proliferation of IP networks that can deliver data bits cost-effectively. VoIP technology has been attracting more and more attention and interest from the industry. VoIP applications such as IP telephony systems involve sending voice transmissions as data packets over private or public IP networks as well as reassembling and decoding at the receiving end. Broadband-based residential customers are also switching to IP telephony due to its convenience and cost-effectiveness. The VoIP architecture pushes intelligence towards the end devices (i.e., PCs, IP phones etc.), giving an opportunity to create many new services that cannot be envisaged using the traditional telephone system.

Recently, there has been an increasing interest in using cellular networks for real-time packet-switched services such as VoIP. The reason behind this increased interest in VoIP is to do with using VoIP in All-IP networks instead of using circuit-switched speech. This would result in cost savings for operators as the circuit-switched part of the core network would not be needed anymore. Similar to the wireline networks, voice in wireless mobile networks started with the circuit-switched networks. Since the first deployment of the commercial wireless systems, wireless mobile networks have evolved from the first generation analog networks to the second generation digital networks. Along with the rapid growth of the wireless voice service, the third generation wireless mobile networks have been deployed offering more efficient circuit-switched services that utilize many advanced techniques to more than double the spectral efficiency of the second generation systems.

VoIP is an attractive choice for voice transport compared to traditional circuit-switching technology for many reasons. These reasons include lower equipment cost, integration of voice and data applications, lower bandwidth requirements, widespread availability of the IP, and the promise of novel, value-added services. In the future, VoIP services will be expected to operate seamlessly over a converged network referred to as the Next Generation Network (NGN), comprising a combination of heterogeneous network infrastructures that will include packet-switched, circuit-switched, wireless and wireline networks.

The *VoIP Handbook* provides technical information about all aspects of VoIP. The areas covered in the handbook range from basic concepts to research grade material including future directions. The *VoIP Handbook* captures the current state of VoIP technology and serves as a source of comprehensive reference material on this subject. The *VoIP Handbook* comprises four sections: Introduction, Technologies, Applications, and Reliability and Security. It has a total of 23 chapters authored by 46 experts from around the world. The targeted audience for the handbook includes professionals who are designers and/or planners for VoIP systems, researchers (faculty members and graduate students), and those who would like to learn about this field.

The handbook is designed to provide the following specific and salient features:

- To serve as a single comprehensive source of information and as reference material on VoIP technology
- To deal with the important and timely topic of an emerging technology of today, tomorrow, and beyond
- To present accurate, up-to-date information on a broad range of topics related to VoIP technology
- To present material authored by the experts in the field
- To present information in an organized and well-structured manner

Although the handbook is not precisely a textbook, it can certainly be used as a textbook for graduate courses and research-oriented courses that deal with VoIP. Any comments from the readers will be highly appreciated.

Many people have contributed to this handbook in their own unique ways. The first and the foremost who deserve an immense amount of appreciation are the highly-talented and skilled researchers who have contributed the 23 chapters of this handbook. All of them have been extremely cooperative and professional. It has also been a pleasure to work with Ms. Nora Konopka, Ms. Jessica Vakili, and Mr. Glenon Butler of CRC Press, and we are extremely grateful for their support and professionalism. Our families, who have extended their unconditional love and support throughout this project, also deserve our very special appreciation.

Syed Ahson
Plantation, Florida

Mohammad Ilyas
Boca Raton, Florida

Editors

Syed Ahson is a senior staff software engineer with Motorola Inc., Plantation, Florida. He has made significant contributions in leading roles toward the creation of several advanced and exciting cellular phones at Motorola. He has extensive experience with wireless data protocols (TCP/IP, UDP, HTTP, VoIP, SIP, H.323), wireless data applications (internet browsing, multimedia messaging, wireless email, firmware over-the-air update) and cellular telephony protocols (GSM, CDMA, 3G, UMTS, HSDPA). Prior to joining Motorola, he was a senior software design engineer with NetSpeak Corporation (now part of Net2Phone), a pioneer in VoIP telephony software.

He has published more than ten books on emerging technologies such as WiMAX, RFID, Mobile Broadcasting and IP Multimedia Subsystem. His recent books include *IP Multimedia Subsystem Handbook* (2008) and *Handbook of Mobile Broadcasting: DVB-H, DMB, ISDB-T and MediaFLO* (2007). He has published several research articles and teaches computer engineering courses as adjunct faculty at Florida Atlantic University, Boca Raton, Florida, where he introduced a course on Smartphone Technology and Applications. He received his BSc in Electrical Engineering from Aligarh University, India, in 1995 and his MS in Computer Engineering in July 1998 at Florida Atlantic University, USA.

Dr. Mohammad Ilyas is associate dean for Research and Industry Relations in the College of Engineering and Computer Science at Florida Atlantic University, Boca Raton, Florida. Previously, he served as chair of the Department of Computer Science and Engineering and interim associate vice-president for Research and Graduate Studies. He received his PhD from Queen's University in Kingston, Canada. His doctoral research involved switching and flow control techniques in computer communication networks. He received his BSc in Electrical Engineering from the University of Engineering and Technology, Pakistan, and MS in Electrical and Electronic Engineering at Shiraz University, Iran.

Dr. Ilyas has conducted successful research in various areas including traffic management and congestion control in broadband/high-speed communication networks, traffic characterization, wireless communication networks, performance modeling, and simulation. He has published over twenty books on emerging technologies and over 150 research articles.

His recent books include *RFID Handbook* (2008) and *WiMAX Handbook—3 Volume Set* (2007). He has supervised 11 PhD dissertations and more than 37 MS theses to completion. He has been a consultant to several national and international organizations. Dr. Ilyas is an active participant in several IEEE Technical committees and activities. Dr. Ilyas is a senior member of IEEE and a member of ASEE.

Contributors

Pedro García Alcaraz   The Directorate of Telematic Services (DIGESET), University of Colima, Colima, Mexico

Albert Banchs   Department of Telematic Engineering, Carlos III University of Madrid, Spain

Qi Bi   Wireless Technologies, Bell Laboratories, Murray Hill, New Jersey

Adrienne Brown   Faculty of Business and Information Technology, University of Ontario Institute of Technology, Oshawa, Ontario, Canada

Hsiao-Hwa Chen   Department of Engineering Science, National Cheng Kung University, Tainan City, Taiwan, Republic of China

Tao Chen   Devices R&D, Nokia, Oulu, Finland

Ángel Cuevas   Department of Telematic Engineering, Carlos III University of Madrid, Spain

Rubén Cuevas   Department of Telematic Engineering, Carlos III University of Madrid, Spain

Swapna S. Gokhale   Department of Computer Science and Engineering, University of Connecticut, Storrs, Connecticut

Tero Henttonen   Devices R&D, Nokia, Helsinki, Finland

Patrick C. K. Hung   Faculty of Business and Information Technology, University of Ontario Institute of Technology, Oshawa, Ontario, Canada

Hun Jeong Kang   Department of Computer Science and Engineering, University of Minnesota, Minneapolis, Minnesota

Markku Kuusela   Devices R&D, Nokia, Helsinki, Finland

David Larrabeiti   Department of Telematic Engineering, Carlos III University of Madrid, Spain

Keonbae Lee   Department of Electronic Engineering, Kyonggi University, Suwon-shi, Gyeonggi-do, Republic of Korea

Jijun Lu   Department of Computer Science and Engineering, University of Connecticut, Storrs, Connecticut

Kejie Lu   Department of Electrical and Computer Engineering, University of Puerto Rico at Mayagüez, Mayagüez, Puerto Rico

Petteri Lundén   Research Center, Nokia, Helsinki, Finland

Zhibin Mai   Department of Computer Science, Texas A&M University, Texas

Esa Malkamäki   Devices R&D, Nokia, Helsinki, Finland

Raymundo Buenrostro Mariscal   The Directorate of Telematic Services (DIGESET), University of Colima, Colima, Mexico

Gianluigi Me   Department of Information Systems and Production, University of Tor Vergata, Rome, Italy

Antonio Nucci   Narus Inc., Mountain View, California

Minseok Oh   Department of Electronic Engineering, Kyonggi University, Suwon-shi, Gyeonggi-do, Republic of Korea

Yi Qian   Advanced Network Technologies Division, National Institute of Standards and Technology, Gaithersburg, Maryland

Jussi Ojala   Devices R&D, Nokia, Helsinki, Finland

Supranamaya Ranjan   Narus Inc., Mountain View, California

Zhen Ren   Department of Computer Science, College of William and Mary Williamsburg, Virginia

Bo Rong   Department of Research, International Institute of Telecommunications, Montreal, Quebec, Canada

Piero Ruggiero   IT Consultant, Accenture Ltd., Rome, Italy

Khaled Salah   Department of Information and Computer Science, King Fahd University of Petroleum and Minerals, Dhahran, Saudi Arabia

Hemant Sengar   Voice and Data Security (VoDaSec) Solutions, Fairfax, Virginia

Dongsu Seong   Department of Electronic Engineering, Kyonggi University, Suwon-shi, Gyeonggi-do, Republic of Korea

Sok-Ian Sou   Department of Electrical Engineering, National Cheng Kung University, Tainan, Taiwan, Republic of China

Barry Sweeney   Computer Science Corporation, Falls Church, Virginia

Akira Takahashi   NTT Service Integration Laboratories, NTT, Musashino, Tokyo, Japan

Manuel Urueña   Department of Telematic Engineering, Carlos III University of Madrid, Spain

Ahmet Uyar   Department of Computer Engineering, Mersin University, Ciftlikkoy, Mersin, Turkey

Miguel Vargas Martin   Faculty of Business and Information Technology, University of Ontario Institute of Technology, Oshawa, Ontario, Canada

Haiming Wang   Devices R&D, Nokia, Beijing, People's Republic of China

Shengquan Wang   Department of Computer and Information Science, University of Michigan-Dearborn, Dearborn, Michigan

Xin Wang   Wireless Technologies, Bell Laboratories, Murray Hill, New Jersey

Xinbing Wang   Department of Electronic Engineering, Shanghai Jiaotong University, Shanghai, People's Republic of China

Duminda Wijesekera   Department of Computer Science, George Mason University, Fairfax, Virginia

Dong Xuan   Department of Computer Science and Engineering, The Ohio State University, Columbus, Ohio

Yang Yang   Wireless Technologies, Bell Laboratories, Murray Hill, New Jersey

Qinqing Zhang   Applied Physics Laboratory and Computer Science Department, Johns Hopkins University, Laurel, Maryland

Zhi-Li Zhang   Department of Computer Science and Engineering, University of Minnesota, Minneapolis, Minnesota

Wei Zhao   School of Science, Rensselaer Polytechnic Institute, Science Center, Troy, New York

Junaid Ahmed Zubairi   Department of Computer Science, State University of New York at Fredonia, Fredonia, New York

Part I

Introduction

1

Deploying VoIP in Existing IP Networks

Khaled Salah

CONTENTS

1.1 Introduction

Many network managers find it attractive and cost effective to merge and unify voice and data networks. A unified network is easier to run, manage, and maintain. However, the majority of today's existing data networks is Ethernet-based and use Internet Protocols (IP).

Such networks are best-effort networks in that they were not designed to support real-time applications such as Voice over Internet Protocol (VoIP). VoIP requires timely packet delivery with low latency, jitter, packet loss, and sufficient bandwidth. To achieve this, efficient deployment of VoIP must ensure that these real-time traffic requirements can be guaranteed over new or existing IP networks.

When deploying a new network service such as VoIP over existing data networks, many network architects, managers, planners, designers, and engineers are faced with common strategic, and sometimes challenging, questions. What are the quality of service (QoS) requirements for VoIP? How will the new VoIP load impact the QoS for currently running network services and applications? Will my existing network support VoIP and satisfy standardized QoS requirements? If so, how many VoIP calls can the network support before it becomes necessary to upgrade any part of the existing network hardware?

Commercial tools can answer some of these challenging questions, and a list of the commercial tools available for VoIP can be found in [1,2]. For the most part, these tools use two common approaches to assess the deployment of VoIP into the existing network. One approach is based on first performing network measurements and then predicting the readiness of the network to support VoIP by assessing the health of network elements. The second approach injects real VoIP traffic into the existing network and measures the resulting delay, jitter, and loss.

There is a definite financial cost associated with the use of commercial tools. Moreover, no commercial tool offers a comprehensive approach to successful VoIP deployment. Specifically, none is able to predict the total number of calls that can be supported by the network, taking into account important design and engineering factors, including VoIP flow and call distribution, future growth capacity, performance thresholds, the impact of VoIP on existing network services and applications, and the impact of background traffic on VoIP. This chapter attempts to address these important factors and lays out a comprehensive methodology to successfully deploy any multimedia application such as VoIP and videoconferencing. Although the chapter focuses essentially on VoIP, it also contains many useful engineering and design guidelines, and discusses many practical issues pertaining to the deployment of VoIP. These issues include the characteristics of VoIP traffic and QoS requirements, VoIP flow and call distribution, defining future growth capacity, and the measurement and impact of background traffic. As a case study, we illustrate how our approach and guidelines can be applied to a typical network of a small enterprise.

The rest of the chapter is organized as follows. Section 1.2 outlines an eight-step methodology to successfully deploy VoIP in data networks. Each step is described in considerable detail. Section 1.3 presents a case study of a VoIP introduced to a typical data network of a small enterprise, using the methods described in the previous section. Section 1.4 summarizes and concludes the study.

1.2 Step-by-Step Methodology

In this section, an eight-step methodology is described for the successful deployment of a VoIP (Figure 1.1). The first four steps are independent and can be performed in parallel. Steps 6 and 7, an analysis and simulation study, respectively, can also be done in parallel. Step 5, however, involves the early and necessary re-dimensioning or modification to the existing network. The final step is pilot deployment.

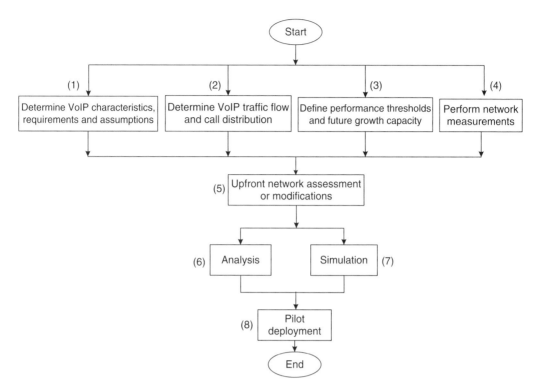

FIGURE 1.1 Flowchart of an eight-step methodology. (*Source:* K. Salah, "On the deployment of VoIP in ethernet networks: Methodology and case study," *International Journal of Computer Communications*, Elsevier Science, vol. 29, no. 8, 2006, pp. 1039–1054. With permission.)

This methodology can be used to deploy a variety of network services other than VoIP, including videoconferencing, peer to peer (p2p), online gaming, internet protocol television (IPTV), enterprise resource planning (ERP), or SAP services. The work in [3,4] show how these steps can be applied to assess the readiness of IP networks for desktop videoconferencing.

1.2.1 VoIP Traffic Characteristics, Requirements, and Assumptions

In order to introduce a new network service such as VoIP, one must first characterize the nature of the traffic, QoS requirements, and the need for additional components or devices. For simplicity, we assume a point-to-point conversation for all VoIP calls with no call conferencing. First, a *gatekeeper* or `CallManager` node, which can handle signaling to establish, terminate, and authorize all VoIP call connections, has to be added to the network [5–7]. Also, a VoIP *gateway* responsible for converting VoIP calls to/from the Public Switched Telephone Network (PSTN) is required to handle external calls. From an engineering and design standpoint, the placement of these nodes in the network is critical (see Step 5). Other hardware requirements include a VoIP client terminal, which can be a separate VoIP device (i.e., IP phone) or a typical PC or workstation that is VoIP-enabled and which runs VoIP software such as IP SoftPhones [8–10].

Figure 1.2 identifies the end-to-end VoIP components from sender to receiver [11]. The first component is the *encoder*, which periodically samples the original voice signal and assigns a fixed number of bits to each sample, creating a constant bit rate stream. The traditional sample-based encoder G.711 uses pulse code modulation (PCM) to generate

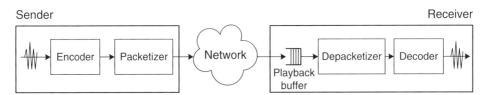

FIGURE 1.2 End-to-end components of VoIP. (*Source:* K. Salah, "On the deployment of VoIP in ethernet networks: Methodology and case study," *International Journal of Computer Communications*, Elsevier Science, vol. 29, no. 8, 2006, pp. 1039–1054. With permission.)

8-bit samples every 0.125 ms, leading to a data rate of 64 kbps [12]. Following the encoder is the *packetizer*, which encapsulates a certain number of speech samples into packets and adds the RTP, UDP, IP, and Ethernet headers. The voice packets travel through the data network to the receiver where an important component called the *playback buffer* is placed for the purpose of absorbing variations or jitter in delay and for providing a smooth playout. Packets are then delivered to the *depacketizer* and eventually to the *decoder*, which reconstructs the original voice signal.

The widely adopted recommendations of the H.323, G.711, and G.714 standards for VoIP QoS requirements are followed here [13,14]. Table 1.1 compares some commonly used International Telecommunication Union-Telecommunication (ITU-T) standard codecs and the amount of one-way delay that they impose. To account for the upper limits and to meet the ITU recommended P.800 quality standards [15], we adopt G.711u codec standards for the required delay and bandwidth. G.711u has a *mean opinion score* (MOS) rating of around 4.4—a commonly used VoIP performance metric scaled from 1 to 5 with 5 being the best [16,17]. However, with little compromise in quality, it is possible to implement different ITU-T codecs that require much less bandwidth per call with a relatively higher, but acceptable, end-to-end delay. This can be accomplished by applying compression, silence suppression, packet loss concealment, queue management techniques, and encapsulating more than one voice packet into a single Ethernet frame [5,11,18–23].

1.2.1.1 End-to-End Delay for a Single Voice Packet

Figure 1.2 illustrates the sources of delay for a typical voice packet. The end-to-end delay is sometimes referred to as mouth-to-ear (M2E) delay [9]. The G.714 codec imposes a maximum total one-way packet delay of 150 ms from end-to-end for VoIP applications [14].

TABLE 1.1

Common ITU-T Codecs and Their Defaults

Codec	Data Rate (kbps)	Datagram Size (ms)	A/D Conversion Delay (ms)	Combined Bandwidth (Bidirectional) (kbps)
G.711u	64.0	20	1.0	180.80
G.711a	64.0	20	1.0	180.80
G.729	8.0	20	25.0	68.80
G.723.1 (MPMLQ)	6.3	30	67.5	47.80
G.723.1 (ACELP)	5.3	30	67.5	45.80

In Ref. [24], a delay of up to 200 ms was considered acceptable. We can break this delay down into at least three different contributing components:

1. encoding, compression, and packetization delay at the sender's end;
2. propagation, transmission, and queuing delay in the network; and
3. buffering, decompression, depacketization, decoding, and playback delay at the receiver's end.

1.2.1.2 Bandwidth for a Single Call

The required bandwidth for a single call, in one direction, is 64 kbps. As the G.711 codec samples 20 ms of voice per packet, 50 such packets need to be transmitted per second. Each packet contains 160 voice samples, which gives 8000 samples per second. Each packet is sent in one Ethernet frame. With every packet of size 160 bytes, headers of additional protocol layers are added. These headers include RTP + UDP + IP + Ethernet, with a preamble of sizes, 12 + 8 + 20 + 26, respectively. Therefore, a total of 226 bytes, or 1808 bits, must be transmitted 50 times per second, or 90.4 kbps, in one direction. For both directions, the required bandwidth for a single call is 100 pps or 180.8 kbps, assuming a symmetric flow.

1.2.1.3 Other Assumptions

We base our analysis and design on the worst-case scenario for VoIP call traffic. Throughout our analysis and work, we assume that voice calls are symmetric and that no voice conferencing is implemented. We also ignore signaling traffic, mostly generated by the *gatekeeper* prior to establishing the voice call and when the call has been completed. This traffic is relatively small compared to the actual voice call traffic. In general, the *gatekeeper* generates no, or very limited, signaling traffic throughout the duration of an already established on-going VoIP call [5].

In this chapter, we implement no QoS mechanisms that can enhance the quality of packet delivery in IP networks. There are a myriad of QoS standards available that can be enabled for network elements and may include IEEE 802.1p/Q, the IETF's RSVP, and DiffServ. Analysis of implementation cost, complexity, management, and benefit must be weighed carefully before adopting such QoS standards. These standards can be recommended when the cost for upgrading some network elements is high, and network resources are scarce and heavily loaded.

1.2.2 VoIP Traffic Flow and Call Distribution

Before further analysis or planning, collecting statistics about the current telephone call usage or volume of an enterprise is important for successful VoIP deployment. The sources of such information are an organization's private branch exchange (PBX), telephone records, and bills. Key characteristics of existing calls can include the number of calls, number of concurrent calls, time and duration of calls, and so on. It is important to determine the location of the call endpoints, that is, the sources and destinations as well as their corresponding path or flow. This will aid in identifying call distribution and the calls made internally or externally. Call distribution must include the percentage of calls made within and outside of a floor, building, department, or organization. As a prudent capacity planning measure, it is recommended that VoIP call distribution plans be based on the busy hour traffic for the busiest day of a week or month. This will ensure support of calls at all times, leading to a high QoS for all VoIP calls. When such current statistics are combined with the projected extra calls, the worst-case VoIP traffic load to be introduced into the existing network can be predicted.

1.2.3 Define Performance Thresholds and Growth Capacity

We now define the network performance thresholds or operational points for a number of important key network elements. These thresholds are to be considered when deploying the new service. The benefit is twofold. First, the requirements of the new service are satisfied. Second, adding the new service leaves the network healthy and capable of growth in the future.

Two important performance criteria are to be taken into account: first, the maximum tolerable end-to-end delay; second, the utilization bounds or thresholds of network resources. The maximum tolerable end-to-end delay is determined by the most sensitive application to be run on the network. In our case, it is 150 ms end-to-end for VoIP. It is imperative that if the network has certain delay-sensitive applications, such delay be monitored when introducing VoIP traffic, so that they do not exceed their required maximum values. As for the utilization bounds for network resources, such bounds or thresholds are determined by factors such as current utilization, future plans, and foreseen growth of the network. Proper resource and capacity planning is crucial. Savvy network engineers must deploy new services with scalability in mind, and ascertain that the network will yield acceptable performance under heavy and peak loads, with no packet loss. VoIP requires almost no packet loss. In the literature, a 0.1% to 5% packet loss was generally considered inevitable [8,23–25]. However, in Ref. [26] the required VoIP packet loss was conservatively suggested to be less than 10^{-5}. A more practical packet loss, based on experimentation, of below 1% was required in [24]. Hence, it is extremely important not to fully utilize the network resources. As a rule-of-thumb guideline for switched, fast, full-duplex Ethernet, the average utilization limit for links should be 190%, and for switched, shared, fast Ethernet, 85% [27].

The projected growth in users, network services, business, and so on, must all be taken into consideration to extrapolate the required growth capacity or the future growth factor. In our study, we reserve 25% of the available network capacity for future growth and expansion. For simplicity, we apply this evenly to all network resources of the router, switches, and switched-Ethernet links. However, it must be kept in mind that, in practice, this percentage is variable for each network resource and may depend on current utilization and required growth capacity. In our methodology, these network resources are reserved upfront, before deploying the new service, and only the left-over capacity is used for investigating the extent of network support available to the new service.

1.2.4 Network Measurements

Network measurements characterize the existing traffic load, utilization, and flow. Measuring the network is a crucial step, as it can potentially affect the results to be used in analytical study and simulation. There are a number of commercial and non-commercial tools available for network measurement. Popular open-source measurement tools include MRTG, STG, SNMPUtil, and GetIF [28]. A few examples of popular, commercially available measurement tools include HP OpenView, Cisco Netflow, Lucent VitalSuite, Patrol DashBoard, Omegon NetAlly, Avaya ExamiNet, and NetIQ Vivinet Assessor.

Network measurements must be determined for elements such as routers, switches, and links. Numerous types of measurements and statistics can be obtained using measurement tools. As a minimum, traffic rates in bits per second (bps) and packets per second (pps) must be measured for links directly connected to routers and switches. To get adequate assessment, network measurements have to be taken over a long period of time, at least 24 h. Sometimes, it is desirable to take measurements over several days or a week.

Network engineers must consider the worst-case scenario for network load or utilization in order to ensure good QoS at all times, including peak hours. The peak hour is different from one network to another and depends totally on the nature of business and the services provided by the network.

1.2.5 Upfront Network Assessment and Modifications

In this step, we assess the existing network and determine, based on the existing traffic load and the requirements of the new service to be deployed, if any immediate modifications are necessary. Immediate modifications to the network may include adding and placing new servers or devices (such as a VoIP gatekeeper, gateways, IP phones), upgrading PCs, and re-dimensioning heavily utilized links. As a good upgrade rule, topology changes need to be kept to a minimum and should not be made unless they are necessary and justifiable. Over-engineering the network and premature upgrades are costly and considered poor design practices.

Network engineers have to take into account the existing traffic load. If any of the network links are heavily utilized, say, 30% to 50%, the network engineer should decide to re-dimension it to a 1-Gbps link at this stage. As for shared links, the replacement or re-dimensioning of such links must be decided on carefully. A shared Ethernet scales poorly, and is not recommended for real-time and delay-sensitive applications. It introduces excessive and variable latency under heavy loads and when subjected to intense bursty traffic [27]. In order to consistently maintain the VoIP QoS, a switched, fast, full-duplex Ethernet LAN becomes necessary.

1.2.6 Analysis

VoIP is bounded by two important metrics: first, the available bandwidth, and second, the end-to-end delay. The actual number of VoIP calls that the network can sustain and support is bounded by those two metrics. Depending on the network under study, either the available bandwidth or the delay can be the key factor in determining the number of calls that can be supported.

1.2.6.1 Bandwidth Bottleneck Analysis

Bandwidth bottleneck analysis is an important step to identify the network element, whether it is a node or a link, that limits the number of VoIP calls that can be supported. As illustrated in Figure 1.3, for any path that has N network nodes and links, the bottleneck network element is the node or link that has the minimum available bandwidth. According to [29], this minimum available bandwidth is defined as follows:

$$A = \min_{i=1,\ldots,N} A_i$$

and

$$A_i = (1 - u_i)C_i,$$

where C_i is the capacity of network element i and u_i is its current utilization. The capacity C_i is the maximum possible transfer or processing rate.

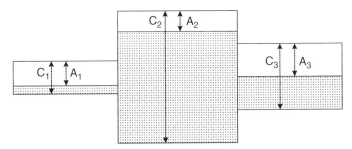

FIGURE 1.3 Bandwidth bottleneck for a path of three network elements. (*Source*: K. Salah, "On the deployment of VoIP in ethernet networks: Methodology and case study," *International Journal of Computer Communications*, Elsevier Science, vol. 29, no. 8, 2006, pp. 1039–1054. With permission.)

Theoretically, the maximum number of calls that can be supported by a network element E_i can be expressed in terms of A_i as

$$MaxCalls_i = \frac{A_i(1 - growth_i)}{CallBW},\tag{1.1}$$

where $growth_i$ is the growth factor of network element E_i and takes a value from 0 to 1. $CallBW$ is the VoIP bandwidth for a single call imposed on E_i. As previously discussed in design Step 2 of Section 2.2, the bandwidth for one direction is 50 pps or 90.4 kbps. In order to find the bottleneck in the network that limits the total number of VoIP calls, the engineer has to compute the maximum number of calls that can be supported by each network element, as in Equation 1.1, and the percentage of VoIP traffic flow passing through this element. The percentage of VoIP traffic flow for E_i, denoted as $flow_i$, can be found by examining the distribution of calls. The total number of VoIP calls that can be supported by a network can be expressed as

$$TotalCallsSupported = \min_{i=1,\dots,N}\left(\frac{MaxCalls_i}{flow_i}\right).$$

1.2.6.2 Delay Analysis

As defined in Section 2.3 for the existing network, the maximum tolerable end-to-end delay for a VoIP packet is 150 ms. The maximum number of VoIP calls that the network can sustain is bounded by this delay. We must always ensure that the worst-case end-to-end delay for all calls is less than 150 ms. Our goal is to determine the network capacity for VoIP, that is, the maximum number of calls that the existing network can support while maintaining VoIP QoS. This can be done by adding calls incrementally to the network while monitoring the threshold or bound for VoIP delay. When the end-to-end delay, including network delay, becomes larger than 150 ms, the maximum number of calls that the network can support becomes known.

As described in Section 2.1, there are three sources of delay for a VoIP stream: sender, network, and receiver. An equation is given in Ref. [26] to compute the end-to-end delay D for a VoIP flow in one direction, from sender to receiver.

$$D = D_{pack} + \sum_{h \in Path} (T_h + Q_h + P_h) + D_{play},$$

where D_{pack} is the delay due to packetization at the source. At the source, there are also D_{enc} and $D_{process}$ where D_{enc} is the encoder delay when converting A/D signals into samples and $D_{process}$ is the information exchanged between the PC of the IP phone and the network, which includes encapsulation. In G.711, D_{pack} and D_{enc} are 20 ms and 1 ms, respectively. Hence, it is appropriate for our analysis to assume a worst-case situation and introduce a fixed delay of 25 ms at the source. D_{play} is the playback delay at the receiver, including jitter buffer delay. The jitter delay is at most 2 packets, that is, 40 ms. If the receiver's delay of $D_{process}$ is added, we obtain a total fixed delay of 45 ms at the receiver. $T_h + Q_h + P_h$ is the sum of delays that occurs in the packet network due to transmission, queuing, and propagation going through each hop h in the path from the sender to the receiver. The propagation delay P_h is typically ignored for traffic within a LAN, but not in a WAN. For transmission delay T_h and queueing delay Q_h, we apply the queueing theory. Hence, any delay in the network, expressed as $\sum_{h \in Path}(T_h + Q_h)$, should not exceed (150–25–45) or 80 ms.

We utilize queueing analysis to approximate and determine the maximum number of calls that the existing network can support while maintaining a delay of less than 80 ms. In order to find the network delay, we utilize the principles of Jackson's theorem to analyze queueing networks. In particular, we use the approximation method of analyzing queueing networks by decomposition, as discussed in Ref. [32]. In this method, a Poisson arrival rate is assumed, and the service times of network elements are exponentially distributed. Analysis by decomposition entails first isolating the queueing network into subsystems, for example, into single queueing nodes. Next, each subsystem is analyzed separately, taking into consideration its own surroundings in the network of arrivals and departures. Then, the average delay for each individual queueing subsystem is found. And finally, all the delays of the queueing subsystems are aggregated to find the average total end-to-end network delay.

For our analysis we assume the VoIP traffic is Poisson. In reality, the inter-arrival time, $1/\lambda$, of VoIP packets is constant, and hence its distribution is deterministic. However, modeling the voice arrival as Poisson gives an adequate approximation, according to Ref. [26], especially when the number of calls is high. More importantly, the network element with a non-Poisson arrival rate makes it difficult to approximate the delay and leads to an intractable analytical solution. Furthermore, analysis by the decomposition method will be violated if the arrival rate is not Poisson.

Figure 1.4 shows queueing models for three network elements: router, switch, and link. The queueing model for the router has two outgoing interfaces: an interface for Switch 1,

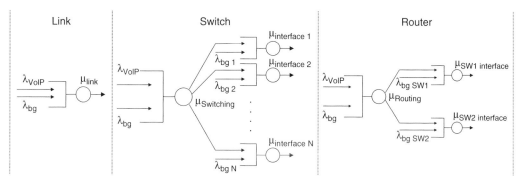

FIGURE 1.4 Queueing models for a network link, switch, and router. (*Source:* K. Salah, "On the deployment of VoIP in ethernet networks: Methodology and case study," *International Journal of Computer Communications,* Elsevier Science, vol. 29, no. 8, 2006, pp. 1039–1054. With permission.)

SW1, and another for Switch 2, SW2. The number of outgoing interfaces for a switch is many, and such a number depends on the number of ports it has. We modeled the switches and the router as *M/M/1* queues. Ethernet links are modeled as *M/D/1* queues. This is appropriate, since the service time for Ethernet links is more deterministic than variable. However, the service times of the switches and the router are not deterministic, as these are all CPU-based devices. According to the datasheet found in Refs. [30,31], the switches and the router (which have been used for our case study in Section 3) have a somewhat similar design, that of a store-and-forward buffer pool with a CPU responsible for pointer manipulation to switch or route a packet to different ports. Gebali [32] provides a comprehensive model of common types of switches and routers. According to Leland et al. [34], the average delay for a VoIP packet passing through an *M/M/1* queue is basically $1/(\mu - \lambda)$, and through an *M/D/1* queue is $[1 - (\lambda/2\mu)]/(\mu - \lambda)$, where λ is the mean packet arrival rate, and μ is the mean network element service rate. The queueing models in Figure 1.4 assume Poisson arrivals for both VoIP and background traffic. In Ref. [26], it was concluded that modeling VoIP traffic as Poisson is adequate. However, in practice, background traffic is bursty in nature and characterized as self-similar with long-range dependence [35]. For our analysis and design, using bursty background traffic is not practical. For one, under the network of queues being considered, an analytical solution becomes intractable when considering non-Poisson arrival. Also, in order to ensure good QoS at all times, we have based our analysis and design on the worst-case scenario of network load or utilization, that is, the peak of aggregate bursts. And thus, in a way, our analytical approach takes into account the bursty nature of traffic.

It is worth noting that the analysis by decomposition of queueing networks in Ref. [33] assumes exponential service times for all network elements including links. But Suri [36] proves that acceptable results, of adequate accuracy, can be obtained even if the homogeneity of service times of nodes in the queueing network is violated, the main system performance being insensitive to such violations. Also, when changing the models for links from *M/D/1* to *M/M/1*, the difference was negligible. More importantly, as will be demonstrated with simulation, our analysis gives a good approximation.

The total end-to-end network delay starts from the outgoing Ethernet link of the sender's PC or IP phone to the incoming link of receiver's PC or IP phone. To illustrate this further, we compute the end-to-end delay encountered for a single call initiated between two floors of a building. Figure 1.5 shows an example of how to compute network delay. Figure 1.5a shows the path of a unidirectional voice traffic flow going from one floor to another. Figure 1.5b shows the corresponding networking queueing model for such a path.

For the model shown in Figure 1.5b, in order to compute the end-to-end delay for a single bi-directional VoIP call, we must compute the delay at each network element: the switches, links, and router. For the switch, $\mu = (1 - 25\%) \times 1.3\,\text{Mpps}$, where 25% is the growth factor. We assume the switch has a capacity of 1.3 Mpps. $\lambda = \lambda_{VoIP} + \lambda_{bg}$, where λ_{VoIP} is the total new traffic added from a single VoIP in packets per second, and λ_{bg} is the background traffic, also in packets per second. For an uplink or downlink, $\mu = (1 - 25\%) \times 100\,\text{Mbps}$, $\lambda = \lambda_{VoIP} + \lambda_{bg}$. Since the service rate is in bits per second, λ_{VoIP} and λ_{bg} too must be expressed in bits per second. Similarly for the router, $\mu = (1 - 25\%) \times 25{,}000\,\text{pps}$ and $\lambda = \lambda_{VoIP} + \lambda_{bg}$. Both λ_{VoIP} and λ_{bg} must be expressed in packets per second. For a single bi-directional VoIP call, λ_{VoIP} at the router and switches for will be equal to 100 pps. However, for the uplink and downlink links, it is 90.4 kbps. One should consider no λ_{bg} for the outgoing link if IP phones are used. For multimedia PCs equipped with VoIP software, a λ_{bg} of 10% of the total background traffic is utilized in each floor. In Figure 1.5, we use multimedia PCs.

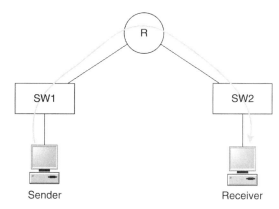

(a) *Unidirectional voice traffic flow path from Floor 1 to Floor 3*

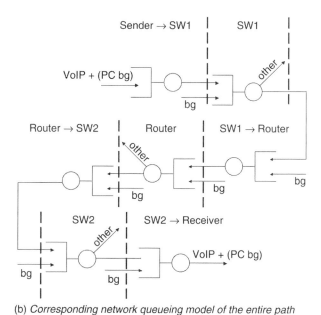

(b) *Corresponding network queueing model of the entire path*

FIGURE 1.5 Computing network delay. (a) Unidirectional voice traffic flow path from Floor 1 to Floor 3. (b) Corresponding network queueing model of the entire path. (*Source:* K. Salah, "On the deployment of VoIP in ethernet networks: Methodology and case study," *International Journal of Computer Communications,* Elsevier Science, vol. 29, no. 8, 2006, pp. 1039–1054. With permission.)

The total delay for a single VoIP call, shown in Figure 1.5b, can be determined as follows:

$$D_{path} = D_{Sender - SW1Link} + D_{SW1} + D_{SW1 - Router\ Link} + D_{Router} + D_{Router - SW2Link} + D_{SW2} + D_{SW2 - Receiver\ Link}$$

In order to determine the maximum number of calls that can be supported by an existing network, while limiting VoIP delay constraint, we devise a comprehensive algorithm that determines network capacity in terms of VoIP calls. Algorithm 1 is essentially the

---

Algorithm 1: Computes the maximum number of calls considering VoIP delay
constraint

Input: n : number of network elements
            $\lambda[1...n]$: background traffic for network elements $1,2,...n$
            $Delay[1...n]$: delay for network elements $1,2,...n$
            P: set of call-flow paths (p) where p is a subset of $\{1,2,...n\}$
Output: MaxCalls: maxmimum number of calls

λ_{VoIP} ← 100pps, or 180.8kbps;
VoIP_MaxDelay ← 80; // network delay for VoIP call in ms
MaxDelay   ← 0;
MaxCalls   ← −1;
$Delay[1..n]$ ← 0;

while MaxDelay < VoIP_MaxDelay do

1. MaxCalls ← MaxCalls + 1
2. Generate a call according to call distribution and let p_c be its
    flow path
3. for each element i in p_c do
            $\lambda_i ← \lambda_i + \lambda_{VoIP}$
            if i is a link then
                    $Delay_i ← (1−\lambda_i/2\mu_i)/(\mu_i−\lambda_i)$
            Else
                    $Delay_i ← 1/(\mu_i−\lambda_i)$
            end if
     end for

4. for each p in P where $p \cap p_c \neq \phi$ do
            PathDelay (p) ← $\sum_i Delay_i$, where i is a network element in
            path p
            if PathDelay (p) > MaxDelay then
                    MaxDelay ← PathDelay (p)
            end if
     end for

end while

---

analytical simulator's engine for computing the number of calls based on delay bounds. Calls are added iteratively until the worst-case network delay of 80 ms has been reached.

It is to be noted that in Step 2 of Algorithm 1, a uniform random number generator is used to generate VoIP calls according to the call distribution. Call distribution must be in the form of values from 1% to 100%. Also, the delay computation for the link in Step 3 is different from that in other network elements such as switches and routers. For links, it is more appropriate to use the average delay formula for $M/D/1$, as the service rate μ is almost constant. However, for switches and routers, it is more appropriate to use the average delay formula for $M/M/1$ as the service rate μ is variable because the routers and switches are CPU-based. For links, the average delay per packet is calculated first using the average bit delay and then multiplying it by the packet size, which is 1808 bits. For this, the link service rate and incoming rate have to be in bits per second. However, for switches and routers, the calculation is done in packets per second. In the algorithm above, the link delay calculation is for a unidirectional link. The total bandwidth that will be introduced as a result of adding one call on the link is 50 pps in one direction, and another 50 pps in the opposite direction. However, for switches and routers, the extra bandwidth introduced per call will be 100pps.

1.2.7 Simulation

The object of the simulation is to verify analysis results of supporting VoIP calls. There are many simulation packages available that can be used, including commercial and open source. A list and classification of such available network simulation tools can be found in Ref. [37]. In our case study in Section 3, we used the popular MIL3's OPNET Modeler simulation package, Release 8.0.C [38]. OPNET Modeler contains a vast number of models of commercially available network elements, and has various real-life network configuration capabilities. This makes its simulation of a real-life network environment close to reality. Other features of OPNET include a GUI interface, a comprehensive library of network protocols and models, a source code for all models, graphical results and statistics, and so on. More importantly, OPNET has gained considerable popularity in academia as it is being offered free of charge to academic institutions. That has given OPNET an edge over DES NS2 in both the market place and in academia.

1.2.8 Pilot Deployment

Before changing any of the network equipment, a pilot project deploying VoIP in a test lab is recommended, to ensure smooth upgrade and transition with minimum disruption of network services. A pilot deployment is done after training of the IT staff. It is the place for the network engineers and support and maintenance teams to get firsthand experience with VoIP systems and their behavior. During this pilot deployment, new VoIP devices and equipment are evaluated, configured, tuned, tested, managed, and monitored. The whole team needs to get comfortable with how VoIP works, how it mixes with other traffic, and how to diagnose and troubleshoot potential problems. Simple VoIP calls can be set up and some benchmark testing can be done.

1.3 Case Study

In this section, we present a case study, a typical IP network of a small enterprise located in a high-rise building. We briefly describe the methodology of successfully deploying VoIP in this network. The network is shown in Figure 1.6. The network is Ethernet-based and has two Layer-2 Ethernet switches connected by a router. The router is Cisco 2621, and the switches are 3Com Superstack 3300. Switch 1 connects Floor 1 and Floor 2 and two servers, while Switch 2 connects Floor 3 and four servers. Each floor LAN is basically a shared Ethernet connecting the employees' PCs with the workgroup and printer servers. The network makes use of VLANs in order to isolate broadcast and multicast traffic. A total of five LANs exist. All VLANs are port-based. Switch 1 is configured such that it has three VLANs. VLAN1 includes the database and file servers.

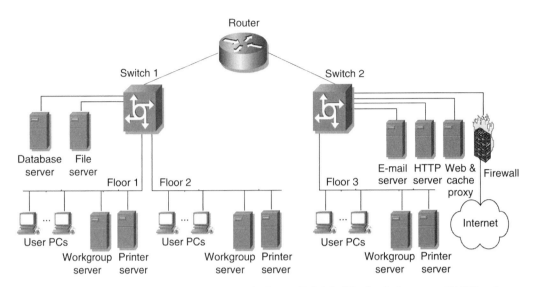

FIGURE 1.6   Topology of a small enterprise network. (*Source:* K. Salah, "On the deployment of VoIP in ethernet networks: Methodology and case study," *International Journal of Computer Communications*, Elsevier Science, vol. 29, no. 8, 2006, pp. 1039–1054. With permission.)

VLAN2 includes Floor 1. VLAN3 includes Floor2. On the other hand, Switch 2 is config-ured to have two VLANs. VLAN4 includes the servers for e-mail, HTTP, Web and cache proxy, and firewall. VLAN5 includes Floor 3. All the links are switched Ethernet 100 Mbps full-duplex, except for the links for Floors 1, 2, and 3, which are shared Ethernet 100 Mbps half-duplex.

For background traffic, we assume a traffic load not exceeding 10% of link capacity. Precise values are described in Ref. [39,40]. The values are those of peak-hour utilization of link traffic in both directions connected to the router and the two switches of the network topology shown in Figure 1.6. These measured results will be used in our analysis and simulation study. For call distributions, thresholds, and projected growth, we used those described in Refs. [39,40].

For an upfront assessment and on the basis of the hardware required to deploy VoIP, as in Step 5, two new nodes have to be added to the existing network: a VoIP *gateway* and a *gatekeeper*. As a network design issue, these two nodes have to be placed appropriately. Since most of the users reside on Floor 1 and Floor 2 and are directly connected to Switch 1, connecting the *gatekeeper* to Switch 1 is practical, and keeps the traffic local. The VoIP *gateway* we connect to Switch 2, in order to balance the projected load on both switches. Also, it is a more reliable and fault-tolerant method to not connect both nodes to the same switch in order to eliminate problems that stem from a single point of failure. For example, currently, if Switch 2 fails, only external calls to/from the network will be affected. It is proper to include the gatekeeper as a member of the VLAN1 of Switch 1, which includes the database and file servers. This isolates the gatekeeper from the mul-ticast and broadcast traffic of Floor 1 and Floor 2. In addition, the *gatekeeper* can locally access the database and file servers to record and log phone calls. On the other hand, we create a separate VLAN for the *gateway* in order to isolate the *gateway* from the multicast and broadcast traffic of Floor 3 and from the servers of Switch 2. Therefore, the network has now a total of six VLANs. Figure 1.7 shows the new network topology after the incorporation of necessary VoIP components. As shown, two new *gateway* and *gatekeeper*

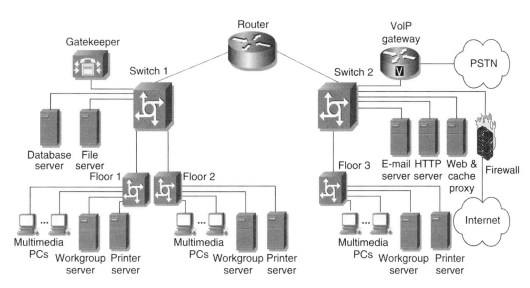

FIGURE 1.7 Network topology with necessary VoIP Components. (*Source:* K. Salah, "On the deployment of VoIP in ethernet networks: Methodology and case study," *International Journal of Computer Communications*, Elsevier Science, vol. 29, no. 8, 2006, pp. 1039–1054. With permission.)

nodes for VoIP were added and the three shared Ethernet LANs were replaced by 100 Mbps switched Ethernet LANs.

For Step 6 of the analysis, there are two implementation options. One uses MATLAB®, and the second, the analytical simulator described in Ref. [41]. In the first option, MATLAB programs can be written to implement the bandwidth and the delay analyses described in Section 2.6. Algorithm 1 was implemented using MATLAB, and the results for the worst-incurred delay have been plotted in Figure 1.8. It can be observed from the figure that the delay increases sharply when the number of calls goes beyond 310. To be more precise, MATLAB results show that the number of calls bounded by the 80 ms delay is 315. From the bandwidth analysis done to compute $MaxCalls_i$ for all network elements, it turns out that the router is the bottleneck. Hence, the *TotalCallsSupported* is 313 VoIP calls. When comparing the number of calls that the network can sustain, based on bandwidth and worst-delay analysis, we find that the number of calls is limited by the available bandwidth more than by the delay, though the difference is small. Therefore, we can conclude that the maximum number of calls that can be sustained by the existing network is 313.

The second option is more flexible and convenient as it avoids using MATLAB. It uses a GUI-based analytical simulator that works on any generic network. The analytical simulator is publicly available, and can be downloaded from http://www.ccse.kfupm.edu.sa/~salah/VoIP_Analytical_Simulator.rar. A complete description of the simulator can be found in Ref. [41]. The simulator has a GUI, using which a network topology can be built (i.e., it is comparable to building a network in OPNET). In other words, the simulator has drag-and-drop features to construct a generic network topology and feed it into the analytical engine. The simulator also allows users to set and configure a variety of settings and parameters related to VoIP deployment. The analytical engine is based on the approach described in Section 2.6. Within seconds, the simulator gives results on how many VoIP calls can be supported: the user can easily tune the network configurations and parameters and determine the results. The results obtained by the simulator and MATLAB were the same.

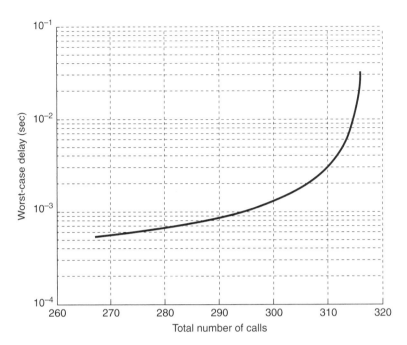

FIGURE 1.8   Worst-incurred delay versus number of VoIP calls. (*Source:* K. Salah, "On the deployment of VoIP in ethernet networks: Methodology and case study," *International Journal of Computer Communications*, Elsevier Science, vol. 29, no. 8, 2006, pp. 1039–1054. With permission.)

Figure 1.9 shows the corresponding network model constructed by the VoIP simulator for the network topology of Figure 1.7. In order to avoid having numerous PC nodes or IP phones on every floor to represent end-users (and thereby cluttering the network topology diagram), floor LANs have been simply modeled as a LAN that encloses an Ethernet switch and three designated Ethernet PCs that are used to model the activities of the LAN users. For example, Floor 1 has three nodes (labeled F1C1, F1C2, and F1C3). F1C1 is a source for sending voice calls. F1C2 is a sink for receiving voice calls. F1C3 is a sink and source of background traffic. This model allows for generating background traffic and also for establishing intra-floor calls or paths from F1C1 and F1C2, and passing through the floor switch of F1SW. The sending and sinking PC nodes of VoIP (e.g., F1C1 and F1C2) have infinite capacity, and there is no limit on how many calls can be added or received by them. As mentioned in Section 1.2.1.3, we ignore the signaling traffic generated by the *gatekeeper*.

Figure 1.10 shows throughput and delay analyses. Figure 1.10a reports the number of calls that can be supported based on bandwidth analysis: 315 calls can be supported for the whole network. In order to identify possible bottlenecks, the report also shows individual calls that can be supported per node and per link. This identifies the router as the bottleneck, and supporting more than 315 calls would definitely require replacement of the router. Figure 1.10b reports the number of calls that can be supported based on network analysis: 313 calls can be supported such that network delay of any of the specified VoIP flows does not exceed the required 80 ms. The figure shows that with 313 calls, a network delay of 16.76 ms is introduced. This means that adding even one more call would lead to network delay, as the maximum of 80 ms was exceeded. Figure 1.10b shows the network delay per flow or path. In our example, there were a total of nine VoIP flows.

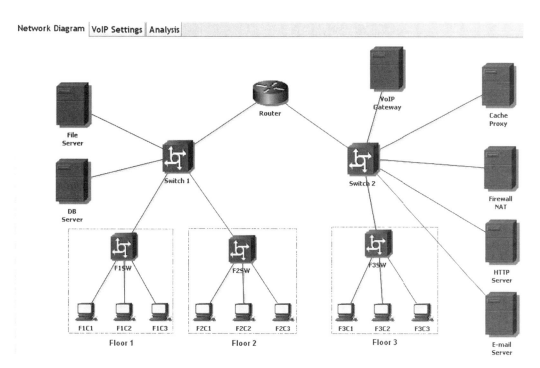

FIGURE 1.9 Corresponding network diagram constructed by analytical simulator. (*Source:* K. Salah, "On the deployment of VoIP in ethernet networks: Methodology and case study," *International Journal of Computer Communications,* Elsevier Science, vol. 29, no. 8, 2006, pp. 1039–1054. With permission.)

As shown, the first triple is for intra-floor flows. The second triple is for inter-floor flows. And the third triple is for external flows. Such information gives insight as to the source of the delays and also identifies the path that causes most of the delays. As the figure shows, the inter-floor flows from F1C1 to F3C2 and F2C1 to F3C2 experience the greatest delays, as they pass through the router.

We chose OPNET to verify our analytical approach. A detailed description of the simulation model, configurations, and results can be found in Ref. [41]. With OPNET simulation, the number of VoIP calls that could be supported was 306. From the results of analysis and simulation, it is apparent that both results are in line and are a close match, as based on the analytic approach, a total of 313 calls can be supported. There is only a difference of seven calls between the analytic and simulation approaches. The difference can be attributed to the degree of accuracy between the analytic approach and OPNET simulation. Our analytic approach is an approximation. Also, the difference is linked to the way the OPNET Modeler adds the distribution of calls. It was found that external and inter-floor calls are added before intra-floor calls. In any case, to be safe and conservative, one can consider the minimum number of calls supported by the two approaches.

The following network design and engineering decisions can be justified from the analytic and simulation perspectives. First, the existing network, with a reserved growth factor of 25%, can safely support up to 306 calls while meeting the VoIP QoS requirements and having no negative impact on the performance of existing network services or applications. Second, the primary bottleneck of the network is the router. If the enterprise under study is expected to grow in the near future, that is, if more calls than 306 are required, the router must be

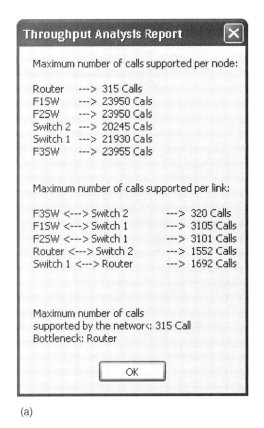

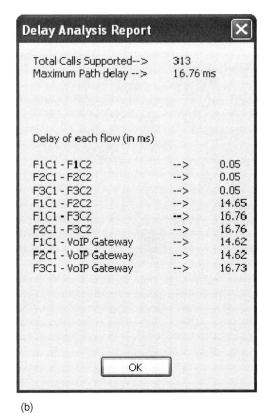

(a)   (b)

FIGURE 1.10   (a) Throughput analysis report. (b) Delay analysis report. (*Source:* K. Salah, "On the deployment of VoIP in ethernet networks: Methodology and case study," *International Journal of Computer Communications,* Elsevier Science, vol. 29, no. 8, 2006, pp. 1039–1054. With permission.)

replaced. The router can be replaced with a popular Layer-3 Ethernet switch, relieving it from routing inter-floor calls from Floor 1 to Floor 2. Before prematurely changing other network components, one has to find out how many VoIP calls can be sustained by replacing the router. To accomplish this, the design steps and guidelines outlined in this chapter must be revisited and re-executed. And finally, the network capacity to support VoIP is bounded more by the network throughput than the delay. This is because the network currently under study is small and does not have a large number of intermediate nodes. The network delay bound can become dominant if there is a large-scale LAN or WAN.

1.4 Summary and Conclusion

This chapter outlined a step-by-step methodology on how VoIP can be successfully deployed in existing IP networks. The methodology can help network designers determine quickly and easily how well VoIP will perform on a network prior to deployment. Prior to the purchase and deployment of VoIP equipment, it is possible to predict the number of

VoIP calls that can be sustained by the network while satisfying the QoS requirements of all existing and new network services and leaving enough capacity for future growth. In addition, many design and engineering issues pertaining to the deployment of VoIP have been discussed. These include characteristics of VoIP traffic and QoS requirements, VoIP flow and call distribution, defining future growth capacity, and measurement and impact of background traffic. A case study was presented for deploying VoIP in a small enterprise network, and the methodology and guidelines outlined in this chapter were applied. Analysis and OPNET simulation were used to investigate throughput and delay bounds for such a network. Results obtained from the analysis and simulation were in line and matched closely. The methodology and guidelines presented in this chapter can be adopted for the deployment of many other network services (other than p2p VoIP). These services may include VoIP conferencing and messaging, videoconferencing, IPTV, online gaming, and ERP.

References

1. M. Bearden, L. Denby, B. Karacali, J. Meloche, and D. T. Stott, "Assessing network readiness for IP telephony," *Proceedings of IEEE International Conference on Communications*, ICC '02, vol. 4, 2002, pp. 2568–2572.
2. B. Karacali, L. Denby, and J. Meloche, "Scalable network assessment for IP telephony," *Proceedings of IEEE International Conference on Communications*, ICC '04, Paris, June 2004, pp. 1505–1511.
3. K. Salah, P. Calyam, and M. I. Buhari, "Assessing readiness of IP networks to support desktop videoconferencing using OPNET," *International Journal of Network and Computer Applications*, Elsevier Science, vol. 31, no. 4, November 2008, pp. 921–943.
4. K. Salah, "An analytical approach for deploying desktop videoconferencing," *IEE Proceedings Communications*, vol. 153, no. 3, 2006, pp. 434–444.
5. B. Goode, "Voice over Internet Protocol (VoIP)," *Proceedings of IEEE*, vol. 90, no. 9, September 2002, pp. 1495–1517.
6. P. Mehta and S. Udani, "Voice over IP," *IEEE Potentials Magazine*, vol. 20, no. 4, October 2001, pp. 36–40.
7. W. Jiang and H. Schulzrinne, "Towards junking the PBX: Deploying IP telephony," *Proceedings of ACM 11th International Workshop on Network and Operating System Support for Digital Audio and Video*, Port Jefferson, NY, June 2001, pp. 177–185.
8. B. Duysburgh, S. Vanhastel, B. DeVreese, C. Petrisor, and P. Demeester, "On the influence of best-effort network conditions on the perceived speech quality of VoIP connections," *Proceedings of IEEE 10th International Conference on Computer Communications and Networks*, Scottsdale, AZ, October 2001, pp. 334–339.
9. W. Jiang, K. Koguchi, and H. Schulzrinne, "QoS evaluation of VoIP end-points," *Proceedings of IEEE International Conference on Communications*, ICC '03, Anchorage, May 2003, pp. 1917–1921.
10. Avaya Inc., "Avaya IP voice quality network requirements," http://www1.avaya.com/enterprise/whitepapers, 2001.
11. A. Markopoulou, F. Tobagi, and M. Karam, "Assessing the quality of voice communications over internet backbones," *IEEE/ACM Transactions on Networking*, vol. 11, no. 5, 2003, pp. 747–760.
12. Recommendation G.711, "Pulse code modulation (PCM) of voice frequencies," ITU, November 1988.

13. Recommendation H.323, "Packet-based multimedia communication systems," ITU, 1997.

14. Recommendation G.114, "One-way transmission time," ITU, 1996.

15. ITU-T Recommendation P.800, "Methods for subjective determination of transmission quality," www.itu.in/publications/main_publ/itut.html.

16. L. Sun and E. C. Ifeachor, "Prediction of perceived conversational speech quality and effects of playout buffer algorithms," *Proceedings of International Conference on Communications*, ICC '03, Anchorage, May 2003, pp. 1–6.

17. A. Takahasi, H. Yoshino, and N. Kitawaki, "Perceptual QoS assessment technologies for VoIP," *IEEE Communications Magazine*, vol. 42, no. 7, July 2004, pp. 28–34.

18. J. Walker and J. Hicks, "Planning for VoIP," NetIQ Corporation white paper, December 2002, http://www.telnetnetworks.ca/products/netIq/whitepapers/planning_for_voip.pdf.

19. Recommendation G.726, "40, 32, 24, 16 kbit/s adaptive differential pulse code modulation (ADPCM)," ITU, December 1990.

20. Recommendation G.723.1, "Speech coders: dual rate speech coder for multimedia communication transmitting at 5.3 and 6.3 kbit/s," ITU, March 1996.

21. Annex to Recommendation G.729, "Coding of speech at 8 kbit/s using conjugate structure algebraic-code-excited linear-prediction (CS-ACELP)." Annex A: "Reduced complexity 8 kbit/s CS-ACELP speech codec," ITU, November 1996.

22. W. Jiang and H. Schulzrinne, "Comparison and optimization of packet loss repair methods on VoIP perceived quality under bursty loss," *Proceedings of ACM 12th International Workshop on Network and Operating System Support for Digital Audio and Video*, Miami, FL, May 2002, pp. 73–81.

23. J. S. Han, S. J. Ahn, and J. W. Chung, "Study of delay patterns of weighted voice traffic of end-to-end users on the VoIP network," *International Journal of Network Management*, vol. 12, no. 5, May 2002, pp. 271–280.

24. J. H. James, B. Chen, and L. Garrison, "Implementing VoIP: A voice transmission performance progress report," *IEEE Communications Magazine*, vol. 42, no. 7, July 2004, pp. 36–41.

25. W. Jiang and H. Schulzrinne, "Assessment of VoIP service availability in the current Internet," *Proceedings of International Workshop on Passive and Active Measurement* (PAM2003), San Diego, CA, April 2003.

26. M. Karam and F. Tobagi, "Analysis of delay and delay jitter of voice traffic in the Internet," *Computer Networks Magazine*, vol. 40, no. 6, December 2002, pp. 711–726.

27. S. Riley and R. Breyer, "*Switched, Fast, and Gigabit Ethernet*," Macmillan Technical Publishing, 3rd Edition, 2000.

28. CAIDA, http://www.caida.org/tools/taxonomy, April 2004.

29. R. Prasad, C. Dovrolis, M. Murray, and K. C. Claffy, "Bandwidth estimation: Metrics, measurement techniques, and tools," *IEEE Network Magazine*, vol. 17, no. 6, December 2003, pp. 27–35.

30. Cisco Systems Inc., "Cisco 2621 modular access router security policy," 2001, http://www.cisco.com/univercd/cc/td/doc/product/access/acs_mod/cis2600/secure/2621rect.pdf.

31. 3Com, "3Com networking product guide," April 2004, http://www.3com.co.kr/products/pdf/productguide.pdf.

32. F. Gebali, *Computing Communication Networks: Analysis and Designs*, Northstar Digital Design, Inc., 3rd Edition, 2005.

33. K. M. Chandy and C. H. Sauer, "Approximate methods for analyzing queueing network models of computing systems," *Journal of ACM Computing Surveys*, vol. 10, no. 3, September 1978, pp. 281–317.

34. L. Kleinrock, *Queueing Systems: Theory*, vol. 1, New York, Wiley, 1975.

35. W. Leland, M. Taqqu, W. Willinger, and D. Wilson, "On the self-similar nature of ethernet traffic," *IEEE/ACM Transactions on Networking*, vol. 2, no. 1, February 1994, pp. 1–15.

36. R. Suri, "Robustness of queueing network formulas," *Journal of the ACM*, vol. 30, no. 3, July 1983, pp. 564–594.

37. K. Pawlikowski, H. Jeong, and J. Lee, "On credibility of simulation studies of telecommunication networks," *IEEE Communications Magazine*, vol. 40, no. 1, January 2002, pp. 132–139.

38. OPNET Technologies, http://www.mil3.com.

39. K. Salah, "On the deployment of VoIP in ethernet networks: Methodology and case study," *International Journal of Computer Communications*, Elsevier Science, vol. 29, no. 8, 2006, pp. 1039–1054.

40. K. Salah and A. Alkhoraidly, "An OPNET-based simulation approach for deploying VoIP," *International Journal of Network Management*, John Wiley, vol. 16, no. 3–4, 2006, pp. 159–183.

41. K. Salah, N. Darwish, M. Saleem, and Y. Shaaban, "An analytical simulator for deploying IP telephony," *International Journal of Network Management*, John Wiley, published online 17 January 2008.

2

Multipoint VoIP in Ubiquitous Environments

Dongsu Seong, Keonbae Lee, and Minseok Oh

CONTENTS

2.1 Introduction

Ubiquitous computing environments, which have recently been receiving much attention, provide communication and computing services to people any place, any time. They are a new paradigm, and their social, cultural, and economic impacts are immense [1]. The multipoint VoIP conference technique is one of the essential instruments that provides services in a ubiquitous environment.

One of the characteristics of multipoint VoIP in a ubiquitous environment is that users having a wireless mobile node are usually collocated in a confined area. For example, firefighters at a fire scene or workers in a construction area may form a wireless local area communication network. In order to communicate with other nodes in a confined area, a mobile node may have to communicate with a fixed conference server located far away from the users. When this configuration, known as a centralized conference scheme [2], is used to set up calls between users close to each other, it may result in poor throughput performance due to the distance the signaling traffic has to travel. When a distributed conference scheme [2], in which user data alone are delivered directly to the other party without going through the conference server, is used to set up a conference call, mobile nodes can exchange users' audio/video data peer-to-peer. This may, however, result in rapid dissipation of battery power, because each node has to perform data compression/decompression and mixing by itself. Therefore, an endpoint mixing scheme, which does not use a conference server, may be an alternative for such a ubiquitous environment as described above.

Another characteristic of multipoint VoIP in a ubiquitous environment is that a large portion of the conference calls are not planned ahead, so that two different conference call groups can be merged into one while they are in session or a conference call group can be split into two while it is in session. When a conference server is involved, the split-and-merge procedure is not easy. In contrast, the endpoint mixing conference scheme [2–4] can manage the split-and-merge operations with relative ease.

However, there are shortcomings with the endpoint mixing conference scheme as well, as when it is applied to a hierarchical, multipoint conference system. The first is that it can cause echoes. Second, when a mobile node leaves a conference call session, it may divide the session into two. Third, it may cause an extra delay in delivering user (audio/video) data due to the hierarchical structure of routing paths. Finally, when too many child nodes are connected to a single parent node, the parent node may experience rapid battery power consumption. This chapter includes solutions to these problems.

The chapter is organized as follows: first, we investigate several multipoint conference systems in a ubiquitous environment and explain why the endpoint mixing conference scheme works effectively. Second, we look into the issues that arise when the endpoint mixing scheme is applied to a ubiquitous environment. Finally, we introduce solutions to these issues and validate them with simulation results.

2.2 Conference Schemes in a Ubiquitous Environment

2.2.1 UFC Terminals in a Ubiquitous Environment

Ubiquitous Fashionable Computers (UFCs) play an important role in a ubiquitous environment where scattered computing devices interact with others and have access to user-centric services. They have initiated the development of a variety of new technologies, such as communication and networking capability, middleware, and applications. They provide a new user-centric lifestyle without time-and-space constraints on the delivery of information. UFC terminals have been developed to provide voice and image recognition, an action–response interface (responsive to movement of a pointer device), portability and durability, which can be achieved through minimal power consumption and size. They also have robust communication capability in a ubiquitous environment [1].

2.2.2 Characteristics of the Multipoint Conference System in a Ubiquitous Environment

As mentioned earlier, users in a ubiquitous environment are most likely to be closely located, in a confined area. A multipoint conference system can be divided into two categories, depending on whether the conference server is used or not. When the conference server is used, it can be categorized further into a centralized conference system and a distributed conference system, depending on whether user data are concentrated in the network or not. In a centralized conference system, call control messages and user data are exchanged through a conference server, as shown in Figure 2.1. In a distributed system, call control messages are delivered via a fixed conference server, but user data are delivered directly to the other party without going through the conference server, as shown in Figure 2.2 [2].

The conference server is usually located in a fixed core network. When a group of closely located users sets up a conference call session, the centralized conference server may have difficulty providing a high quality audio/video service, since the traffic from several users is gathered into the server and distributed through the server back to the users; that is, too much traffic is concentrated around the conference server. When a distributed conference server is used, traffic is exchanged directly between users, which may relieve data concentration, but may result in high power-consumption in the mobile node battery because each mobile node has to receive traffic from all other users and process it within itself. We use a scheme without a conference server as an alternative.

Schemes that do not require a conference server include the pseudo-centralized conference scheme, the pseudo-distributed conference scheme, and the endpoint mixing conference scheme. The pseudo-centralized conference scheme shown in Figure 2.3

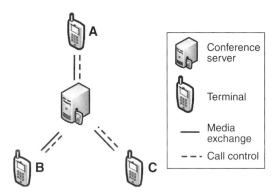

FIGURE 2.1 Centralized conference system.

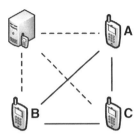

FIGURE 2.2   Distributed conference system.

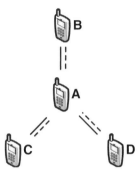

FIGURE 2.3   Pseudo-centralized conference system.

selects a mobile node for processing call control messages as well as relaying user data [3]. But the selected mobile node inevitably experiences a shortage of resources, such as CPU usage, memory availability, or battery power, since all traffic is delivered through the selected mobile node. Therefore, tasks such as audio/video compression, decompression, and mixing may not be completed within the required time frame. The pseudo-distributed conference scheme shown in Figure 2.4 chooses a mobile node only for call control messages. Data are now exchanged directly between users, but since each mobile node processes and communicates user data, the battery power consumption may be significantly high, similar to that in the distributed conference system.

As the number of nodes increases, the selected mobile node gets proportionally exhausted, both in a pseudo-centralized system and in a pseudo-distributed system. As an alternative, to relieve overloading of a mobile node, we consider the endpoint mixing

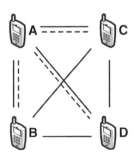

FIGURE 2.4   Pseudo-distributed conference system.

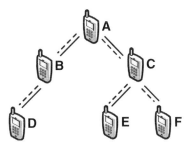

FIGURE 2.5 Endpoint mixing conference system.

conference system shown in Figure 2.5. This is similar to the pseudo-centralized conference scheme in that the exchange of call control messages and user data for child nodes is performed via a parent node. Since the mobile nodes are connected hierarchically, a single mobile node is not responsible for every node's signaling and user data as in the pseudo-centralized system [2–4]. The parent mobile node takes care of call control message- and user-data exchanges solely for its child nodes. In other words, a parent node functions as a conference server for its child nodes. The endpoint mixing conference scheme can alleviate battery power consumption since user data and delivery processing is distributed hierarchically to all parent nodes. In Figure 2.5, when mobile node *D* requests a connection to node *B*, node *D* becomes a child node of node *B*, and node *B*, which accepts the connection request, becomes a parent node of node *D*. Once a hierarchical structure is built using the endpoint mixing scheme, node *A*, which has child nodes and no parent node, is denoted a root node. In contrast, nodes *D*, *E*, and *F*, which have a parent node and no child nodes, are denoted leaf nodes. All nodes except root and leaf nodes will have a parent and a child node at the same time.

Other characteristics of the multipoint conference system in a ubiquitous environment include that a large portion of the conference calls are not preplanned, two different conference call groups can be merged into one while they are in session, or a conference call group can be split into two while it is in session. Since in the centralized conference system the call control messages and user data are exchanged through the conference server, the capacity of the server may limit the number of users being serviced through it. Therefore, the centralized conference system is suitable for a preplanned conference call. For calls not planned earlier, the server may not be available for additional users. This can be resolved by moving the call control and user data service from the current server to another, but that will take up unnecessary time and resources. However, when the endpoint mixing scheme is used, a session can be established through a hierarchical structure, virtually without limit on the number of users.

In an *ad hoc* conference call session, two conference call groups can be merged into one, or a conference call group can be divided into two while they are in session. For example, when construction workers are having a conference call in a construction area, a director or other staff from the headquarters may join or leave the session before it ends. In order for two conference call groups, say conference group 1 and conference group 2, to be merged (suppose conference server 1 continues to serve the group after merging), neither the centralized conference scheme nor the distributed conference scheme, which requires a conference server, can merge the two instantly. Conference call group 2 ends first, and then the participants in this group request a connection to the server to allow them to join conference call group 1.

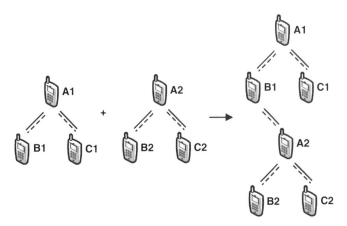

FIGURE 2.6 Combining two conference groups.

When a conference call group has to be divided into two, that is, conference call group 1 and conference call group 2, both the centralized conference scheme and the distributed conference scheme let the participants who want to form a separate conference call group leave the current session first and then form a new group, that is, they form conference call group 2.

However, the procedures can be simplified if we use the endpoint mixing conference scheme, in which the conference call session has a hierarchical structure, a tree topology. Merging two conference call groups can be interpreted as merging two hierarchical structures into one. The simplest case of merging will be that a root node of group 2 joins as a child node in group 1, as shown in Figure 2.6. Then group 1 becomes the resultant merged conference call group.

In the endpoint mixing scheme, a conference session can easily be separated, forming two tree structures from one. In an *ad hoc* environment, it is common for two conferences to be merged into one and later separated, making them as they were before. The root mobile node of the conference call group being separated from the main tree, which is connected to a parent node in the current conference group, has to be simply dismantled, as shown in Figure 2.7.

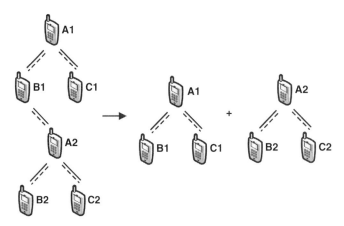

FIGURE 2.7 Separating a conference group.

2.2.3 Dialogs and Transactions Within a Session Initiation Protocol User Agent in the Endpoint Mixing Conference System

In an endpoint mixing conference system, when a third user joins a conference between two users, it becomes a multipoint conference group as shown in Figure 2.8a, and the parent node, *Kim* in Figure 2.8b, which is responsible for mixing user data, creates two dialogs [5]. The first dialog is for the conversation between *Kim* and *Lee*, and the second dialog is for that between *Kim* and *Park*. Note that *Lee* and *Park* have one dialog for each. Transactions can be created and deleted as needed within dialogs.

2.2.4 User Data Processing in the Endpoint Mixing Scheme

2.2.4.1 User Data Processing Unit in Two-User Conference

Figure 2.9 shows a block diagram of a user data processing unit in a two-user conference. The unit compresses data received from a camera unit and microphone. The compressed data is transmitted to other nodes through a Real-time Transport Protocol (RTP) [6]. The node, which receives the compressed data through a Real Time Receiver (RTR), decompresses it using the data decompressing unit and delivers it to the data output unit.

2.2.4.2 Data Processing Unit

The endpoint mixing scheme allows a multipoint conference without a conference server. Figure 2.10 shows a data processing unit that can process data coming from up to three users. It mixes its own data with the data coming in from other users and delivers the

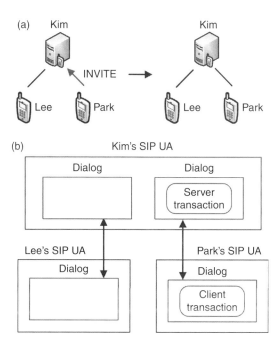

FIGURE 2.8 Endpoint mixing scheme in multipoint conference system: (a) joining an existing session, (b) dialogs and transactions within SIP UA.

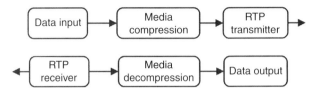

FIGURE 2.9   Data processing unit in two-user conference.

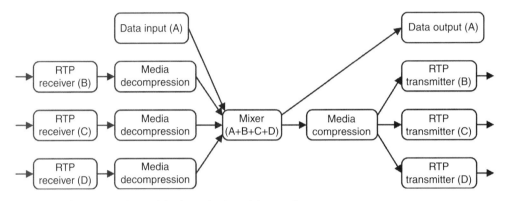

FIGURE 2.10   Data processing unit in the endpoint mixing conference system.

combined data to itself and to other users. The data processing unit for *n* users can be easily obtained by expanding the unit of Figure 2.10. For *n* users there will be $n - 1$ RTP receivers, $n - 1$ data decompressors, $n - 1$ RTP transmitters, one mixer, one data input unit, one data output unit, and one data compressor.

2.3 Problems of an Endpoint Mixing Conference Scheme in a Ubiquitous Environment

The endpoint mixing conference scheme has certain shortcomings when it is built hierarchically. These include an echo effect, unwanted release of a mobile node from a conference call group during the separation process, delay caused by long hierarchical paths, and rapid battery power consumption in a parent mobile node that carries many child nodes.

2.3.1 Duplicate Reception of User Data

Echoes can occur when a hierarchical tree topology is formed using the endpoint mixing conference scheme. A parent mobile node's data processing unit receives user data from its neighbor nodes, mixes them with its own user data, and transmits the combined user data to its neighbor nodes.

Figure 2.11 shows a snapshot of user data streams where nodes are equipped with the data processing unit depicted in Figure 2.10. The arrow indicates user data flow. For example, mobile node *A* receives data $b + c + d$ (assume that *b*, *c*, and *d* denote the data

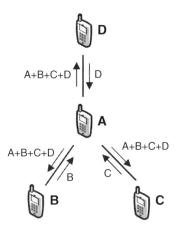

FIGURE 2.11 Conference system equipped with the data processing unit depicted in Figure 2.10.

transmitted from mobile nodes *B*, *C*, and *D*, respectively) and sends the combined data ($a + b + c + d$, data received from *B*, *C*, *D* plus data generated at *A*) to mobile nodes *B*, *C*, and *D*. The shortcoming of this scheme is that the mixed data contains the mixing node's user data as well. This scheme is easy to implement, but causes echoes. For example, node *B* delivers user data *b*, but the mixed user data received from *A* includes user data *b* as well.

Figure 2.12 shows another example of a hierarchical multipoint conference system, where each node is equipped with the data processing unit depicted in Figure 2.10. When one of

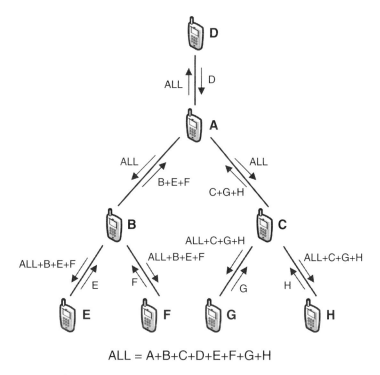

FIGURE 2.12 Another conference system equipped with the data processing unit depicted in Figure 2.10.

nodes *B*, *C*, and *D*, of which the data mixing mode is inactive, is requested for a connection, the conference system is transformed into a hierarchical multipoint conference system. For example, new users *E* and *F* ask for connections to *B* and *B* accepts them, and the configuration changes into the one in Figure 2.12. Then, node *B*'s user data mixing capability is activated.

Mobile node *A* receives data $b + e + f$ from mobile node *B* and sends data ALL $(a + b + c + d + e + f + g + h$: data received from *B*, *C*, *D* plus data generated at *A*) to mobile node *B*. Note that mobile node *A* functions as a root node and collects user data from all the nodes, including itself, due to its central position in the topology. Mobile node *E* will eventually receive ALL $+ b + e + f$ $(a + b + c + d + e + f + g + h + b + e + f$: data received from *A*, *E*, *F* plus data generated at *B*), where data *b*, *e*, and *f* are received twice. This only illustrates the early part of data stream duplication and as time progresses, the echoing will get worse.

2.3.2 Separation of Conference Group

In a hierarchical multipoint conference system, when a certain mobile node leaves an in-session conference call group, it may cause some mobile nodes to be kicked off from the conference call. This can occur when a mobile node that mixes the user data leaves the conference, so that all the child mobile nodes under it are separated from the original conference call group. For example, when mobile node *A* in Figure 2.12 leaves the conference call group, it segregates the connection between nodes *B*, *C*, *D* and then divides the conference call group into three, as shown in Figure 2.13.

2.3.3 User Data Delay and Resource Shortage of a Parent Mobile Node

In a hierarchical multipoint conference system, user data is delivered through a hierarchical route, which may result in longer delay due to its tree structure. For example, in Figure 2.14, in order for node *G* to communication with node *C*, the traffic must traverse through nodes *D*, *B*, and *A*.

The delay may be even greater if the two nodes are located deep in two different tree branches. When a parent mobile node carries many child nodes, which results in a great

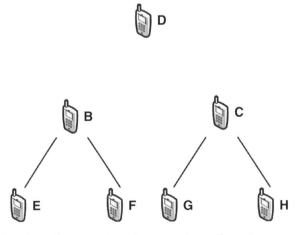

FIGURE 2.13 Segregation of a conference system when a user leaves the system.

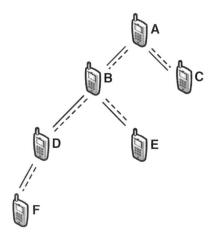

FIGURE 2.14 Problems in user data delay and resource shortage.

deal of call control signaling and user data processing in the parent node, the parent node may experience a shortage of resources such as CPU usage and memory availability. Most importantly, this may lead to rapid battery power consumption. In Figure 2.14, node *B* has the most child nodes, so its resources will be depleted sooner than that of any other node.

2.4 Resolving Problems in an Endpoint Mixing Conference Scheme

2.4.1 Resolving Duplicate Reception of User Data in a Hierarchical Multipoint VoIP

Figure 2.15 shows a data processing unit that can prevent user data being received twice in a hierarchical multipoint conference system. It can process user data from up to three

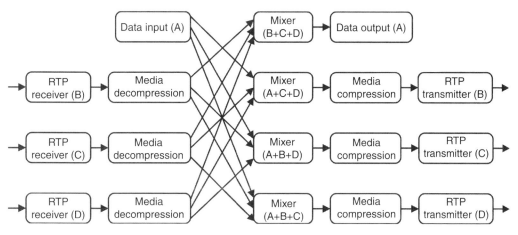

FIGURE 2.15 Data processing unit supporting endpoint mixing scheme.

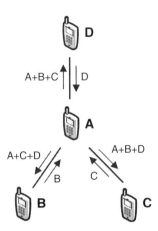

FIGURE 2.16 Conference scheme using the data processing unit depicted in Figure 2.15.

users. The unit receives data, selectively mixes it with its own user data, and then sends it to other users, including itself. In this case, the combined user data transmitted through RTP transmitter (b) is the combined data received from all users except the one through the RTP receiver (b), that is, $a + b + d$. If extended to accommodate n users, the unit will be composed of n-1 RTP receivers, n-1 data decompressors, n-1 RTP transmitters, and n mixers, one data input unit, and one data output unit.

Figure 2.16 shows user data streams where node A is equipped with the data processing unit described in Figure 2.15. Node A selectively mixes incoming user data and its own user data so that echoes do not occur, and sends the combined user data to the other nodes (B, C, D). Each node will receive combined user data, excluding its own user data. For example, node B will receive user data $a + c + d$ and node C will receive $a + b + d$. The scheme increases the complexity of the user data processing unit.

Figure 2.17 shows another example of a multipoint conference scheme using this data processing unit. The figure shows that each node does not receive its own user data twice. For example, when node E receives combined user data from node B, the data does not include e. This resolves the echo problem in a hierarchical conference system. In addition, since the endpoint mixing scheme functions separately from the session initiation protocol (SIP), there will be no need to change the SIP. Therefore, the endpoint mixing scheme using the newly proposed data processing unit can be used effectively in a hierarchical multipoint conference system.

2.4.2 Resolving Segregation of a Conference Group

In the endpoint mixing scheme, when a node responsible for user data mixing leaves the session, it segregates the session into two. This can be resolved by using the REFER command right before or after the node that wants to leave the conference call group sends the BYE command. Figure 2.18 shows the signal flow when node A leaves the session. Node A asks node B to make a connection to node C (using REFER) before it initiates its own departing procedure (using BYE). Node B and C will continue to be in session without stopping or having to reset the connection.

Figure 2.19 shows the flowchart for the procedure described in Figure 2.18. Whereas in the figure, the REFER command is transmitted after the BYE command, it can also be sent before the BYE.

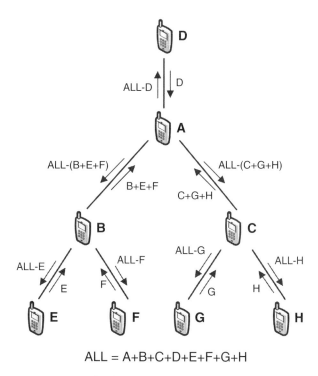

FIGURE 2.17 Conference system using the data processing unit depicted in Figure 2.15.

This scheme works in a hierarchical multipoint conference system such that the session is not divided when a node executes the user data mixing operation. Figure 2.20 shows a signal flow when node *A* leaves a session. Node *A* asks nodes *B* and *C* to make a connection to node *D* (using REFER) before it initiates its own departing procedure (using BYE) and nodes *B* and *C* send a connection request to node *D* using INVITE. Then nodes *B*, *C*, and *D* will construct a new formation as shown in Figure 2.21 without stopping or resetting the session.

2.4.3 Resolving Media Delay and Terminal Resource Deficiency

Delivering user data traffic, whether audio or video, through hierarchical paths may cause delay. This can be resolved by forming a hierarchical tree such that the number of nodes

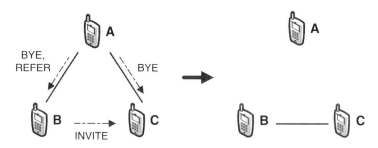

FIGURE 2.18 Preventing segregation of a session using REFER.

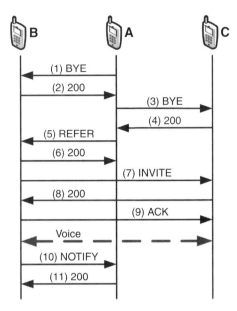

FIGURE 2.19   Flowchart for Figure 2.18.

that data traffic has to go through is reduced. This reduction can be done by carrying as many as child nodes as possible under a parent node so that the tree depth does not increase.

However, if a node carries too many neighbors, including a parent node and child nodes, the amount of data it processes will increase and the resources required for processing will

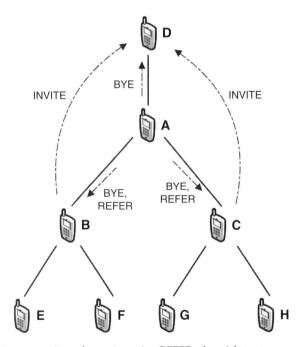

FIGURE 2.20   Preventing segregation of a session using REFER when A leaves.

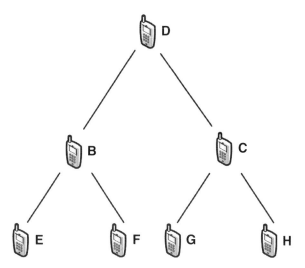

FIGURE 2.21 A newly formed conference session after A leaves.

deplete quickly. Its main functions, such as media compression/decompression and media mixing, may not be executed properly. It will also cause early battery power shortage due to the overloading process. This can be resolved by limiting the number of child nodes for a parent node based on the current state of the parent node's resources, such as CPU power, memory usage, and battery power. The parent node should be able to accept or reject a connection request according to its resource status.

We want to minimize the tree depth to reduce delay in transmitting user data, as shown in Figure 2.22, and also minimize the number of child nodes to save resources in parent nodes, as shown in Figure 2.23. There is a trade-off between the reduction in media delay and reduction in the resource shortage of a node. In order to achieve both, we should keep to a minimal tree depth, restricting the number of child nodes under a parent node.

In the existing method [2], when a call request is received, the decision to accept or reject the call is made by the node receiving the request. As shown in *call assignment algorithm 1*,

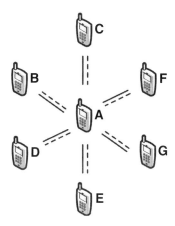

FIGURE 2.22 The optimal configuration to minimize the user data delay.

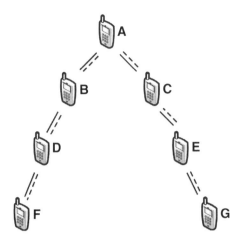

FIGURE 2.23   The optimal configuration to conserve resources of nodes.

the system accepts the call request only if the parent node has enough resources to hold the call [3]. This method does not take into account the problem of delay when accepting a call request. We propose *call assignment algorithm 2*, which takes into account both resource depletion and user data delay.

Call assignment algorithm 2 works as follows. Suppose a new participant requests a connection to node N_i. Then check whether N_i is a leaf node or a parent node. If N_i is a leaf node, check the number of the neighbor nodes of the parent nodes in the system. If a parent node P_i is found, which has a lesser number of neighbor nodes than is allowed, then N_i rejects the request and hands the request over to P_i. If all parent nodes have more neighbor nodes than are allowed, then measure the depth of each leaf node. If a leaf node, say T_i, has depth less than N_i, reject the connection request and hand the request over to leaf node T_i. If N_i is a leaf node and N_i satisfies the two conditions mentioned above, that is, all parent nodes carry more neighbor nodes than allowed and all leaf nodes have a depth greater than that of N_i, N_i accepts the connection request.

If N_i is a parent node, check the number of its neighbor nodes. If it has fewer neighbor nodes, it accepts the connection request, otherwise it rejects the request and hands it over to another node. How does the rejecting node, in our case, N_i, select another node as a possible candidate to take the call connection? It checks all parent nodes except itself. If there is a parent node, say P_i, with less-than-allowed neighbor nodes, it hands the request over to P_i. If there is none, the request is handed over to the leaf node having the least depth.

The algorithm increases the tree depth by one when the number of child nodes in every parent within the tree reaches the maximum number of the allowed neighbor nodes minus one. For example, as shown in Figure 2.24, when node M asks for a connection to node F, node F, with the maximum number of neighbor nodes four, rejects the request and transfers it to node C, which has fewer than four neighbor nodes. In Figure 2.25, when node H requests a connection to node F, node F rejects it and transfers the connection to leaf node C, which has less depth than node F. In order to relieve the resource shortage, when the number of child nodes connected to a parent node reaches the maximum, the parent node can reject the connection request. In Figure 2.26, when node M requests a connection to node B, node B rejects the request and transfers it to node C, which has child nodes less than or equal to four. In Figure 2.27, when node H requests a connection to node B, node B rejects the request and transfers it to node C, which has the least depth.

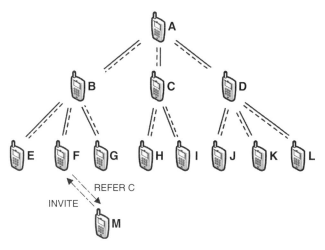

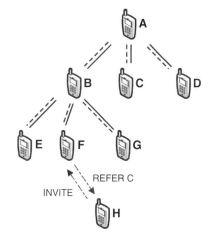

FIGURE 2.24 Example 1 of rejecting request to relieve user data delay when neighbor nodes are 4.

FIGURE 2.25 Example 2 of rejecting request to relieve user data delay when neighbor are 4.

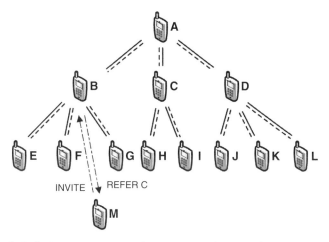

FIGURE 2.26 Example 1 of rejecting request to relieve resource shortage when neighbor nodes are 4.

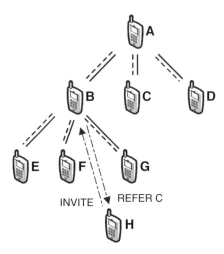

FIGURE 2.27   Example 2 of rejecting request to relieve resource shortage when neighbor nodes are 4.

Call Assignment Algorithm 1

```
(Considering only the resource depletion problem)
if (no. of neighbor nodes of node receiving a request < max. no. of
allowed neighbor nodes)
      accept request;
else
      reject request;
```

Call Assignment Algorithm 2

```
(Considering both the resource depletion and user data delay problems)
if (node N_i receiving a request is a leaf node)
      for (parent node P_i ∈ all the parent nodes in the system)
            if (no. of neighbor nodes of node P_i < max. no. of allowed
neighbor nodes)
                  hand the request to another P_i;
                  exit;
      for (leaf node T_i ∈ all the leaf nodes except N_i in the system)
            if (tree depth of node N_i receiving a request > tree depth of
leaf node T_i)
                  hand the request to another T_i;
                  exit;
      accept request;
else
      if (no. of neighbor nodes of parent node P_i < max. no. of allowed
neighbor nodes)
            accept request;
      else
            for (parent node P_i ∈ all the parent nodes except N_i in the
system)
                  hand it to another P_i;
                  exit;
      hand the request to a leaf node with the least tree depth;
```

It is true that the algorithm for both resource shortage and user data delay will be more complex than the algorithm for resource shortage alone. But compared with the complexity of the SIP itself, it is still negligible.

2.5 Simulation Results

The UFC VoIP has been implemented in a main module, which is attached to a jacket. The main module has an ARM9 CPU and is operated on Linux OS. It supports IEEE802.11, Bluetooth, and Zigbee [1]. The UFC VoIP has been implemented according to IETF RFC3261 and is composed of the follow modules: SIP, session description protocol (SDP) [7,8], real-time transport protocol (RTP)/real-time transport control protocol (RTCP) [6], the audio processing, multipoint processor, and multipoint controller, as shown in Figure 2.28. The SIP header module creates and analyzes SIP messages. The SDP module creates and analyzes SDP messages and exchanges user data. The RTP/RTCP and the audio processing modules handle the real-time transmission of user data. The multipoint processor mixes user data coming from up to three users with its own data and delivers the combined user data to up to three users. The multipoint controller controls SIP calls for up to three users.

The total bandwidth usage for each conference system is as follows. The distributed conference system shown in Figure 2.2 requires 45 connections in total for user data transmission when there are 10 users in a session. For N users, the total number of connections required will be $\sum_{k=2}^{N}(k-1)$. As the number of users increases, the total number of connections increases rapidly, which results in high bandwidth usage. In the centralized conference system shown in Figure 2.1, since each user maintains a connection to the server, the total number of connections for N users will be N. In the endpoint mixing conference system, the nodes that do not perform a mixing operation require only one connection, and the nodes that do, up to three connections. When there are 10 users in a session, the total number of connections will be 9, as depicted in Figure 2.29. In an endpoint mixing conference system, the total number of connections will be $N-1$ for N users. This indicates that the endpoint mixing scheme requires the least number of connections among the three. Table 2.1 shows the number of connections and the bandwidth usages for the three conference schemes. We assume that user data require a 6.3 kbps audio stream, and since the exchange is bidirectional, a single connection requires 12.6 kbps.

Multipoint processor			Multipoint controller
Audio (G.729, G.723.1, G711)	Audio (G.729, G.723.1, G711)	Audio (G.729, G.723.1, G711)	SIP (RFC 3261)
RTP/RTCP (RFC 3550)	RTP/RTCP (RFC 3550)	RTP/RTCP (RFC 3550)	SDP (RFC 3264)
TCP/IP			

FIGURE 2.28   Protocol stack of UFC VoIP.

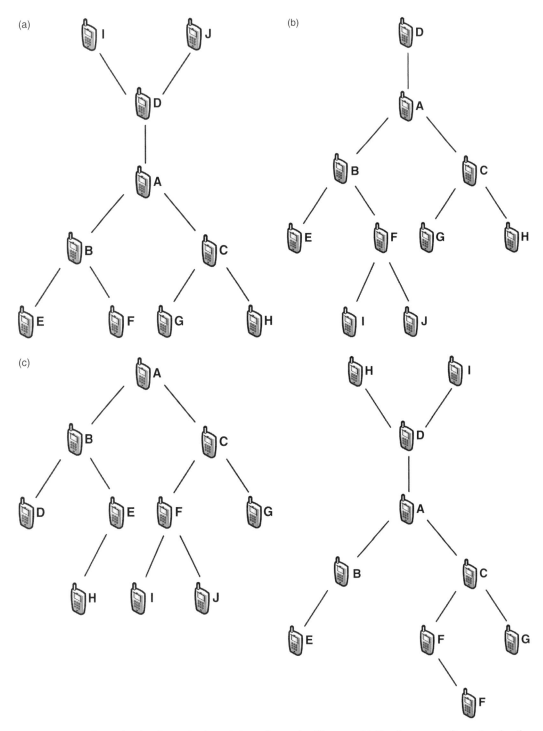

FIGURE 2.29 Examples for the endpoint mixing scheme for 10 users: (a) Conference configuration by the proposed algorithm, (b) example 1 of conference configuration by existing algorithm, (c) example 2 of conference configuration by existing algorithm, and (d) Example 2 of conference configuration by existing algorithm.

TABLE 2.1

Total Bandwidth Usage for Three Conference Schemes

Number of Users		2	3	4	5	6	7	8	9	10
Distributed	Connections	1	3	6	10	15	21	28	36	45
	Bandwidth (kbps)	12.6	37.8	75.6	126	189	264.6	352.8	453.6	567
Concentrated	Connections	2	3	4	5	6	7	8	9	10
	Bandwidth (kbps)	25.2	37.8	50.4	63	75.6	88.2	100.8	113.4	126
Endpoint mixing	Connections	1	2	3	4	5	6	7	8	9
	Bandwidth (kbps)	12.6	25.2	37.8	50.4	63	75.6	88.2	100.8	113.4

Table 2.2 shows the maximum power usage and maximum user data delay for each node in the endpoint mixing scheme.

High bandwidth usage means high power usage, and a greater number of hops means high user data delay. The above results can be interpreted in terms of power usage and user delay as follows.

When up to three neighbor nodes are allowed for each parent node:

- the maximum power usage decreases but the user delay increases, compared with when up to four neighbor nodes are allowed for each node;
- the maximum user delay decreases but the maximum power usage increases, compared with when up to two neighbor nodes are allowed for each node.

Therefore the number of maximum neighbor nodes should be chosen based on the maximum user data delay acceptable and optimum power usage.

The proposed algorithm is designed such that the maximum delay experienced in transmitting user data in the system is minimized compared to that in the existing endpoint mixing schemes [2,3,5], and it has been verified through a series of simulations. Figure 2.29 shows examples of endpoint mixing schemes when there are 10 users in the session.

TABLE 2.2

Maximum Bandwidth and Maximum User Data Delay with Varying Neighbor Nodes

Number of Users		2	3	4	5	6	7	8	9	10
Maximum number of neighbor nodes: 2	Maximum bandwidth (kbps)	12.6	25.2	25.2	25.2	25.2	25.2	25.2	25.2	25.2
	Maximum number of hops	1	2	3	4	5	6	7	8	9
Maximum number of neighbor nodes: 3	Maximum bandwidth (kbps)	12.6	25.2	37.8	37.8	37.8	37.8	37.8	37.8	37.8
	Maximum number of hops	1	2	2	3	3	4	4	4	4
Maximum number of neighbor nodes: 4	Maximum bandwidth (kbps)	12.6	25.2	37.8	50.4	50.4	50.4	50.4	50.4	50.4
	Maximum number of hops	1	2	2	2	3	3	3	4	4

TABLE 2.3

Maximum Distance for Each Participant in Figure 2.29a

	A	B	C	D	E	F	G	H	I	J
Maximum distance (hops)	2	3	3	3	4	4	4	4	4	4

TABLE 2.4

Maximum Distance for Each Conference of Figure 2.29

	(a)	(b)	(c)	(d)
Maximum distance (hops)	4	5	6	5

An example for the proposed algorithm is shown in Figure 2.29a and the other three show examples of the existing endpoint mixing scheme. Table 2.3 shows the maximum distance between users in a conference session for the proposed algorithm depicted in Figure 2.29a, which is defined as the greatest number of hops between any two users in a session. The average maximum distance is defined as the average for the maximum distance between every pair of nodes in a session. Table 2.3 shows that the maximum distance in Figure 2.29a is 4 and the average maximum distance is 3.5. Table 2.4 shows the maximum distance in a session depicted in Figures 2.29a–d, and Table 2.5 shows the average maximum distance in a session for Figure 2.29a–d. The tables show that the proposed algorithm performs best.

The proposed connection assignment algorithm has shown less user data delay in actual experiments using UFC terminals. But to make a systematic assessment, we performed a series of simulations. The simulations were run varying the number of participants from 1 to 20. For each simulation, we established 50,000 conference sessions. When a new node enters a conference in session, the call connection request is assigned to an arbitrary node, and the node receiving the request determines whether to accept it or hand it over to another node. We established 1,000,000 (20 × 50,000) conference call sessions for the existing and proposed algorithms, and compared the results.

Figure 2.30 shows the average of hops between two nodes having the greatest hops, that is, the farthest in the system, varying the number of nodes when the number of the maximum allowable neighbor nodes is set to three. Figure 2.31 shows that the average is over 50,000 runs for the average hops between nodes in the system. Simulation results show that both the number of the greatest hops and the average hops start to decrease when the number of nodes is greater than three. The proposed algorithm reduces the number of the greatest hops by 27%, as shown in Figure 2.32, and the average hops by 22.2%, as shown in Figure 2.33. Figure 2.34 shows the average of the hops between two nodes having the greatest hops between them when the number of the maximum allowable neighbor nodes is set to four, and Figure 2.35 shows the average over 50,000 runs for the average hops between nodes, varying the number of nodes in the system. The proposed algorithm

TABLE 2.5

Average of Maximum Distance for Users of Each Conference in Figure 2.29

	(a)	(b)	(c)	(d)
Maximum distance (hops)	3.5	4.2	4.9	4.2

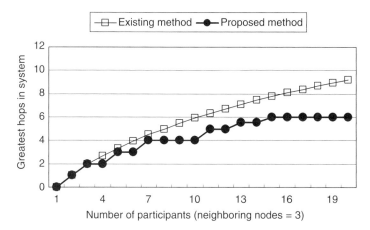

FIGURE 2.30 Number of the greatest hops when the maximum allowable neighbor node is set to 3.

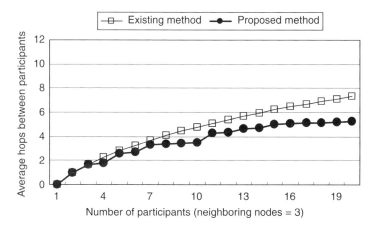

FIGURE 2.31 Average hops when the maximum allowable neighbor nodes are set to 3.

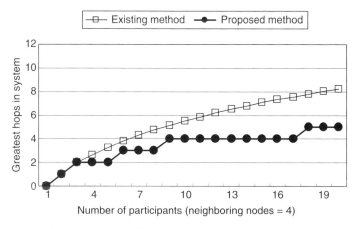

FIGURE 2.32 Reduction rate of the number of the greatest hops in the system.

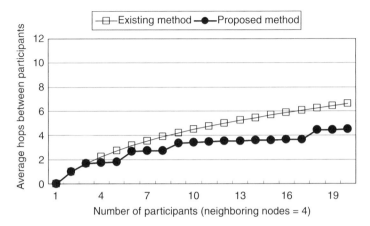

FIGURE 2.33 Average hops for varying number of nodes in the system.

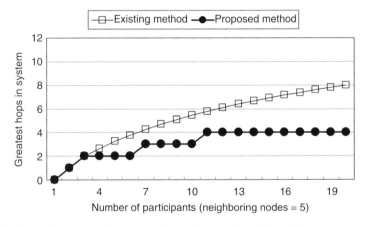

FIGURE 2.34 Number of the greatest hops when the maximum allowable neighbor nodes are set to 4.

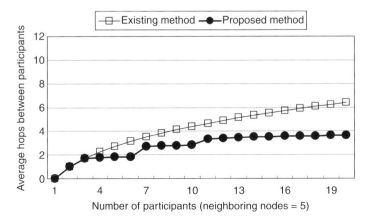

FIGURE 2.35 Average hops when the maximum allowable neighbor nodes are set to 4.

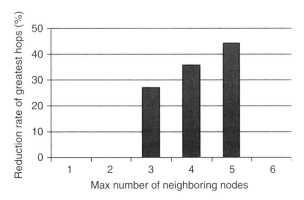

FIGURE 2.36 Number of the greatest hops when the maximum allowable neighbor nodes are set to 5.

reduces the number of the greatest hops by 35.8%, as shown Figure 2.32, and the average hops by 30.3%, as shown in Figure 2.33. Figure 2.36 and Figure 2.37 show the results when the number of the maximum allowable neighbor nodes is set to five. The proposed algorithm reduces the number of the greatest hops by 44.2%, as shown in Figure 2.32, and the average hops by 37.6%, as shown in Figure 2.33. The reduced number of the greatest hops and the average hops are closely related to the reduction of user data delay.

2.6 Conclusion

Ubiquitous environments target systems that provide a means of communication and computing service to people any time, any place. Multipoint VoIP is one of the essential techniques that provides services in a ubiquitous environment. In this chapter, we compared various conference systems for their efficiency and described why the endpoint mixing scheme works better than the rest. However, the endpoint mixing conference scheme too has drawbacks when applied in a ubiquitous environment, and we proposed solutions for these issues. By resolving these problems, various conference systems can be built in a ubiquitous environment, and application QoS (AQoS) improvement can also be achieved in a hierarchical conference system. The multipoint VoIP service will help in

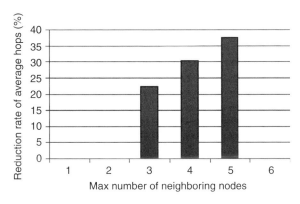

FIGURE 2.37 Average hops when the maximum allowable neighbor nodes are set to 5.

enhancing various application services, and improving the lifestyle of the ubiquitous terminal user. The multipoint conference system is considered to be an important area for further research given the proliferation of ubiquitous application services.

References

1. Kyu Ho Park, Seung Ho Lim, and Dae Yeon Park, "UFC: A ubiquitous fashionable computer," *Next Generation PC 2005 International Conference*, 2005, pp. 142–147.
2. J. Rosenberg, "A framework for conferencing with the session initiation protocol," Draft-IETF-Sipping-Conferencing-Framework-05, 2005.
3. Sung-min Lee, Ki-yong Kim, Hyun-woo Lee, Pyung-su Kim, Dong-su Seong, and Keon-bae Lee, "Multipoint MoIP in ubiquitous fashionable computer," *21st International Technical Conference on Circuits/Systems, Computers and Communication*, 2006, pp. 285–288.
4. M. Irie, K. Hyoudou, and Y. Nakayama, "Tree-based mixing: a new communication model for voice over IP conferencing systems," *Proceedings of the 9th IASTED International Conference on Internet and Multimedia Systems and Applications*, 2005, pp. 353–358.
5. J. Rosenberg, H. Schulzrinne, G. Camarillo, A. Johnston, J. Peterson, R. Sparks, M. Handley, and E. Schooler, "SIP: Session initiation protocol," *IETF*, RFC 3261, 2002.
6. H. Schulzrinne, S. Casner, R. Frederick, and V. Jacobson, "RTP: a transport protocol for real-time applications," *IETF*, RFC 3550, 2003.
7. M. Handley and V. Jacobson, "SDP: session description protocol," *IETF*, RFC 2327, 1998.
8. J. Rosenberg and H. Schulzrinne, "An offer/answer model with session description protocol," *IETF*, RFC 3264, 2002.

3

VoIP in a Wireless Mobile Network

Qi Bi, Yang Yang, and Qinqing Zhang

CONTENTS

3.1 Introduction

3.1.1 Evolution of Circuit Voice Services in Mobile Networks

Voice over Internet Protocol (VoIP) has rapidly become popular in wireline networks. The evolution of voice service from circuit-switched voice to VoIP has been due to the proliferation of IP networks that can deliver data bits cost-effectively. Similar to the wireline networks, voice in wireless mobile networks started with circuit-switched networks. Since the deployment of the first commercial wireless system, wireless mobile networks have evolved from first-generation analog networks to second-generation digital networks. Third-generation wireless mobile networks that are currently in use offer a highly efficient circuit-switched service and have double the spectral efficiency of second-generation systems. After the deployment of the third-generation wireless voice system, however, the number of voice subscribers, having reached a saturation point in many parts of the world, no longer increased. Consequently, the focus on wireless system design shifted to wireless data applications. This resulted in the emergence of high-speed packet-switched data (HSPD) networks, which have been gradually deployed alongside circuit-switched voice networks worldwide.

The growth of the HSPD networks has made it even more desirable to offer a voice service using VoIP together with data to support highly diversified multimedia services. When HSPD and VoIP are deployed together, a separate circuit-switched network need not be dedicated for voice services. Besides, as most of the HSPD networks were designed using the most up-to-date technology, the spectral efficiency of wireless mobile VoIP networks can be greatly improved over that of packet data services designed for third-generation voice networks. This makes the wireless mobile VoIP service attractive to service providers who face constraints on the availability of the frequency spectrum; they also have the advantage of operating an integrated network offering both voice and data services.

3.1.2 Challenges of VoIP Over Wireless

Although VoIP has become very popular and successful in wireline systems, it is still in its infancy in the wireless system, and many technical challenges remain. Listed below are some of the major challenges.

- *Delay and jitter control.* In wireline systems, channels are typically clean and end-to-end transmission can be almost error-free, requiring no retransmissions. However, a wireless channel could be unfavorable, resulting in bit errors and corrupted packets. Packets may have to be retransmitted multiple times to ensure successful reception, and the number of retransmissions depends on the dynamic radio frequency (RF) conditions. This could introduce significant delay and delay variations. Further, unlike the circuit channels, which have a dedicated fixed bandwidth for continuous transmission, packet transmissions are typically bursty and share a common channel that allows multiplexing for efficient channel utilization. This operation also results in loading-dependent delay and jitter.

- *Spectral efficiency.* In a wireline VoIP system, bandwidth is abundant, and it is often used to trade-off a shorter delay. In fact, more bandwidth-efficient circuit-switched transmissions have been abandoned in favor of the flexibility of packet-switched transmissions, even though packet transmissions incur extra overheads. In wireless systems, however, the spectrum resource is generally regarded as the most expensive resource in the network, and high-spectral efficiency is vitally important for service providers. Therefore, wireless VoIP systems must be designed such that they can control delay and jitter without sacrificing spectral efficiency. Packet transmission overheads must also be kept to a minimum over the air interface.

- *Mobility management.* In many HSPD systems, mobility management has been designed mainly for data applications. When mobile users move among cell sites, the handoff procedure follows the break-before-make principle. This leads to a large transmission gap when the mobile is being handed off from one cell site to another. While a transmission gap is often acceptable in data applications, it is unacceptable for voice. To support VoIP applications, the handoff design must be optimized so that the transmission gap during the handoff is minimized and does not impact voice continuity.

- *Transmission power and coverage optimization.* With the packet overheads, a higher power is needed to transmit the same amount of voice information, which results in smaller coverage. In addition, bursty packet transmissions also cause a higher peak-to-average transmission power ratio, which in turn may lead to a higher power requirement in the short term and degraded performance in the outer reaches of the areas covered by the network. Therefore, more advanced techniques must be adopted to compensate for these shortcomings.

3.1.3 3G/4G HSPD System Overview

The two dominant bodies that define wireless mobile standards worldwide are 3GPP and 3GPP2, which have produced their respective third generation standards: UMTS terrestrial radio access (UTRA) and CDMA2000. Their respective fourth generation evolution systems, namely, long-term evolution (LTE) and ultra mobile broadband (UMB), are currently being standardized. The technologies used by the two standard systems are similar; therefore we use the 3GPP2 standards as the example in this chapter.

On the 3G side, CDMA2000 standards include two components: 3G1x, focusing on circuit switched voice service, and the CDMA2000 evolution data optimized (EVDO) standard dedicated to high-speed packet data services. For our discussion of VoIP, we focus on the EVDO system.

The EVDO air interface design adopts a dynamic time division multiplexing (TDM) structure on the forward link with a small time-slot structure (600 time-slots per second) to enable fast and flexible forward packet scheduling. Each active user feeds the desired channel rate back in real time. To improve the transmission efficiency under dynamic RF-fading conditions, a hybrid automatic repeat request (HARQ) technique is employed with incremental redundancy (IR). In EVDO Revision0 (Rev0) [1], 12 nominal data rates were defined, ranging from 38.4 kbps to 2.4 Mbps. A scheduler at the base transceiver station (BTS) decides which user's packet will be transmitted in each open time slot, and the packet must be transmitted with the matching data rate requested by the terminal.

On the reverse link, the design of EVDO Rev0 is similar to that of 3G1x. CDMA technology is used for data traffic as well as real-time feedback (overhead) information transmissions. Each user can transmit packets at any time with data rates from 9.6 kbps up to 153 kbps. The packet transmission duration is 16 time-slots. The reverse data rates are controlled jointly by each mobile and the BTS in a distributed fashion. The BTS monitors the aggregate interference level and broadcasts the overload information via a dedicated control channel on the forward link. The information is used by each mobile to adjust the reverse traffic channel rate upwards or downwards, subject to the available power headroom. When there is no packet to be sent, the reverse data channel is gated off, whereas the pilot and other overhead channels are continuously transmitted.

In EVDO RevisionA (RevA) [2], more forward data rates are defined up to 3.1 Mbps, which are further augmented, up to 4.9 Mbps, by EVDO RevisionB (RevB) [3]. The BTS has the flexibility of transmitting at either the data rate requested by the terminal or at one of several compatible data rates lower than the requested data rate; it can also multiplex packets from multiple users onto a single time slot.

The reverse link operation is significantly upgraded in EVDO RevA. Twelve encoder packet sizes are defined with effective data rates ranging from 4.8 kbps to 1.8 Mbps. The HARQ technique is used with up to four transmissions: each transmission or subpacket lasts four time-slots. Reverse data rate control is done via a comprehensive resource management scheme that adjusts the allowable terminal transmission power, which in turn determines the transmission data rate that can be achieved.

Besides the channel operation-related improvements, EVDO RevA also specifies the quality of service (QoS) framework that enables the radio access network (RAN) to differentiate between applications with different QoS expectations. On the air interface, the forward link scheduler applies different scheduling policies and priorities to traffic flows with different QoS requirements either across users or to the same user. The reverse resource management scheme allows multiplexing data from different flows that share the same physical channel, and flow QoS requirements are taken into consideration to determine resource distribution.

UMB [4,5] is an evolution technology of 1x EVDO. It is the fourth generation (a.k.a. 4G) broadband high-speed data system developed by the 3GPP2 standard body. It operates on a much wider carrier bandwidth, ranging from 1.25 MHz to 20 MHz, and uses orthogonal frequency division multiple access (OFDMA) technology for both forward and reverse link data transmissions. OFDMA enjoys advantages over CDMA in several aspects. The OFDM transmission symbol duration is much longer than that of the CDMA chip, which requires less stringent time synchronization between transmitter and receiver to combat inter-symbol interference in multi-path fading environments. As compared with CDMA, orthogonal transmission between OFDMA subcarriers improves reverse link performance, as it eliminates intra-cell interference. The RF transmission can be scheduled in time as well as frequency dimensions, hence improving scheduling efficiency.

In UMB, a carrier bandwidth is divided into a set of subcarriers with a spacing of 9.6 kbps. The basic transmission unit is defined as a tile that consists of 16 subcarriers in frequency and 8 OFDM symbols in time, which lasts about 1 ms depending on the cyclic prefix length configuration. Each tile can deliver an instantaneous data rate between 75 kbps and 630 kbps under different coding and modulation combinations. Up to six HARQ transmissions are allowed for a packet, which helps it to exploit the time diversity gain. In a 5 MHz carrier, this leads to data rate support over a wide range: from 0.4 Mbps to 17 Mbps. Further, the UMB standard specifies multiple types of advanced antenna technologies that can improve spectral efficiency, which includes:

- Multiple input multiple output (MIMO) technology, where multiple data streams are transmitted to and from a single user simultaneously via multiple transmitting and receiving antennae. Depending on the antenna configuration, MIMO can double or even quadruple spectral efficiency in a multi-path rich environment and also under relatively benign RF conditions. With MIMO, the user peak data rate in UMB is increased to 35 Mbps for a 2 × 2 antenna configuration and to 65 Mbps for a 4 × 4 antenna configuration.

- The spatial diversity multiple access (SDMA) scheme, where the BTS simultaneously transmits data streams to different users using the same frequency resources, given these users are located at distinctively different locations and the signals of those streams cause minimum mutual interference via multiple antenna beam-forming techniques. With SDMA, the aggregate RF transmission efficiency can be significantly improved.

OFDMA is used in UMB for both forward- and reverse-link data transmissions. The orthogonal nature of OFDMA significantly improves the spectral efficiency on the reverse link, as compared with the CDMA-based EVDO system. However, to optimize mobility management, UMB employs a hybrid CDMA + OFDMA structure on the reverse link operation. That is, a certain portion of the tile resources are configured to operate in CDMA fashion, and are used by each user to feedback information, such as the change requests of the forward/reverse serving sector. CDMA and OFDMA transmissions maintain separate pilots. When there is no data transmission, the strong OFDMA pilot is not transmitted, whereas the weak CDMA pilot is, which the BTSs continue to monitor to track the reverse channel conditions and provide channel quality feedback information to the terminal. The hybrid operation and the split of pilots not only facilitate a fast handoff process, but also improve the mobile battery power performance.

We shall now discuss the various techniques that have been designed in the EVDO and UMB systems to address the VoIP service challenges described earlier.

3.2 Techniques to Improve VoIP Transmission Efficiency

3.2.1 Speech Codec and Silence Suppression

Due to the limited bandwidth over the air interface, the speech codec used in cellular systems is often in the category of a narrowband, low bit-rate speech coding. In CDMA2000 systems, the enhanced variable rate codec (EVRC) [6] for low bit-rate speech is used to generate a voice frame every 20 ms. There are four different frame types defined in the EVRC (also called EVRC-A), that is, full rate: 171 bits (equivalent to 8.55 kbps), 1/2 rate: 80 bits (equivalent to 4 kbps), 1/4 rate: 40 bits (equivalent to 2 kbps), and 1/8 rate: 16 bits (equivalent to 0.8 kbps), although the 1/4 rate frame is actually not used in the EVRC voice coder (vocoder). The percentage of frames with each rate varies depending on the talk spurt structure and voice activity. The EVRC vocoder has been widely deployed in 3G1x voice networks.

Recently, a new speech codec, EVRC-B [7], has been standardized and implemented as the next generation speech codec for CDMA2000 networks in 3GPP2. The main design consideration of EVRC-B is to provide a smooth and graceful tradeoff between network capacity and voice quality. The EVRC-B vocoder uses all the four frame types listed above. It defines eight modes (encoder operating points) with different source data rates, and utilizes a reduction rate mechanism to select a mode, which trades average encoding rate for the network capacity.

TABLE 3.1

EVRC-B Target Rates for Different Modes

	Mode							
	0	1	2	3	4	5	6	7
Active source rate (kbps)	8.3	7.57	6.64	6.18	5.82	5.45	5.08	4.0

Table 3.1 shows the estimated average source data rates for active voice encoding under different modes [7]. Mode0 of EVRC-B, like EVRC, does not use the 1/4 rate frame. It has a data rate similar to that of EVRC, but is not compatible with EVRC and, in general, offers better voice quality than does EVRC. Other modes produce lower data rates than EVRC, and show a gradual degradation of voice quality. Mode7 of EVRC-B uses only the 1/4 and 1/8 rate frames. The different modes provide a mechanism for service providers to dynamically adjust and prioritize voice capacity and quality in their networks.

Among the different data rate frames, the 1/8 rate frames usually represent background noise during silent periods and during the speech gaps within talk spurts. They do not carry any actual speech signals and hence have minimum impact on the receiving voice quality. Therefore, most of the 1/8 rate frames can be dropped at the encoder side and need not be transmitted. Only a very small percentage of the 1/8 rate frames are transmitted to assist noise frame reconstruction at the receiver end in order to generate a more natural background noise environment. This technique is called silence detection and suppression, and it greatly improves voice transmission efficiency over the air interface by eliminating a significant portion of the packets containing 1/8 frames when packets are transmitted for a VoIP call.

3.2.2 Header Compression

Header compression, which works by exploiting redundancy in headers, is essential to providing a cost-effective VoIP service in wireless networks. A VoIP packet is generally carried over the RTP/UDP/IP protocol stack. The uncompressed header is about 40 bytes per voice packet. In a full-rate EVRC voice frame of 22 bytes, the RTP/UDP/IP header adds an overhead of more than 180% over the voice information. Without any header compression, the overhead would become dominant and have a significant impact on the air interface capacity. RTP, UDP, and IP headers across the packets of a media stream have significant redundancy. The fields in a header can be classified into multiple categories:

- Static or known fields, which do not need to be sent with every packet. For instance, the source and destination address in the IP header, the source and destination port in the UDP header.
- Inferred fields, which can be inferred from other fields and thus do not need to be sent with every packet; for example, the time stamp in the RTP header, which increases by a fixed amount with every increase in the sequence number. The time stamp hence need not be sent during a talk spurt once the sequence number has been sent. Only at the beginning of a new voice spurt after a silence suppression will the time stamp have to be sent to the receiver again.
- Changing fields that can be sent in compressed form to save bandwidth.

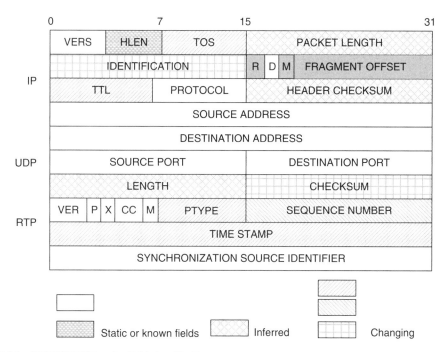

FIGURE 3.1 RTP, UDP, IP header field classification.

Figure 3.1 shows the header field classification in the RTP/UDP/IP headers.

Header compression and suppression have been specified in both 3GPP and 3GPP2 standards. The most popular header compression scheme is the robust header compression (ROHC) [8], which is the Internet Engineering Task Force (IETF)'s standard specification for a highly robust and efficient header compression scheme. The ROHC protocol specifies several operating modes, each requiring different level of feedback information. The selection of the mode depends on the type of the underlying communication channel between the compressor and the decompressor that may have different capability of providing such feedback information. With ROHC compression, RTP/UDP/IP headers are reduced from 40 down to 1–4 bytes. Furthermore, ROHC design is also resilient to packet errors, making it suitable for use with the error-prone wireless link transmissions. As a result, ROHC greatly improves the transmission efficiency of VoIP packets over the air interface. The performance of ROHC in the EVDO radio access network has been analyzed and evaluated in Ref. [10].

It should be emphasized that ROHC is typically not used as an end-to-end header compression scheme, as the full IP header information is still needed for general routing purposes. It is mostly used in the radio access network. A typical implementation of ROHC is shown in Figure 3.2. The ROHC compressor and decompressor are located in the mobile handset and radio network controller (RNC) or packet data service node (PDSN).

3.3 Techniques to Support VoIP in EVDO RevA and RevB

As mentioned earlier, EVDO RevA and RevB standards have been enhanced to support QoS applications such as VoIP. In particular, the low volumes and directionally symmetric

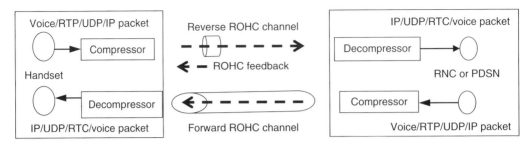

FIGURE 3.2 ROHC implementation architecture.

nature of VoIP traffic do not align well with the traffic assumptions in the original design of the EVDO Rev0, which was designed for Internet-oriented packet data services. As a result, a set of novel features, specially tailored to improve VoIP performance and capacity, have been introduced into the standard. This section describes some of the salient techniques used in EVDO RevA and RevB to support VoIP services.

3.3.1 QoS Enforcement in EVDO RevA/RevB

When a VoIP service is offered together with the general packet data service over the same air interface, the issue of QoS immediately comes up. The original EVDO design focused on supporting general data services, and lacked supporting mechanisms for QoS applications. All data from and to a user were mixed, with no means of differentiating between them. EVDO RevA fundamentally changed the picture by classifying various applications targeted to different users or even the same user. A QoS application traffic is mapped to a radio link protocol (RLP) flow, which bears a specific traffic characteristic and performance target. An abbreviated approach to negotiating QoS parameters is provided via the standardized FlowProfileID, which identifies the type of application, delay sensitivity, and the required data rates.

EVDO RevA and RevB allow a user to open multiple flows over the same air interface connection, and multiplex the traffic from different flows on the same physical traffic channel. The radio link packet header identifies the flow the packet belongs to, so that the mobile and the RAN can treat the packet appropriately, based on flow QoS requirements. For the VoIP service, QoS enforcement is focused on expedited packet delivery. Therefore, a VoIP packet is usually assigned a higher priority than a non-real-time data packet while scheduling traffic throughout the RAN. The specific approaches to the air interface QoS enforcement are different between the forward link and the reverse link.

On the forward link, the air interface scheduler is the central entity that evaluates the priority of the packet and determines the air interface transmissions for all users served by the sector. The specific QoS scheduling algorithm used in the BTS is typically proprietary to the infrastructure vendor, while the commonly used scheduling metrics, though not limited to, include:

- the QoS requirements of the flow, such as packet latency and packet loss requirements, flow throughput expectation, and so on;
- the current flow performance;
- the data backlog of the flow;

- the dynamic RF condition of the mobile user;
- the fairness among flows to different users with similar QoS requirements.

Since the objective of QoS enforcement is to satisfy the flow performance regardless of the user's RF condition, it often leads to conflict with the RF efficiency optimization, and it is up to the scheduler to strike a balance between the two objectives.

On the reverse link, the QoS enforcement is provided via the enhanced reverse traffic channel MAC (RTCMAC) protocol in EVDO RevA. The RTCMAC protocol supports multiple MAC layer flows corresponding to flow of different applications. Each MAC flow can be configured as a high capacity (HiCap) or low latency (LoLat) flow. A LoLat flow always has transmission priority over a HiCap flow, although data from both flows can be multiplexed on the same physical layer packet. A LoLat flow typically enjoys boosted traffic power to reduce the transmission latency via early termination of HARQ. On the aggregation side, the shared RF resource on the reverse link is the interference received by the BTS. It is measured using the rise over thermal (RoT) characteristic, which is defined as the total power received by the BTS normalized by the thermal noise level. RoT needs to be controlled within a target operating point to ensure the stability of CDMA power control. To that end, the BTS monitors the RoT level and compares it with the target operating point. Once the RoT exceeds the target, the BTS broadcasts an "overload" indication to the sector. The resource management scheme embedded in the terminal determines the distribution of traffic power for each active MAC flow based on multiple factors such as:

- the overload indications received from the BTS;
- the flow type, HiCap or LoLat;
- the traffic demand and the performance of the flow

In addition, the terminal also considers other constraints such as the available power headroom and the recent history of the packet transmission patterns, so that it does not produce an abrupt surge of interference to other users. Once the transmission power is determined, the attainable physical transmission data rate is determined.

Interested readers can find more information about the QoS framework within the EVDO system in Ref. [11].

3.3.2 Efficient Packet Format Design for VoIP

Compared with typical Internet application data packets, VoIP packets are much smaller in payload size. For example, the full-rate EVRC vocoder generates a 22 byte voice frame, each of 20 ms interval, during a talk-spurt. Using the header compression techniques described earlier, the VoIP packet size presented to the RAN is usually brought down to within 26 bytes. In order to transmit the small VoIP packets efficiently over the air interface, EVDO RevA defines a new RLP packet format which enables an ROHC compressed full-rate EVRC VoIP packet to be placed perfectly into an RL physical packet of size 256 bits without any segmentation or padding, so that the air interface transmission overhead is minimized.

On the forward link, a different technique is used to improve the VoIP transmission efficiency. In the original design, a physical packet carries information targeted to only a single user. To support VoIP traffic, significant padding would have to be appended to the physical packet leading to poor RF efficiency. The EVDO forward link is partitioned into

600 time-slots per second and it would become a serious resource limit if each time slot carried only single-user packets. To overcome these issues, EVDO RevA defines a new MAC packet format on the forward link, namely, multiuser packet (MUP), where up to eight RLP packets from different users can be multiplexed within a single physical packet. This not only improves the packet transmission efficiency greatly, but also effectively removes the time-slot resource as a capacity constraint.

3.3.3 Expediting VoIP Packet Delivery within EVDO RevA/RevB RAN

Besides the QoS enforcement that in general enables the EVDO RevA/RevB RAN to expedite VoIP packet delivery, there are certain techniques used on the packet delivery method to match the VoIP traffic characteristics.

VoIP packets arrive in 20 ms intervals, hence it is desirable to deliver the VoIP packets within a 20-ms interval, or within 12 time-slots. The HARQ operation of the EVDO RevA/RevB reverse link allows up to four subpacket transmissions, which takes 16 time-slots. To match VoIP requirements, a termination target is introduced, which can control the number of subpackets it takes to transmit a packet under a given packet error rate (PER). For VoIP flow, the termination target is set to three subpackets with a 1% PER. This eliminates reverse link queuing delay, and expedites VoIP packet delivery. Figure 3.3 shows the worst-case (with 1% PER) time-line for the VoIP transmission on EVDO RevA reverse link.

Beyond the air interface, EVDO RevA/RevB allows out-of-order RLP packet delivery to minimize VoIP transmission delay within the RAN. That is, the RLP receiver in the mobile or the network submits a VoIP packet to the upper layer as soon as it is received, even if there are RLP packets with smaller sequence numbers yet to be received. This is based on the consideration that a wireless VoIP client design typically includes a de-jittering algorithm that is able to tolerate a certain degree of varying packet delay, and some can even make use of the information contained in an out-of-order arrived packet. The support of out-of-order packet delivery supports the client design and helps optimizing the end-to-end VoIP application performance.

3.3.4 Smooth Mobility Support

One of the major challenges of supporting VoIP service in a mobility environment is to ensure service continuity when users are moving within the network. EVDO supports soft-handoff in the reverse link as part of the CDMA operation, thereby providing a smooth make-before-break transition when the user moves across cell boundaries.

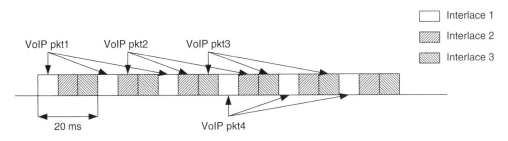

FIGURE 3.3   VoIP transmission time-line in EVDO RevA.

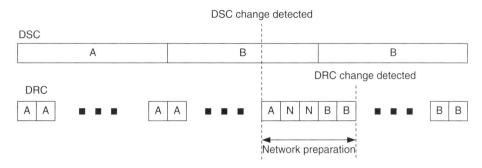

FIGURE 3.4   Serving sector switch time-line in EVDO RevA/RevB.

On the forward link, though, EVDO operates in the simplex mode, wherein the terminal only communicates with a single sector (a.k.a. the serving sector) at any moment of time. The terminal indicates the serving sector along with the desired channel rate through the reverse link data rate control (DRC) channel. If the terminal detects a stronger signal from another sector, it changes the serving sector indication on the DRC channel and switches immediately to the new serving sector in order to receive packets. In EVDO Rev0, the network detects the serving-sector change based on DRC channel decoding, and re-routes the packets if the new serving sector resides in a different BTS so that the user can be served again. This process is similar to the break-before-make transition incurred in a hard hand-off, which creates a long traffic gap in the range of ~50–100 ms. For VoIP applications, such long gaps can cause a short-term voice break and impact the service quality.

In order to minimize the service gap during serving sector change, especially during the serving BTS change, EVDO RevA introduces a new reverse control channel, namely, data source control (DSC) channel, dedicated to forward link handoff. The terminal indicates the serving BTS on the DSC channel. Whenever the terminal determines that it has to change the serving BTS, it indicates the change on the DSC channel, while continuing to point to the current serving sector on the DRC channel for a predefined duration. This staggered operation provides the crucial time required by the network to detect the imminent serving-BTS change and arrange for the data to be re-routed accordingly. When the terminal finally switches to the new serving sector, the new serving BTS is ready to serve the user. In this way, the traffic gap is minimized, often to within 20 ms, a range undetectable by the VoIP application. From the service perspective, it effectively achieves a seamless handoff. Figure 3.4 illustrates the process.

3.3.5   Other VoIP Performance Improvement Techniques Used in EVDO

3.3.5.1   *Mobile Receiver Diversity*

Mobile receiver diversity (MRD) is a well-known technique to improve signal quality. Since the EVDO forward link operates in the simplex mode, MRD maintains the RF link quality at or above the acceptable level leading to continuous voice packet transmissions within the network coverage. With MRD, the overall mobile received signal quality can improve significantly, up to 3 dB on average with balanced and uncorrelated antennae. It also greatly reduces the possibility of the simplex RF connection becoming extremely poor due to deep channel-fading conditions.

3.3.5.2 Interference Cancellation on EVDO Reverse Link

The EVDO reverse link air interface capacity is generally interference limited due to its CDMA operation, so naturally interference mitigation or avoidance techniques have been widely investigated to improve the capacity. One of the interference mitigation solutions, namely, successive interference cancellation (SIC), has been considered in the latest EVDO BTS design. The concept works like this: the BTS first tries to decode all users' data. If some users' data can be successfully decoded, the BTS reconstructs the received signals and removes those signals from the total received signal. The BTS then tries again to decode the remaining users' data. With the removal of a part of the interference from those signals that have been successfully decoded, the effective signal quality of the remaining users' data improves, and with it, the possibility of being successfully decoded as well. Since the strongest interference source is typically from the transmissions within the same sector, SIC is able to eliminate a significant fraction of the interference from the decoding process. As a result, the same data can be transmitted with a reduced power while maintaining the same decoding performance, which in turn lowers the interference received by the BTS. Under the same operational target of interference level, SIC provides an effective means to improve the overall RF capacity on the EVDO reverse link. Interested users can refer to Ref. [9] for more information.

3.3.5.3 Reverse Link Discontinuous Transmission

Each active EVDO terminal maintains an RF connection with the BTS. In EVDO Rev0 and RevA, the terminal continuously transmits through the pilot channel to facilitate channel estimation in the BTS receiver. The reverse feedback channel signals are also transmitted continuously to assist forward link operations. For applications that involve large amounts of data transmissions, the transmission power overheads are minor. However, for VoIP applications, which generate a low volume of user data, the overheads become high both in terminal power consumption as well as the interference generated. Given the nature of the VoIP application, and the enabling of silence suppression, almost no VoIP packets are sent over the air interface when a user is not talking, resulting in low reverse traffic channel activity. Under this condition, due to the continuous transmission of the pilot channel and all the overhead channels the power overheads are quite heavy on the VoIP terminal.

To address the issue, EVDO RevB allows discontinuous transmission (DTX) on the reverse link. Figure 3.5 illustrates the DTX reverse link channel transmission. All feedback channels for a forward link operation are gated with a 50% duty cycle. If there is no data traffic during a subpacket interval, the terminal also gates off the pilot and the associated reverse rate indication (RRI) channel transmissions with a 50% duty cycle. To maintain the performance, the gated overhead channels are transmitted with boosted power. Meanwhile, to conserve the terminal power, the terminal also shuts off the forward signal reception during gated periods. From a VoIP capacity perspective, pilot gating using a DTX operation reduces interference to other users and hence improves the overall VoIP capacity on the reverse link.

With all the techniques described above, the VoIP capacity in the EVDO RevA is estimated as 35 Erlang per sector-carrier with the EVRC vocoder [12]. The EVDO RevB system provides further improvement by about 35%, yielding a significantly higher capacity than the circuit-based voice capacity of the existing CDMA2000 network, which operates at a target capacity of about 26 Erlang per sector-carrier.

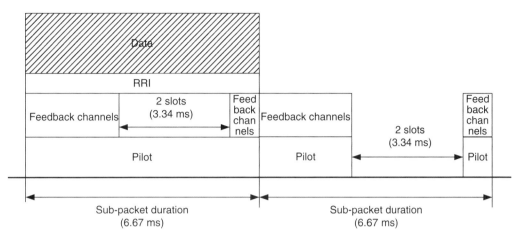

FIGURE 3.5   EVDO RevB reverse channels with DTX operation.

3.4 Techniques and Performance to Support VoIP in UMB

Unlike EVDO, the support for multi-media applications, including VoIP, is an inherent part of the UMB standard. Many techniques have been specified to optimize the performance of VoIP over the air interface.

3.4.1 Fine Resource Allocation Unit

In UMB, the resource allocation unit for a packet transmission is a tile, which consists of a group of 16 contingent subcarriers for one frame duration. Each subcarrier occupies a bandwidth of 9.6 kbps, and each frame lasts around 1 ms. The small VoIP packet size generally occupies only a single tile resource assigned for packet transmission. A straightforward calculation shows that in a 5 MHz carrier, more than 1000 VoIP users can be accommodated solely from the tile resource perspective, which fundamentally lifts the tile resource as a potential bottleneck when evaluating VoIP capacity. Through the HARQ operation, the tile resource can be traded off against transmission power resource.

3.4.2 Flexible RL Frame Durations for Extended Coverage

In order to improve RF transmission efficiency, UMB, like EVDO, employs HARQ on both the forward and reverse links. The HARQ process consists of up to six transmissions, each lasting a single frame duration for most packet formats. For VoIP applications, however, continuous network coverage and reliable packet transmission are essential, as the real-time nature of the service precludes the use of application layer re-transmissions, which usually incurs long delayed. Due to the limited power of transmission in the mobile terminal, under poor RF conditions, reverse link performance can become a matter of concern. To address this, UMB designs a special set of packet formats that use extended frame structures, where each HARQ transmission lasts three frames instead of one frame. This effectively reduces the transmission power requirements to 1/3 for the same amount of information, greatly improving the VoIP network coverage. The tradeoff is that more tile resources will be used for extended frame packet transmissions, which is another example of the tile versus power resource tradeoff.

3.4.3 Persistent Assignment to Minimize Signaling Overhead

The UMB air interface packet transmissions are determined by the BTS scheduler for both the forward and reverse links. In each open frame, the scheduler decides which subset of users will transmit or receive packets, together with the tile allocations, power allocations, and packet format selections. The scheduling decisions are notified to the target users via the forward link control channel. For general packet data applications that can tolerate relatively long latency and large jitter, the scheduling decision can be made on a packet-to-packet basis, optimizing the aggregate spectral efficiency by analyzing RF dynamics. On the other hand, VoIP packets have a much more stringent delay requirement, and the VoIP traffic is composed of a stream of small but frequent packets. If each VoIP packet has to be explicitly scheduled and the scheduling decision has to be explicitly notified to the user on the forward control channel, it can lead to an excessively high signaling overhead when the number of VoIP users increases, degrading the overall RF performance. As the packets arrive periodically during a talk-burst, it is much more efficient to preallocate a sequence of RF resource to the user throughout the talk-burst as long as the mobile's RF condition does not change significantly. This is the idea behind the persistent assignment specified in UMB. The resource assignment decisions by the scheduler are categorized into two:

1. Non-persistent assignment, which applies to a single packet transmission. This type of assignment is highly suited to non-periodical bursty data applications that have relatively loose delay requirements, and allows the scheduler to select the best user for packet transmission.
2. Persistent assignment, which applies to all subsequent packet transmissions unless overridden by new assignments or implicitly nullified by packet transmission failures. This type of assignment is well suited to real-time applications such as VoIP, which have recognized packet arrival patterns and relatively stable packet sizes.

With persistent assignment, the scheduler signaling overhead is greatly reduced. For example, for a talk-burst composed of 100 VoIP packets, nonpersistent assignment would cause up to 100 assignment messages to be sent to the mobile, and persistent assignment would, in most cases, need only one assignment and one possible deassignment message.

3.4.4 Seamless Mobility Support for VoIP

In UMB the mobile terminal still maintains an active set with the network as in a CDMA system. The active set represents the top set of the sectors radiating strong signals to the mobile. The terminal transmits CDMA pilot and some other feedback channel information to the active set members in a dedicated CDMA control segment allocated in both the frequency and time domains. The embedded CDMA operation allows the terminal to communicate with multiple BTSs and determines which RF connectivity has the best quality for data communications, as the packet transmissions in UMB operate in simplex mode on both forward and reverse links. The terminal receives packets from only one of the sectors within the active set, that is, the forward link serving sector (FLSS), and the terminal transmits packets to only one of the sectors within the active set, namely, the reverse link serving sector (RLSS). Typically, the FLSS and RLSS are the sectors with the best forward and reverse channel qualities, respectively. At any moment of time, the FLSS and RLSS can be the same, or they can be different if a significant link imbalance occurs. When the user moves across different cell or sector boundaries, the FLSS and RLSS must change to match the user's new RF conditions. If the user is engaged in an active VoIP call during this period, the process of changing FLSS and/or RLSS must be done quickly and reliably to ensure voice continuity.

In UMB, the terminal determines the desired FLSS (DFLSS) based on the strength of the forward pilot signal of each sector within the active set, and the desired RLSS (DRLSS) based on the reverse pilot quality indications fed back by each sector within the active set. Since packet transmission on the forward link requires feedback support from the reverse link channels, and vice versa, the new sector serving in one direction must have reasonable channel quality in the other direction to maintain acceptable performance of the feedback channels. To that end, UMB specifies the following criterion to determine switching for the serving sector.

> The reverse pilot quality indicated by the FLSS must not be poorer than the best reverse pilot quality indicated by the active set by more than a preconfigured margin. Otherwise the terminal must find a different DFLSS that complies with this requirement. The rule applies to RLSS and DRLSS as well.

Once the terminal has determined a new serving sector, it sends a handoff indication to the new sector. Meanwhile all the feedback channels for data transmissions still communicate to the existing serving sector, so that it can continue to serve the user during the handoff. Once the new serving sector detects the handoff request, it coordinates with the existing serving sector and other network components to re-arrange the data path for the user. When it is ready to serve the user, the new serving sector notifies the mobile on the forward control channel. The mobile then switches the data transmission and reception to the new serving sector. This ensures that data transmission interrupts due to the handoff is minimized. Figure 3.6 illustrates the process.

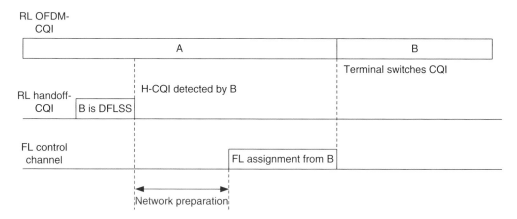

FIGURE 3.6 Serving sector switch time-line in UMB.

3.4.5 Other Performance Improvement Techniques for VoIP in UMB

3.4.5.1 Spatial Time Transmit Diversity

Besides receiving diversity as employed in the EVDO system, UMB also introduces spatial time transmit diversity (STTD) on the forward link to further improve performance for users located at outer reaches of the network coverage. STTD is a technique to transmit multiple transformed versions of a data stream across multiple antennae to improve the reliability of data transmission. The data transmitted on each antenna is a certain combination of multiple temporal stream data points based on a preconfigured combination matrix. The transmission antennae are widely spaced toprovide spatial channel diversity.

For example, Ref. [4] specifies that $A = \begin{bmatrix} s_i & -s_{i+1}^* \\ s_{i+1} & s_i^* \end{bmatrix}$ can be used as a combination matrix for

STTD with two transmission antennae.

3.4.5.2 Spatial Diversity Multiple Access

When two users are at different locations, such as when the propagation paths between the BTS and the users are sufficiently separated, spatial diversity multiple access (SDMA) can be used to improve the RF capacity. That is, the BTS can transmit different data to the users on the same channel resource using differently steered antenna beams that cause little interference between data streams. The aggregate spectral efficiency can thus be increased multiple fold.

In SDMA operations, a steered antenna beam is created by transmitting the data over multiple antennae with different phases. The antennae are placed close together to create a narrow beam pointing in the desired direction when fed with a certain combination of antenna gains and phases. The available set of steered beams is specified by a predefined or downloadable set of precoding matrices. Figure 3.7 illustrates the SDMA operation on the forward link traffic channel. The terminal measures the difference between channel qualities of different beams. It reports the channel qualities, together with the preferred precoding matrix and the selected beam index. Based on feedback from the terminal, the BTS selects a set of users to be superposed on the same channel resource. Adaptive modulation and coding (AMC) and power allocation are performed for each selected user, based on the respective channel condition. Precoding is applied based on the beam index reported by the terminal within the precoding matrix, and the beam-formed data streams are sent to the transmitting antennae.

3.4.5.3 Fractional Frequency Reuse

Like CDMA and EVDO systems, UMB systems operate with universal frequency reuse, that is, each sector in the network is configured to operate in the same carrier frequency. Since the traffic loading in different sectors are expected to fluctuate, the interference between

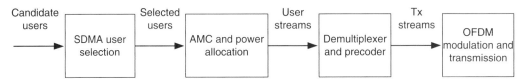

FIGURE 3.7   SDMA operation for forward link in UMB.

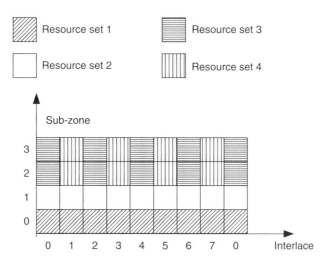

FIGURE 3.8 Resource set concept in UMB.

sectors is dynamic as well. In general, the OFDMA system capacity is other-cell interference limited; hence, the system's performance can be improved if the data transmissions from different sectors are coordinated in the frequency domain to minimize co-channel interference. Fractional frequency reuse (FFR) is a technique used to achieve this goal.

When FFR is enabled, the entire RF resource is partitioned into multiple resource sets, defined as a set of two-dimensional resource pairs: subzone in frequency and interlace in time. Figure 3.8 depicts the concept of the resource set. The resource set partition is broadcast to the sector, and each terminal reports the RF conditions observed in each resource set. This enables the BTS to preferentially allocate the resource sets to different users. For example, a resource set can be preferentially allocated to users under good RF conditions by every BTS, since these users generally require small channel transmission power and hence cause minimum interference to other cells. Some other resource sets can be preferred allocations to users close to cell boundaries. From the feedback given by the terminal, the BTS is able to sense which resource set is experiencing a high interference level from cells adjacent to the user's location, and avoid allocating the same resource set to the user at that moment. In this manner, data transmissions to and from different BTSs are effectively coordinated to minimize the cross-cell interference among adjacent cells, especially when the traffic condition does not demand full band transmission simultaneously across all cells. For real network deployments, the traffic situations typically vary across cells and rarely become fully loaded at the same time, hence the FFR is able to improve the overall performance of the system significantly.

A recent performance study [13] has shown that the OFDMA-based UMB system is able to support over 200 Erlang of VoIP calls within the 5 MHz spectrum. As compared to EVDO RevB, it offers yet another significant improvement on VoIP capacity.

3.5 Summary and Conclusion

In this chapter, we presented the challenges of supporting VoIP service in wireless mobile data networks, and the various techniques that can be used to meet these challenges. With careful design and optimization, the 3G/4G mobile data networks are able to provide VoIP service with not only a similar service quality as in the existing 2G circuit voice network, but also a significantly higher RF capacity for the application. By supporting voice service using VoIP techniques, the 3G/4G mobile data networks offer an integrated solution that has high spectral efficiency and high mobility for multimedia applications.

A final note: while the discussions are based on 1x EVDO and UMB systems, most of the principles and many of the techniques also apply to high speed packet applications and long term evolution networks, since the technologies of the two wireless networks are quite similar.

References

1. 3GPP2 C.S0024-0_v4.0, "cdma2000 high rate packet data air interface specification," version 4.0, October 2002.
2. 3GPP2 C.S0024-A_v1.0, "cdma2000 high rate packet data air interface specification," version 1.0, March 2004.
3. 3GPP2 C.S0024-B_v2.0, "cdma2000 high rate packet data air interface specification," version 2.0, March 2007.
4. 3GPP2 C.S0084-001.0, "Physical layer for ultra mobile broadband (UMB) air interface specification," version 2.0, August 2007.
5. 3GPP2 C.S0084-002.0, "Medium access control layer for ultra mobile broadband (UMB) air interface specification," version 2.0, July 2007.
6. 3GPP2 C.S0014-A_1.0, "Enhanced variable rate codec, speech service option 3 for wideband spread spectrum digital systems," April 2004.
7. 3GPP2 C.S0014-B_1.0, "Enhanced variable rate codec, speech service option 3 and 68 for wideband spread spectrum digital systems," version 1.0, May 2006.
8. RFC 3095 "Robust header compression (ROHC): Framework and four profiles: RTP, UDP, ESP, and uncompressed," July 2001.
9. J. Soriaga, J. Hou, and J. Smee, "Network performance of EV-DO CDMA reverse link with interference cancellation," GLOBECOM '06.
10. Q. Zhang, "Performance of robust header compression for VoIP in 1× EVDO system," GLOBECOM '06.
11. P. Chen, R. Da, C. Mooney, Y. Yang, Q. Zhang, L. Zhu, and J. Zou, "Quality of service support in 1× EV-DO revision A systems," *Bell Labs Technical Journal*, vol. 11, no. 4, pp. 169–184.
12. Q. Bi, P. Chen, Y. Yang, and Q. Zhang, "An analysis of voip service using 1× EV-DO revision A system," *IEEE JSAC*, vol. 24, no. 1, 2006, pp. 36–45.
13. Q. Bi, S. Vitebsky, Y. Yuan, Y. Yang, and Q. Zhang, "Performance and capacity of cellular OFDMA systems with voice over IP traffic," *IEEE Trans. on Vehicular Technology*, in press.

4

SIP and VoIP over Wireless Mesh Networks

Bo Rong, Yi Qian, and Kejie Lu

CONTENTS

In recent times, the wireless mesh network (WMN) has become a popular choice in Internet connectivity offers, since it allows fast, easy, and inexpensive network deployment. In WMNs, it is important to support high quality multimedia service, such as voice over IP (VoIP), in a flexible and intelligent manner. The ultimate goal of VoIP is to deliver reliable, high quality voice service comparable to that provided by traditional circuit-switching networks. However, due to the limitation on bandwidth and the deficiency in control mechanisms, the delivery of VoIP through WMN-accessed Internet often suffers from unpredictable delay/jitter and packet loss. This is because both the Internet and the WMNs were originally designed for data communications.

To address this issue, this chapter studies the session initiation protocol (SIP) for wireless VoIP applications. We especially investigate the technical challenges in WMN VoIP systems, and propose to design an enhanced SIP proxy server to overcome them. An analysis of the signaling process has shown the advantages of our proposed approach.

4.1 Introduction

Providing a WMN is the first step toward providing broadband wireless access with large network coverage [1–3]. Through several years of study, the research community has made significant progress in developing the WMN as a fast, easy, and inexpensive solution for wireless service providers. However, there still exist many challenges in the design of WMNs. One of the most important issues is how to efficiently support wireless VoIP applications, as they are expected to become one of the predominant applications of wireless networks.

Many researchers advocate SIP as a feasible signaling solution for VoIP applications [4,5]. SIP is a signaling protocol for initiating, managing, and terminating voice and video sessions across packet networks. SIP sessions involve one or more participants and can use unicast or multicast communication. In the year 2000, SIP was selected by the Third Generation Partnership Project (3GPP) as the call control protocol for 3G IP-based mobile networks [6]. In this chapter, we address the deployment of SIPs in WMNs in order to support quality of service (QoS)-guaranteed multimedia communication. In particular, we assume that a WMN is connected through a gateway to the IP core network, which employs multi-protocol label switching (MPLS) technology [7].

To deploy SIP in such a network architecture, we have to deal with many new technical challenges that have never been faced in wired networks. These new challenges arise from the inherent combination of wireless infrastructure, user mobility, and heterogeneous network computing [8]. In this chapter, we mainly address the following three issues. First, the interaction between a WMN and the IP core network can increase signaling complexity and cause a long delay in call setup. Second, bandwidth access requirements change from time to time in WMNs, due to the mobility of users and the variation in wireless channel conditions, making it necessary to design a dynamic access bandwidth prediction and reservation scheme. Third, a call admission control (CAC) mechanism has to be implemented, as there may be a distinction between the actual access bandwidth requirements and the predicted/reserved access bandwidth condition.

To overcome these challenges, we have also designed an enhanced SIP proxy server that, using a common open policy service (COPS), reserves some access bandwidth from the IP core network for all the SIP terminals in a WMN. COPS is a protocol that distributes clear-text policy information from a centralized policy decision point to a set of policy enforcement points in the Internet. Moreover, the enhanced SIP proxy server contains two special modules to deal with traffic prediction and CAC problems.

The remainder of this chapter is organized as follows. We first introduce the background of WMNs and SIP-based VoIP. We then discuss the challenges of deploying SIP in WMN and develop an enhanced SIP proxy server to deal with these challenges. Finally, we evaluate the performance of our proposed approach and conclude.

4.2 VoIP in WMNS

4.2.1 Overview of WMNs

WMN is a promising solution for large open areas, both indoors and outdoors, where network cabling does not exist and is costly to install. As the pioneers, the cities of Philadelphia and Tempe are currently planning to install WMNs to provide government employees, residents, and visitors with citywide wireless access to the Internet. Moreover, many universities across the world are considering the deployment of WMNs to provide campus-wide coverage for faculty and students.

Figure 4.1 shows that a WMN consists of two types of nodes: mesh routers and mesh clients. The mesh routers form a backbone infrastructure for mesh clients. In general, mesh routers have minimal mobility and operate exactly like a network of fixed routers, except that they are connected by wireless links through wireless technologies such as IEEE 802.11. We can see from Figure 4.1 that the WMN can access the Internet through a gateway mesh router connected to the IP core network with physical wires.

In this study, we assume that the IP core network employs MPLS technology, which has emerged as the answer to the management challenges of carrier-class IP networks. In MPLS networks, the packets are labeled at the edge of the network and are routed through the network based on these simple labels. This enables them to be explicitly routed and differentiated even as the core routers are kept simple. Although MPLS development was begun with the goal of expediting packet forwarding, the main benefit in the current network environment stems from its traffic-engineering capabilities. The ability of MPLS

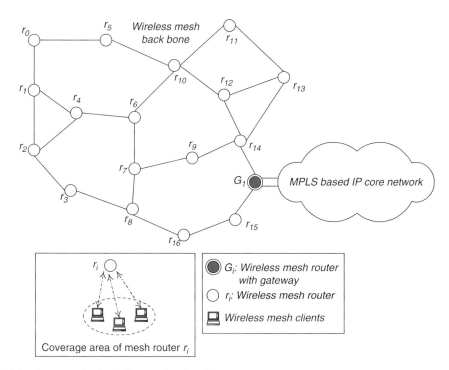

FIGURE 4.1   An example of wireless mesh network.

to provide constraint-routed load-switched paths (LSPs) across MPLS domains enables sophisticated load balancing, QoS, and virtual private network (VPN) deployments over single multipurpose networks. Many researchers recommend MPLS as a reliable way to provide QoS-guaranteed services [7].

In WMNs, every mesh router is equipped with a traffic aggregation device (similar to an 802.11 access point) that interacts with individual mesh clients. The router relays the clients' aggregated data traffic to and from the IP core network. Typically, a mesh router has multiple wireless interfaces to communicate with other mesh routers, and each wireless interface corresponds to one wireless channel. Each of these channels has different characteristics, inherent in the nature of the wireless environment. In practice, wireless interfaces usually run on different frequencies and are built on either the same or different wireless access technologies, such as IEEE 802.11a/b/g/n. It is also possible that directional antennas are employed on some interfaces to establish wireless channels over long distances.

4.2.2 Overview of SIP-based VoIP

VoIP is a technology that enables the transmission of voice over an existing IP network, such as the Internet or a local network. Simply speaking, VoIP is a way of converting the analog signals of the sender to a digital format, transmitting the data across an IP network, and then converting the digital format back into analog for the receiver. Currently, there are two main competing VoIP standards: H.323 and SIP.

H.323 has been developed by the International Telecommunications Union for the transmission of audio, video, and data across IP-based networks, including the Internet [9]. H.323 consists of a series of protocols for signaling, for the exchange of device functionalities and status information, and for the control of dial-up and data flow. The most important protocols of the H.323 standard are H.225, H.245, and H.450.x. H.225 describes signaling protocols like remote access service (RAS) and call signaling, H.245 is the control protocol for multimedia communication, and H.450.x contains additional features, for example, for mapping the capability characteristics between ISDN and IP.

SIP is defined in RFC 2543 as an Internet Engineering Task Force (IETF) standard for multimedia conferencing over IP. It is an ASCII-based, application-layer control protocol that can be used to establish, maintain, and terminate calls between two or more end points. Like other VoIP protocols, SIP is designed to provide the functions of signaling and session management within a packet telephony network. Signaling allows call information to be carried across network boundaries. Session management provides the ability to control the attributes of an end-to-end call.

As compared to H.323, SIP is a much more streamlined protocol, developed specifically for IP telephony [5]. SIP is simpler and more efficient than H.323, and it takes advantage of existing protocols to handle certain parts of the process. For example, the media gateway control protocol (MGCP) is used by SIP to establish a gateway connecting to the PSTN system.

Figure 4.2 shows the deployment of SIP-based VoIP in WMNs. Here, the SIP proxy server is an intermediate device that receives SIP requests from a client and then forwards the requests on their behalf. Basically, proxy servers receive SIP messages and forward them to the next SIP server in the network. Proxy servers can provide functions such as routing, reliable request retransmission, authentication, authorization, and security. Moreover, each WMN is connected to the MPLS-based IP core network through a label edge router (LER) that operates at the edge of an MPLS network, uses routing information to assign labels to datagrams, and then forwards them into the MPLS domain.

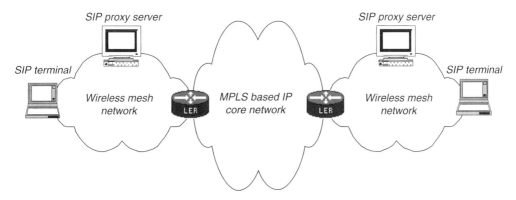

FIGURE 4.2 The deployment of SIP based VoIP in WMNs.

4.3 Technical Challenges of SIP Deployment in WMNs

We now address the issue of how to use SIP to support wireless VoIP in WMN-accessed IP networks. According to the original design, SIP only sets up and tears down sessions, and focuses only minimally on the management of active sessions. To deploy SIP in WMNs, we have to face many new and challenging issues caused by the instability of the wireless environment and the frequent movement of users. Here, we focus on three technical challenges in WMN SIP deployment: call setup delay, access bandwidth prediction/reservation, and CAC.

4.3.1 Call Setup Delay

In the real world, a WMN usually serves as an access network to the Internet. To provide QoS-guaranteed service, IP core networks are built with wired technologies such as MPLS. Moreover, most VoIP applications are intended to go out of their own local WMNs to their counterparts in the Internet. Therefore, when SIP is used to setup a VoIP session, it has to face a heterogeneous network environment, which increases signaling complexity, leading in turn to a long delay in call setup.

Without loss of generality, we study the scenario where a WMN is connected to the MPLS-based IP core network. We assume that the MPLS network has traffic-engineering capability, which is essential to achieving high efficiency. In traffic engineering-enabled MPLS networks, a constraint-based routing label distribution protocol, CR-LDP, or a resource reservation protocol with traffic engineering extensions, RSVP-TE, is employed to set up an LSP dynamically for a connection with QoS requirements. As a result, if an SIP client in a WMN wants to communicate with its counterpart in another through the MPLS-based IP core network, the total delay in setting up the VoIP call will be the sum of the time taken by SIP and MPLS signaling.

4.3.2 Access Bandwidth Prediction and Reservation

When designing a SIP architecture for WMNs, we have to note two facts: (1) users in WMNs are free to move anywhere at any time; (2) wireless channel conditions may vary from time to time. Clearly, these two facts can result in varying access bandwidth require-ments in WMNs.

To accommodate this variation, it is best to let the WMN gateway mesh routers dynamically reserve access bandwidth from the IP core network, since the fixed band-width reservation approach is not efficient in this scenario. For example, there may be two straightforward ways to reserve the fixed access bandwidth for variable requirements. The first is called the optimal user satisfaction scheme, which reserves the maximum band-width that a WMN would ever require. The second is called the optimal cost scheme, which reserves the minimum bandwidth that a WMN would ever require. However, both these methods have their shortcomings. The first is not economical, although it can always provide enough access bandwidth for WMN users, and the second may not satisfy all users, although it is able to reduce WMN operator costs.

Dynamic access bandwidth reservation requires the prediction of outgoing traffic load in WMNs. Since there always exists a distinction between the exact and the predicted access bandwidth requirements, the CAC mechanism must be implemented.

4.3.3 Call Admission Control

The CAC mechanism must be employed when the predicted and reserved access band-width is different from the real bandwidth. The CAC is used to accept or reject connection requests based on the QoS requirements of these connections and the system state information.

In a traditional telephony network, users access the system via the CAC. However, most current IP networks have no admission control and can only offer a best-effect service. In other words, new traffic may continue to enter the network even beyond its capacity. Consequently, both the existing and new flows suffer packet delay and losses. To prevent these, CAC mechanisms should be in place. The CAC mechanism complements the capa-bilities of QoS tools to protect audio/video traffic from the negative effects of other audio/ video traffic and to keep excessive audio/video traffic away from the network. CAC can also help WMNs to provide different types of traffic load with different priorities by manipulating their blocking probabilities.

The most important criterion for an admission controller to admit a new VoIP flow is the availability of the requested resource. Once VoIP calls are accepted, a medium access control (MAC) protocol capable of service differentiation and rate control is required to efficiently coordinate the activities of different types of traffic that coexist in a wireless environment, so that VoIP calls ultimately delivered provide a high QoS. The current fun-damental access method in the IEEE 802.11 MAC protocol is the distributed coordination function [10], which is contention-based and lacks a priority mechanism to guarantee the QoS of VoIP packets. In this mechanism, all stations compete randomly for channel capacity. The work in [11] proposes a rate control mechanism to dynamically adjust the transmission rate of the coexisting best-effort traffic so that it only uses the residual band-width left by the VoIP traffic. VoIP traffic is given the highest priority in the ongoing queue, and it accesses channel capacity before the best-effort traffic. The MAC layer dynamically monitors and updates the free channel capacity, which is then evenly divided among all stations that have best-effort traffic to send. The best-effort traffic will only transmit with the rate given, so that channel capacity is efficiently utilized and VoIP QoS will be guaranteed.

4.4 Enhanced SIP Proxy Servers for Wireless VoIP

4.4.1 Framework of Enhanced SIP Proxy Servers

Conventionally, a proxy server is an optional SIP component that handles routing of SIP signaling but does not initiate SIP messages. Proxy servers can also provide a few auxiliary functions such as authentication, authorization, reliable request retransmission, and security. To overcome the technical challenges in wireless VoIP deployment, we have developed an enhanced SIP proxy server with the framework shown in Figure 4.3.

In particular, the enhanced SIP proxy server utilizes COPS messages to negotiate with the MPLS LER about the overall access bandwidth requirement, on behalf of all SIP terminals in a WMN (not only on behalf of one SIP terminal). Then, the LER exchanges traffic engineering signaling with other routers inside the MPLS core network, thereby setting up the corresponding LSPs before SIP calls are made. As a result, the SIP call setup delay in the MPLS network is decreased significantly.

We use a set of time marks $\{t_0, t_1, t_2, \ldots, t_{n-1}, t_n, t_{n+1}, \ldots\}$ to distinguish the time instances of the system. If the enhanced SIP proxy server knows that the overall access bandwidth requirement of its WMN during time $[t_{n-1}, t_n]$ is $B_{n-1,n}$, then at time t_{n-1}, the enhanced SIP proxy server negotiates with the MPLS network to get $B_{n-1,n}$ outgoing bandwidth using COPS messages. In time, if the overall bandwidth requirement of the WMN during $[t_n, t_{n+1}]$ changes to $B_{n,n+1}$, then at time t_n, the enhanced SIP proxy server should renegotiate with the MPLS network to increase/decrease the bandwidth requirement to $B_{n,n+1}$.

However, it is impossible for the enhanced SIP proxy server to know the exact value of $B_{n-1,n}$ before time t_{n-1}. Usually, the enhanced SIP proxy server can only use a certain bandwidth prediction algorithm to give an approximate value of $B_{n-1,n}$, which can be defined as $B'_{n-1,n}$.

If $B'_{n-1,n} < B_{n-1,n}$ during $[t_{n-1}, t_n]$, the WMN does not have enough outgoing bandwidth to accommodate all SIP calls, and the enhanced SIP proxy server has to utilize a CAC mechanism to decline some of the call requests. In contrast, if $B'_{n-1,n} > B_{n-1,n}$ during $[t_{n-1}, t_n]$, some

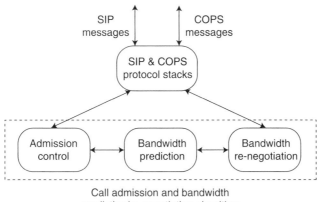

Call admission and bandwidth
prediction/re-negotiation algorithm

FIGURE 4.3 The framework of enhanced SIP proxy server.

outgoing bandwidth resource of the WMN would be wasted. As the preceding discussion shows, the algorithms for access bandwidth prediction and for running the CAC on the enhanced SIP proxy server are critical in our approach.

4.4.2 Access Bandwidth Prediction in Enhanced SIP Proxy Servers

Predicting the behavior of network traffic is always an important issue in designing pro-active management schemes for communication networks. In the literature, extensive studies have been conducted on access bandwidth prediction. Given a history of past values, current prediction approaches construct stochastic models to forecast subsequent time series values. As a result, our proposed enhanced SIP proxy server can directly inherit these research results.

The simplest linear prediction models are the autoregressive (AR) and the autoregressive moving average (ARMA) models [12], which are used in cases of stationary time series. The parameters of these models can be estimated by the least-mean square (LMS) and the recursive-least square (RLS) algorithms [13].

Other than the AR and ARMA models, some researchers suggest employing artificial neural networks for prediction [14]. Studies show that neural networks can naturally account for the practical issues arising in real data: they are often nonlinear, nonstationary, and non-Gaussian.

The authors of [15] have specially developed a multiresolution finite-impulse-response neural-network-based learning algorithm for access bandwidth prediction, using the maximal overlap discrete wavelet transform (MODWT) method. The authors demonstrated that the translation-invariant property of MODWT allows alignment of events in a multiresolution analysis with respect to the original time series and, therefore, preserves the integrity of some transient events. This algorithm is a good choice for the enhanced SIP proxy server, because it has a satisfactory tradeoff between prediction accuracy and computational complexity.

4.4.3 CAC in Enhanced SIP Proxy Servers

The CAC plays an important role in QoS provisioning in terms of the signal quality, call blocking and dropping probabilities, packet delay and loss rate, and transmission rate. CAC in wireless networks has been receiving a great deal of attention during the last two decades due to the growing popularity of wireless communications networking. As pointed out in [16], using CAC can benefit the wireless network in the following areas.

Signal Quality CAC is essential to guarantee the quality of the signal in CDMA networks or other interference-limited wireless networks that have a soft capacity limit. In such cases, a CAC scheme can be used to admit users only if the wireless system is able to provide them with a minimum signal quality.

Call-Dropping Probability In bandwidth-limited wireless networks, CAC is often employed to reduce the handoff failure probability by reserving some resources exclusively for handoff calls, because dropping an active call is usually more annoying than blocking a new call.

Packet-Level Parameters In wireless networks, traffic overloading can cause unacceptable excessive packet delay and/or delay jitter. Therefore, the CAC could be used to limit the traffic load entering the network, in order to guarantee packet-level QoS parameters.

Transmission Rate As in wireline networks, CAC schemes are used in data-service wireless networks to guarantee a minimum transmission rate.

Revenue-Based CAC In general, service providers expect the network to produce the maximal revenue. Considering that different classes of traffic have different revenue rates, CAC can help to achieve this goal by increasing the probability of high revenue rate connections accessing the network. Moreover, if we view revenue rate as the priority factor, then revenue-based CAC can grant differentiated priorities to different users.

Fair Resource-Sharing CAC is an importance measure to guarantee fairness among different users, if the wireless network is required to allocate network resources equally.

In the enhanced SIP proxy server, the CAC is committed to a session layer. More complicated than revenue-based CAC, the CAC policy for an enhanced SIP proxy server can be formulated as an optimization problem, where the demands of both, the WMN service provider and user, are taken into account. To solve this optimization problem, we proposed a utility-constrained optimal revenue policy in [17], which can be easily implemented.

4.5 Analysis of the Signaling Process

Figure 4.4 shows the signaling flow in the proposed SIP architecture containing the enhanced SIP proxy server. The call setup starts with a standard SIP INVITE message sent from the caller to the local enhanced SIP proxy server in a WMN. This message carries the call receiver's URL in the SIP header, and the QoS requirements of the SIP call in the session description protocol (SDP) body.

After identifying the caller ID, QoS requirements, and remaining outgoing bandwidth in the local WMN, the enhanced SIP proxy server decides whether to admit the SIP call request. If the call request is admitted, the enhanced SIP proxy server will forward the original INVITE message to the callee; otherwise, it simply sends the caller a DECLINE message to drop the call.

Furthermore, regardless of whether the call is admitted or not, it would be registered in the enhanced SIP proxy server for the purposes of access bandwidth prediction and CAC in the future. If the access bandwidth has to change, the enhanced SIP proxy server uses COPS messages to negotiate with MPLS-based IP core networks in order to set up new LSPs with the required bandwidth. The COPS protocol is part of the IP suite as defined by the IETF's RFC 2748. COPS specifies a simple client–server model for supporting policy control over QoS signaling protocols (e.g., RSVP). As shown in Figure 4.4, the enhanced SIP proxy server uses COPS REQ and COPS DEC messages to negotiate on-demand bandwidth with the MPLS core network.

These discussions clearly show that the enhanced SIP proxy server approach can considerably reduce call setup delay, because the LSPs needed by SIP telephony are set up in MPLS-based IP core network even before SIP calls begin.

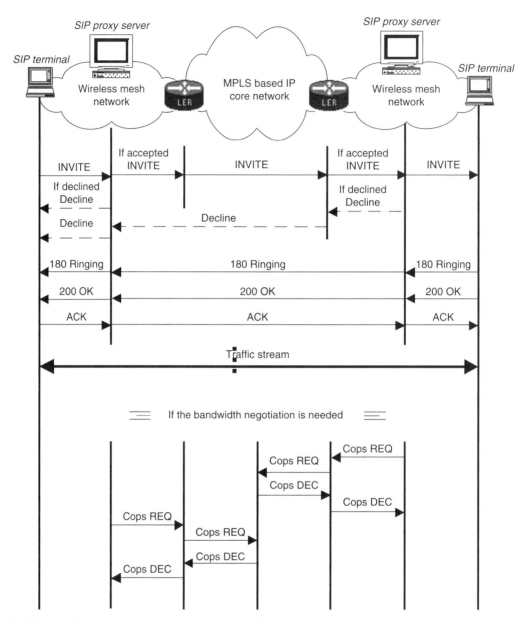

FIGURE 4.4   The signaling flow with the enhanced SIP proxy server.

4.6 Conclusion

One of the key capabilities of next-generation wireless networks is to carry VoIP and support seamless wireless data and voice communications. In this chapter, we investigated the deployment of SIP-based VoIP in WMNs and discussed the technical challenges in wireless VoIP systems, such as call setup delay, access bandwidth prediction and reservation, CAC, and so on. In general, these technical issues arise from the instability of the wireless environment and the frequent movement of users.

To overcome these problems in wireless VoIP systems, we proposed the use of a novel enhanced SIP proxy server that utilizes COPS messages to negotiate the overall access bandwidth requirement with the IP core network on behalf of all SIP terminals in a WMN. Moreover, an extensive analysis of the signaling exchange between the enhanced SIP proxy server and the MPLS-based IP core network was done in order to demonstrate the advantages of our proposed approach.

References

1. I. F. Akyildiz and X. Wang, "A survey on wireless mesh networks," *IEEE Communications Magazine*, vol. 43, no. 9, September 2005, pp. S23–S30.
2. A. Raniwala and T. Chiueh, "Architecture and algorithms for an IEEE 802.11-based multi-channel wireless mesh network," *IEEE INFOCOM 2005*, vol. 3, March 2005, pp. 2223–2234.
3. H. Jiang, W. Zhuang, X. Shen, A. Abdrabou, and P. Wang, "Differentiated services for wireless mesh backbone," *IEEE Communications Magazine*, vol. 44, no. 7, July 2006, pp. 113–119.
4. J. Rosenberg, H. Schulzrinne, G. Camarillo, J. Peterson, R. Sparks, M. Handley, and E. Schooler, "SIP: Session initiation protocol," *IETF*, RFC 3261, June 2002.
5. U. Black, *Voice Over IP*. Prentice-Hall, New Jersey, 2000.
6. 3GPP, "Technical specification group services and system aspects: Network architecture (Release 5)," Technical Report TS23.002, 3GPP, March 2002.
7. T. Li, "MPLS and the evolving Internet architecture," *IEEE Communications Magazine*, vol. 37, no. 12, December 1999, pp. 38–41.
8. E. A. Wan, "Finite impulse response neural networks with applications in time series prediction," PhD dissertation, Stanford University, 1993.
9. ITU-T Recommendation H.323, "Infrastructure of audiovisual services—systems and terminal equipment for audiovisual services," Series H: Audiovisual and multimedia systems, 1999.
10. IEEE, "Part 11: Wireless medium access control (MAC) and physical layer (PHY) specifications," IEEE Standard 802.11, 1999.
11. H. Zhai, X. Chen, and Y. Fang, "A call admission and rate control scheme for multimedia support over IEEE 802.11 wireless LANs," *Proceedings of First International Conference on Quality of Service in Heterogeneous Wired/Wireless Networks* (QShine), 2004.
12. G. E. P. Box and G. M. Jenkins, *Time Series Analysis: Forecasting and Control.* San Francisco, CA, Holden-Day, 1976.
13. S. Haykin, *Adaptive Filter Theory, 3rd ed.*, Prentice-Hall, Upper Saddle River, NJ, 1996.
14. A. S. Weigend and N. A. Gershenfeld, *Time Series Prediction: Forecasting the Future and Understanding the Past,* Addison-Wesley, Reading, MA, 1994.
15. V. Alarcon-Aquino and J. A. Barria, "Multiresolution FIR neural-network based learning algorithm applied to network traffic prediction," *IEEE Transactions on Systems, Man and Cybernetics, Part C: Applications and Reviews*, vol. 36, no. 2, March 2006, pp. 208–220.
16. M. H. Ahmed, "Call admission control in wireless networks: a comprehensive survey," *IEEE Communications Surveys & Tutorials*, vol. 7, no. 1, First Qtr. 2005, pp. 49–68.
17. Bo Rong, Yi Qian, and Kejie Lu, "Integrated downlink resource management for multiservice WiMAX networks," *IEEE Transactions on Mobile Computing*, vol. 6, no. 6, June 2007, pp. 621–632.

Part II

Technologies

5

Compression Techniques for VoIP Transport over Wireless Interfaces

Yang Yang and Xin Wang

CONTENTS

In general, one of the major challenges for wireless communication is the capacity of wireless channels, which is especially limited when a small delay bound is imposed, for example, for voice service. A Voice over Internet Protocol (VoIP) packet, on the other hand, has large overheads compared with circuit-switched voice. VoIP signaling packets are also typically large, which in turn could cause a long signaling delay when transmitted over wireless networks. In order to use the wireless channel capacity efficiently and make VoIP services economically feasible, it is necessary to apply compression techniques to reduce the overheads in the VoIP bearer and signaling packets. In this chapter, we discuss the following compression techniques for VoIP: robust header compression (RoHC) for bearer paths and signaling compression (SigComp) for signaling paths. RoHC and SigComp are the most favored compression techniques for VoIP in wireless communication today.

5.1 Robust Header Compression

The main specification for RoHC is contained in RFC 3095 [1], which was published by the Internet Engineering Task Force (IETF) in July 2001. It has since been clarified in RFC 3759 [2] and RFC 4815 [3] in April 2004 and February 2007, respectively. In this section, the material is based primarily on the above RFCs. In particular, we focus here on the following three aspects:

- Why is RoHC required for VoIP over wireless?
- How is compression efficiency achieved?
- How is robustness achieved?

RoHC has different profiles that target different application flows such as profile for uncompressed IP flow, Profile for User Datagram Protocol (UDP)/IP compression, and so on. In this section, we focus on RoHC for a VoIP application, and therefore the Real-time Transport Protocol (RTP)/UDP/IP compression or RoHC profile 1, is the main subject of discussion.

5.1.1 Motivation

To make VoIP over wireless an economically viable solution, a number of technical challenges must be addressed. One of the most critical is the bandwidth efficiency. For mobile wireless communications, bandwidth remains a scarce resource and it is of vital importance that it be used efficiently. If today's circuit-switched cellular system can support X number of users per cell per carrier, when converted to VoIP, the system must be able to support more than X number of VoIP users; otherwise, the spectrum and deployment costs would be prohibitive.

Having said that bandwidth efficiency is critical for VoIP over wireless, one immediate problem is the large VoIP header overheads. Speech data for IP telephony is typically carried by the RTP protocol. A VoIP packet will have an IP header of 20 octets, a UDP header of 8 octets, and an RTP header of 12 octets. This makes a total of 40 octets for the RTP/UDP/IP header. If IPv6 is used, the IP header alone will be 40 octets and the total RTP/UDP/IP header will be 60 octets. This is in addition to the air interface protocol overhead as well as a possible link layer framing overhead. On the other hand, the size of the speech payload is typically as small as 2 to about 20 octets. As a result, the header overhead ratio is enormous. It is thus imperative to compress the RTP/UDP/IP header in order to utilize the wireless spectrum efficiently.

There have been a number of header-compression techniques prior to RoHC, such as the one specified in RFC 2508 [4]. One of the main differences between RoHC and the other header-compression schemes is robustness, which is critical for maintaining the VoIP service quality. Mobile wireless communications typically operate at high packet error rates in order to achieve spectral efficiency. The packet error rate over wireless channels can be as high as a few percentage points. The loss of a packet must not cause extensive decompression errors or failures in subsequent packets as this could lead to severe impairment of the speech quality.

Another characteristic of the mobile wireless channel is the long round-trip time (RTT), which can be as high as 100–200 ms. In RFC 2508, the header compression scheme uses signaling messages from the decompressor to the compressor to indicate that the context is out-of-sync. The RTT of the wireless link limits the speed of this context-repair mechanism. Each packet lost over the link causes several subsequent packets also to be lost since the context is out-of-sync during at least one link RTT. In VoIP, such bursts of packet drops can cause a perceivable muting of speech and result in voice quality degradations.

The RTP/UDP/IP header compression, therefore, must be both efficient and robust. We now describe in more detail how this can be achieved with RoHC.

5.1.2 Achieving Efficiency While Preserving Robustness

5.1.2.1 Characteristics of Header Fields

In the packet flow of RTP/UDP/IP headers, significant correlations can be found both for the same header field across different packets and for some different header fields within the same packet. By sending static field information only initially, and utilizing dependencies and predictability for other fields, the header size of most packets can be significantly reduced.

Figure 5.1 shows that most header fields never, or seldom, change, so that they can be easily compressed. Only 5 fields, or about 10 octets in total, need more elaborate mechanisms. These fields are:

- IPv4 Identification (IP-ID), 16 bits
- UDP Checksum, 16 bits
- RTP Marker (M-bit), 1 bit
- RTP Sequence Number (SN), 16 bits
- RTP Timestamp (TS), 32 bits

Typically, the SN field increments by one for each packet sent by the RTP source. The TS and IP-ID fields can usually be predicted from the SN. The M-bit is usually the same. The UDP Checksum should not be predicted and is sent as-is when enabled.

Given these characteristics of the header fields, RoHC compression first establishes functions from the SN to the TS and IP-ID fields. Once functions are established, the compressor need only reliably communicate the SN. If a function from the SN to another field changes, additional information is sent to update the parameters of that function.

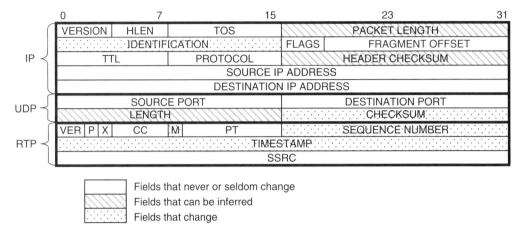

FIGURE 5.1   RTP/UDP/IP header fields.

Next, we describe how SNs are compressed while preserving robustness. If the TS and IP-ID fields need to be updated, for example when establishing the functions initially or when the functions change, the same compression approach is followed.

5.1.2.2 Basic Compression Approach

As mentioned earlier, the packet drop rate in wireless communications is not trivial. In addition, there is the possibility that packets can become out of order. Therefore, although the SN field typically increments by one in successive packets at the RTP source, packet drop and reordering can alter such a pattern at the RoHC compressor or decompressor. The compression approach must therefore communicate the field successfully to the decompressor even when such a pattern is broken. This calls for reliability in addition to compression efficiency.

The basic approach that achieves this is least significant bits (LSB) encoding. With LSB encoding, the k least significant bits of the field value are transmitted instead of the original field value, where k is a positive integer. After receiving k bits, the decompressor derives the original value using a previously received value as a reference (v_ref).

The idea is that even if the wireless link drops or reorders packets, the value to be compressed should have a high likelihood of remaining in the vicinity of the reference value (v_ref) at the decompressor. The parameter k represents the range within which the value to be compressed might fall. The wider the range, the larger the k needed, and the less the compression efficiency. This range is called the interpretation interval (for the k LSB). The interpretation interval can be expressed as $[v\_ref - p, v\_ref + (2^k - 1) - p]$, where p is an integer. It can be seen that by specifying the k LSB, a unique value in the interpretation interval is identified. If the parameters k and p are selected properly, this unique value will be the original field value. Some examples are shown below.

Example 1

With $k = 4$, $p = -1$ and $v\_ref = 0 \times 001f$, the compressor sends bits 1111 for field value $0 \times 002f$. The decompressor finds $0 \times 002f$ to be the unique value in the interpretation interval $[0 \times 0020, 0 \times 002f]$, thus decompressing the field correctly. Note that the decompressor can decompress correctly even if packets containing field values $0 \times 0020, 0 \times 0021, . . ., 0 \times 002e$ are lost. Hence damage propagation is prevented and robustness achieved.

Example 2

With $k = 4$, $p = 6$ and $v\_ref = 0 \times 001f$, suppose packets become disordered and a packet with a field value 0×0019 is received at the decompressor after packet $0 \times 001f$. The decompressor can successfully decompress the bits 1001 as 0×0019, since the interpretation interval is now $[0 \times 0019, 0 \times 0028]$.

From these examples, it can also be seen that with a given k value, p provides a tradeoff between the capability to handle out-of-order packets and the capability to handle packet loss. The value p is fixed for the compression of a particular field. However, different fields may use different p values depending on specific needs. The p value is known to both compressor and decompressor, and is not explicitly communicated.

The value k is dynamically selected at the compressor for each compressed field value. It is implicitly communicated from compressor to decompressor through Packet Type, which is described later. The choice of the k value should be large enough that the field value is highly likely to fall into the interpretation interval at the decompressor, and it should be as small as possible to maintain efficiency.

In practice, the compressor maintains a sliding window, which contains the possible values that the decompressor may use as v_ref. The exact value of v_ref cannot be determined by the compressor due to possible packet drops, reordering, and damage in the communication link between the compressor and decompressor. The compressor then calculates the k value such that no matter which v_ref in the sliding window the decompressor uses, the original value is covered by the resulting interpretation interval, thus guaranteeing successful decompression. By using feedback and/or by making reasonable assumptions based on the characteristics of the communication link, the compressor can limit the size of the sliding window. Compression efficiency is thereby achieved and robustness preserved.

This compression approach is referred to as Window-based LSB encoding, or W-LSB encoding. In Example 1, assuming the compressor uses a sliding window of $[0 \times 001f, 0 \times 0028]$ when compressing the field value $0 \times 002f$, the compressor will determine that $k = 4$ is sufficient and necessary to ensure that the value $0 \times 002f$ is covered by the interpretation interval at the decompressor, as long as the reference value v_ref used by the decompressor is in the compressor's sliding window.

5.1.2.3 Special Treatments of the TS Field

In VoIP, the RTP TS field typically exhibits certain properties, using which the TS field can be compressed more efficiently.

First, from packet to packet in a VoIP flow, the increments of the TS field are usually in multiples of a certain unit. Such a unit is denoted as TS_STRIDE. For example, the sampling rate for voice is typically 8 kHz and one voice frame is 20 ms long. Furthermore, each voice frame is often carried in one RTP packet. In this case, the TS field always increments in multiples of 160. Hence, here we have TS_STRIDE = 160. Note that silence periods have no impact on this characteristic, as the sample clock at the source normally keeps running during silence periods without changing either the frame rate or the frame boundaries.

So for RTP TS field compression, the TS is usually first downscaled by a factor of TS_STRIDE before compression. The resulting scaled TS (TS_SCALED) and the original TS have the following relation:

$$TS = TS\_SCALED * TS\_STRIDE + TS\_OFFSET$$

The parameter TS_STRIDE is explicitly communicated from the compressor to the decompressor. The TS_OFFSET is implicitly communicated to the decompressor by sending several unscaled TS values. After the TS_STRIDE and TS_OFFSET are known to the decompressor and the relation given above is established, the compressor need only send TS_SCALED from then on.

Next, we describe in some detail timer-based compression of the RTP TS. This is very important to maintain RoHC compression efficiency when silence suppression is used.

With silence suppression, RTP packets are not sent, or are sent only occasionally, by the RTP source during silence periods. As a result, at the beginning of a new speech spurt following a silence period, the TS field value of the RTP packet could jump significantly, compared with that of the previously transmitted RTP packet. The TS field value increment depends on the duration of the silence interval or on when the last update was sent. Note that the SN field only increments by one at the RTP source. Hence, TS values must be sent to the decompressor to update the function from the SN to the TS field. The problem here is that a large increase in TS value will require a large k value when using the W-LSB encoding to compress the TS value. By using certain special properties of RTP TS fields, timer-based compression is able to compress the TS field using a small k value, thus improving compression efficiency.

At the RTP source of a VoIP flow, the RTP TS closely follows a linear function of the time when the RTP packet is generated. At the RoHC decompressor, where the compressed packet is received, the RTP TS is approximately a linear function of the time when the packet is received. Clearly, this linear function at the RoHC decompressor will have an offset compared with the linear function at the RTP source, due to the delay from the RTP source to the RoHC decompressor. More importantly, at the RoHC decompressor, there will be more deviations from the linear function because of the delay jitter between the RTP source and the RoHC decompressor. However, during normal operation, the delay jitter is bounded in order to meet the real-time conversational requirements of the VoIP service. Hence, by using a local clock at the RoHC decompressor, the decompressor can obtain an approximation of the TS value from the packet's arrival time. The approximation can then be refined with the k LSBs of the (scaled) TS transmitted by the RoHC compressor. The value of k required to ensure correct decompression is a function of the jitter between the RTP source and the RoHC decompressor.

For the RoHC compressor, the main task associated with timer-based compression of the RTP TS is to determine the value of k. The compressor can estimate the jitter introduced prior to the compressor by measuring the packet arrival times at the compressor and comparing them against the RTP TS. Then, if the compressor knows the potential jitter that could be introduced between compressor and decompressor (it typically makes some assumptions on the jitter, based on the link characteristics between compressor and decompressor), it will be able to derive the jitter all the way from the RTP source to the RoHC decompressor and thereby determine the k value required to ensure a successful decompression.

TIME_STRIDE is defined as the time interval equivalent to one TS_STRIDE. For VoIP, with a fixed sampling rate of 8 kHz, 20 ms in the time domain is equivalent to an increment of 160 in the unscaled TS domain. Therefore, if TS_STRIDE = 160, the corresponding TIME_STRIDE is 20 ms. TIME_STRIDE is explicitly communicated from the RoHC compressor to the decompressor. When the compressor is confident that the decompressor has received the TIME_STRIDE value, it can switch to timer-based compression.

5.1.3 RoHC Packet Formats

Depending on which fields have to be updated, how many bits are required, and which mode RoHC is operating in (U, O, or R modes, which we describe in more detail in later

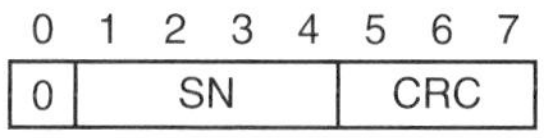

```
0  1  2  3  4  5  6  7
[0|   SN   |  CRC  ]
```

FIGURE 5.2 UO-0 packet format.

sections), there are many packet formats for RoHC. RoHC can compress the 40-byte RTP/ UDP/IP header down to even 1 byte.

RoHC uses three packet types to identify compressed headers. Since the format of a compressed packet also depends on the mode, packet format is expressed in the form:

⟨modes format is used in⟩ − ⟨packet type number⟩ − ⟨some property⟩.

For example, packet format UO-1-TS indicates this format can be used in U or O mode, is of packet type 1, and provides update on TS field.

Next, we briefly describe the packet types and show some examples of packet formats.

5.1.3.1 Packet Type 0

Packet type 0 includes three packet formats: R-0, R-0-CRC, and UO-0. This packet type is used when the decompressor knows all the SN-function parameters, and the header to be compressed adheres to these functions. Thus, only the W-LSB encoded RTP SN have to be communicated. Among the three packet types, packet type 0 provides the least amount of update, and is the most efficient. Figure 5.2 shows what the UO-0 packet format looks like.

With RoHC operating in U or O mode, the first bit "0" of a compressed header indicates that this is a UO-0 packet. The SN field is W-LSB encoded to 4 bits. The compressed header also contains a 3-bit cyclic redundancy check (CRC). The CRC is computed at the RoHC compressor based on the original header, which allows the decompressor to check (with high probability) whether the decompressed header is correct. Hence, the CRC helps prevent context damage at the RoHC decompressor.

In the example of the UO-0, the RoHC decompressor first decompresses the SN field based on the reference value, or in other words, the context, of the SN, as well as the received 4-bit W-LSB encoded SN value. Then, the decompressor derives the other field values such as TS and IP-ID from the decompressed SN value based on the established functions. When decompression is complete, the decompressor must compute a 3-bit CRC over the reconstructed header, and compare it against the CRC sent by the compressor. If the CRC matches, the decompression is successful. The decompressed packet is delivered to the application, and also the context of the SN updated for decompression of subsequent packets. If the CRC does not match, the decompressor may attempt a context repair mechanism. If it still fails, the decompressed packet is discarded and the context of the SN not updated.

5.1.3.2 Packet Type 1

Packet type 1 includes the following packet formats: R-1, R-1-ID, R-1-TS, UO-1, UO-1-ID, and UO-1-TS. This packet type is used when the number of bits required for the SN update exceeds those available in packet type 0, or when the parameters of the SN-functions for TS or IP-ID field change. Figure 5.3 shows a UO-1-TS packet.

With RoHC operating in U or O mode, the first three bits "101" of a compressed header indicate that it is a UO-1-TS packet. The SN field is W-LSB encoded to 4 bits, and the TS field is W-LSB encoded to 5 bits. The RTP Marker (M bit) is also transmitted, and it takes 1 bit. Similarly to UO-0, the UO-1-TS packet format uses a 3-bit CRC to protect against context damage.

```
 0  1  2  3  4  5  6  7
┌──┬──┬──┬──────────────┐
│1 │0 │1 │      TS      │
├──┼──┴──┴────┬─────────┤
│M │   SN     │   CRC   │
└──┴──────────┴─────────┘
```

FIGURE 5.3   UO-1-TS packet format.

If the UO-1-TS packet is decompressed successfully, as verified by CRC, the contexts of both SN and TS are updated at the decompressor.

When timer-based compression of TS is employed, packet type 0 cannot be used since TS must be updated to address the delay jitter between the RTP source and the decompressor. Hence, packets of packet type 1, such as the UO-1-TS packets, are often used at the beginning of talk spurts, after the silence periods. After the function for the TS field is successfully updated at the decompressor, no further update is required, and the compressor can then switch to a more efficient packet format, such as UO-0 of packet type 0.

5.1.3.3 Packet Type 2

Packet type 2 includes packet formats UOR-2, UOR-2-ID, and UOR-2-TS. This packet type can be used to change the parameters of any SN function, or when the number of bits required for the update exceeds those available in packet type 0 or packet type 1. The decompressor updates its context for the fields updated through packet type 2.

Packet type 2 is the least efficient of the three packet types for compressed headers. However, it is occasionally required to establish or update the SN function; for example, when the full 32-bit TS value has to be sent, and sometimes for mode transitions, to be described later.

In addition to the three packet types for compressed headers, there are two for initialization/refresh. They are the initialization and refresh (IR) packet, and initialization and refresh-dynamic part (IR-DYN) packet. The IR packet communicates the static part of the context, and the IR-DYN packet communicates the dynamic part.

5.1.4 RoHC Operation

As mentioned earlier, RoHC retains the relevant information from past packets in a context, and uses it to compress/decompress subsequent packets. The compressor and decompressor update their contexts upon certain events. Communication impairment may lead to inconsistencies between the contexts of the compressor and decompressor, which in turn may cause incorrect decompression. As a robust header compression scheme, RoHC provides mechanisms for avoiding context inconsistencies and for making the contexts consistent when they are out-of-sync.

First, RoHC operates in one of three modes: unidirectional (U-mode), bidirectional optimistic (O-mode), and bidirectional reliable (R-mode). Both compressor and decompressor must be in the same mode.

Under each mode, RoHC can operate in a number of states, which determines what packet type can be sent and received by the compressor and decompressor, respectively. Under different modes, the state transitions, as well as the actions that can be performed in each state, are different.

Next, we describe in more detail the three RoHC operation modes, and how to transition among them. Using the O-mode as an example, we then describe the states under the O-mode, as well as when state transitions occur.

5.1.4.1 ROHC Operation Modes

In the U-mode, packets are sent in only one direction: from the compressor to the decompressor. Due to lack of feedback, the compressor has no knowledge of whether packets have been successfully received at the decompressor. Therefore, to avoid propagation of context damage, periodic refreshing of the context is needed in the U-mode. This is the least efficient of the three operation modes, often used only in the initial stage, as compression must start with it.

In the O-mode, a feedback channel is used to send error recovery requests from the decompressor back to the compressor. The O-mode aims at maximizing compression efficiency with sparse use of the feedback channel. As a result, the frequency of context invalidation may be higher than for the R-mode, especially when bursts of errors occur.

The R-mode is similar to the O-mode, but it makes a more intensive use of the feedback channel. It aims at maximizing robustness against damage propagation, that is, it minimizes the probability of context invalidation, even under bursty error conditions.

The optimal mode in which to operate depends on the characteristics of the communication channel, such as whether a feedback channel is available, the packet error rate, burstiness, and the like.

There are primarily three types of feedback to the compressor: ACK, NACK, and STATIC-NACK.

- ACK acknowledges successful decompression of a packet, which indicates to the compressor that the context at the decompressor is very likely in-sync.
- NACK indicates that the dynamic context of the decompressor is out-of-sync. NACK is generated when several successive packets have failed to decompress correctly.
- STATIC-NACK indicates that the static context of the decompressor is not valid or has not been established.

The feedback packets not only provide error recovery requests, but can also carry a mode parameter indicating the desired compression mode: U, O, or R. This is important for mode transitions, which we see in the next section.

5.1.4.2 RoHC Mode Transitions

The RoHC compressor and decompressor always start from the U mode, and can transition to other modes depending on what the other end supports. The decision to move from one compression mode to another is taken by the decompressor. The possible mode transitions are shown in Figure 5.4, where Feedback (X) denotes a feedback packet with the desired compression mode set to X, with $X \in$ {U, O, R}.

Next, we show two specific mode transition examples, to give an idea how mode transition is done. Mode transitions must be smooth: that is, the decompressor must be able to decompress packets sent by the compressor during the transition period. RoHC achieves this by imposing restrictions on the compressor and decompressor during the transition phase.

Transition from U- to O-modes

When a feedback channel is available, the decompressor may at any moment decide to initiate transition from the U- to the O-modes. Any feedback packet carrying a CRC and

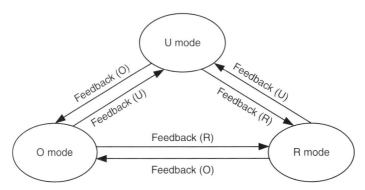

FIGURE 5.4   RoHC mode transitions.

with the desired mode set to O can be used to initiate the transition. CRC is essential for the feedback packet, as it prevents false transition at the compressor in case the feedback packet is erroneously received at the compressor. The decompressor can then directly start working in O-mode, because all packets generated in U-mode can be decompressed in O-mode (such as UO-0, UO-1, UO-1-ID, UO-1-TS, UOR-2, UOR-2-ID, UOR-2-TS, IR, and IR-DYN). The compressor transits from the U- to the O-mode as soon as it receives a feedback packet that has the desired mode set to O and that passes the CRC check. The transition is then complete.

Transition from O- to R-modes

Transition from O- to R-modes is permitted only after at least one packet has been correctly decompressed, which means that at least the static part of the context is established. The decompressor sends an ACK or NACK feedback packet carrying a CRC, and with the desired mode set to R to initiate the mode transition. The compressor must not use packet types 0 or 1 during the transition period. For example, if the compressor sends UO-0, UO-1, UO-1-ID, or UO-1-TS packets and the decompressor has already transited into R mode, it will not be able to decompress these packets correctly. In other words, if the decompressor is still receiving packets of type 0 or 1 after initiating the transition, the decompressor cannot transition into R mode.

When the compressor receives the ACK or NACK packet that has the desired mode set to R and that passes the CRC check, it can transit into the R-mode. However, there are restrictions on what packet types the compressor can send at this point, because the decompressor is still in the O-mode. The only packets generated in the R-mode but that can be decompressed in O-mode are UOR-2, IR, or IR-DYN packets. These packets can carry a mode parameter indicating the mode in which they have been generated. In this case, the compressor sets the mode parameter to R, which indicates to the decompressor that the compressor has already transitioned into the R-mode. The decompressor, upon receiving a UOR-2, IR, or IR-DYN packet with its mode parameter set to R, transits into the R-mode. Meanwhile, the compressor continues to send the UOR-2, IR, or IR-DYN packets until it is sure that the decompressor has transited into the R-mode, the indication of which is an ACK from the decompressor for a UOR-2, IR, or IR-DYN packet that was sent with its mode parameter set to R. After the compressor receives such an ACK, the mode transition is considered complete at the compressor side, and the compressor can start sending type 0 or type 1 packets (such as R-0 and R-1). When the decompressor receives packets of type 0 or 1, the mode transition is considered complete at the decompressor side.

It can be seen during mode transitions that larger packets such as UOR-2, IR, or IR-DYN packets are sometimes needed. The impact on compression efficiency should be quite small, as mode transitions do not happen often during the life of a traffic flow. These packets help to achieve robustness, which is critical for mode transitions.

In the next section, we take O-mode as an example and describe its operations, including what packet types can be sent or received in each state, and when state transitions should occur. The state machines at the compressor and decompressor sides help ensure that the contexts at the two sides are in-sync and that packets can be decompressed successfully.

5.1.4.3 Operations in O-Mode

As mentioned earlier, the O-mode provides good compression efficiency as well as robustness through sparse use of feedback when the packet error rate is not very high. In the following, we describe in more detail the operations in the O-mode, and see how this is achieved.

Compressor states and logic

At the compressor, there are three states: initialization and refresh (IR), first order (FO), and second order (SO). The compressor starts in the lowest compression state, IR, and moves gradually to the higher, more efficient, states. The compressor always operates in the highest possible compression state, under the constraint that it is sufficiently confident that the decompressor has the information necessary to decompress headers compressed according to that state. Figure 5.5 shows the state machine for the compressor in the O-mode.

- **The IR state** In the IR state, the compressor initializes, or refreshes after failure, the static part of the context at the decompressor. The compressor sends the complete header information, including both the static and dynamic fields, in uncompressed form. The compressor stays in the IR state until it is fairly confident that the decompressor has received the static information successfully.

- **The FO state** The purpose of the FO state is to communicate efficiently the irregularities in the packet stream. When operating in this state, the compressor rarely sends information about all dynamic fields, and the information sent is usually at least partially compressed. Only a few static fields can be updated.

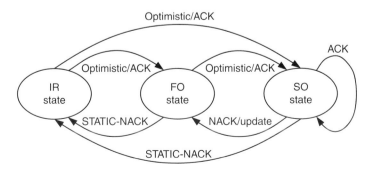

FIGURE 5.5 Compressor state machine in O-mode.

The compressor enters this state from the IR state, and also from the SO state, whenever the headers of the packet stream do not conform to their previous pattern. It stays in the FO state until it is confident that the decompressor has acquired all the parameters of the new pattern. Changes in fields that are always irregular (such as UDP checksums) are communicated in all packets and are therefore part of what is considered a uniform pattern.

All packets sent in the FO state carry context updating information. It is very important to detect corruption of such packets to avoid erroneous updates and context inconsistencies. Strong CRC protection, as available in UOR-2, IR, or IR-DYN packets, meets this requirement.

- **The SO state** Compression is most efficient in the SO state. The compressor enters the SO state when the header to be compressed is completely predictable from the RTP SN, and it is sufficiently confident that the decompressor has acquired all the parameters of the functions from SN to other fields. As long as the decompressor can correctly decompress SN, it will be able to successfully reconstruct the packet headers sent in the SO state.

  The compressor returns to the FO state from SO state when the header no longer conforms to the uniform pattern.

- **Upward transition, optimistic approach** Transition to a higher compression state in the O-mode can be carried out according to an optimistic approach principle. This means the compressor transits to a higher compression state when it is fairly confident that the decompressor has received enough information to correctly decompress the packets sent according to the higher compression state. The compressor normally obtains its confidence by sending several packets with the same information in the lower compression state. If the decompressor receives any of these packets, it will be in-sync with the compressor.

- **Upward transition, optional acknowledgments** Alternatively, positive feedback (ACKs) may be used for UOR-2 packets in the O-mode. Upon reception of an ACK for an updating packet, the compressor knows that the decompressor has received the acknowledged packet and the transition to a higher compression state can be carried out immediately. Hence, if the optional acknowledgment approach is used, it generally provides better compression efficiency and higher robustness compared with the optimistic approach.

- **Downward transition, need for updates** The compressor immediately transits back to the FO state when the header to be compressed does not conform to the established pattern, thus requiring updates to be sent for the parameters.

- **Downward transition, negative acknowledgments (NACKs)** Upon receipt of a NACK, the compressor transits back to the FO state and sends updates (UOR-2, IR-DYN, or possibly IR) to the decompressor. NACKs carry the SN of the last successfully decompressed packet, and this information can help the compressor to determine the fields that require updating.

  Similarly, reception of a STATIC-NACK packet makes the compressor transit back to the IR state.

- **Choosing packet formats** The compressor chooses the smallest possible packet format that can communicate the desired changes and has the required number of bits for W-LSB encoded values.

Decompressor states and logic

The decompressor also has three states: no context (NC), static context (SC), and full context (FC). The decompressor starts in its lowest state, NC, and gradually transits to higher states. The decompressor state machine normally never leaves the FC state once it enters. The decompressor never attempts to decompress headers in the NC or SC state unless sufficient information is included in the packet itself (e.g., IR packet for NC state, and IR, IR-DYN, or type 2 packet for SC state).

Initially, when the decompressor is in an NC state, the decompressor would not yet have successfully decompressed a packet. Once a packet has been decompressed correctly (for example, upon reception of an IR packet), the decompressor can transit all the way to the FC state. The decompressor stays in the FC state unless there are repeated decompression failures, which will make the decompressor transit back to lower decompression states. When that happens, the decompressor first transits back to the SC state. There, successful reception of any packet sent in the FO state from the compressor is usually sufficient to enable transition to the FC state again. Only when decompression of several packets sent in the FO state continually fails in the SC state, will the decompressor return all the way to the NC state.

As mentioned earlier, the decompressor uses feedback to request context updates, and it is essential for the state transitions of the compressor under the O mode. In the following, we describe when feedback is sent by the decompressor in the O mode. The feedback logic depends on the state in which the decompressor is. Figure 5.6 shows the state machine for the decompressor in O-mode.

If the decompressor is in an NC state, under each of the conditions given here, it acts as follows:

- when an IR packet passes the CRC check, the decompressor will send an ACK;
- when receiving a type 0, 1, 2, or IR-DYN packet, or when an IR packet fails the CRC check, the decompressor may send a STATIC-NACK.

If the decompressor is in SC state, under each of the conditions given here, it acts as follows:

- when an IR packet passes the CRC check, the decompressor will send an ACK;
- when a type 2 or IR-DYN packet is successfully decompressed, the decompressor may optionally send an ACK;

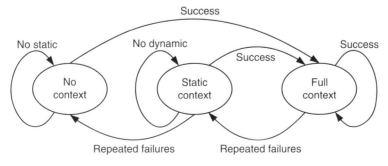

FIGURE 5.6 Decompressor state machine in O-mode.

- when a type 0 or type 1 packet is received, the decompressor will treat it as a failed CRC and may send a NACK;
- when a type 2, IR-DYN, or IR packet fails the CRC check, the decompressor may send a STATIC-NACK.

If the decompressor is in FC state, then under each of the conditions given here, it acts as follows:

- when an IR packet passes the CRC check, the decompressor will send an ACK;
- when a type 2 or IR-DYN packet is successfully decompressed, the decompressor may optionally send an ACK;
- when a type 0 or type 1 packet is successfully decompressed, no feedback is sent;
- when any packet fails the CRC check, the decompressor may send a NACK.

5.1.5 Summary

With RoHC compression, the RTP/UDP/IP header of VoIP packets can usually be compressed down to 1 byte (with packet type 0) or 2 bytes (with packet type 1, used especially at the beginning of talk spurts, with timer-based compression of RTP TS), assuming the UDP checksum is not enabled. With the UDP checksum enabled, 2 more bytes are added to the compressed header.

A few packet drops in the communication link do not necessarily cause context damage and it is highly likely that the RoHC decompressor can still successfully decompress the subsequent packet. In the rare cases when extensive bursts of packet drops happen or when an erroneous packet is delivered to the RoHC decompressor due to undetected errors at lower layers, decompression failure or context damage may occur. However, RoHC is robust enough to prevent damage propagation, and the RoHC state machine is able to bring the contexts at the compressor and the decompressor back into sync.

5.2 Signaling Compression

5.2.1 Motivation

VoIP applications often use session initiation protocol (SIP) [5] as the signaling protocol. SIP is a text-based protocol, and typical SIP messages are rather large, usually in the range of a few hundred bytes to two thousand bytes or more. For example, typical SIP INVITE messages are more than 600 bytes. When transmitting large SIP messages over wireless channels, not only are precious wireless channel resources consumed, but more importantly, long message transmission delays are induced, which can adversely impact the call setup latency performance and other signaling-related performance metrics. Hence, it is highly desirable to compress the signaling messages before sending them over the wireless channel.

Signaling compression was specified in IETF RFC 3320 [6] in January 2003, and was further clarified in RFC 4896 [7] in June 2007. SigComp provides a method to compress messages generated by application protocols such as SIP. Different compression algorithms can be used in SigComp, such as LZ77 [8] and DEFLATE [9]. SigComp standardizes the decompressor, the so-called universal decompressor virtual machine (UDVM), which can be programmed to understand the output of many well-known compressors.

5.2.2 Compression Algorithms

To choose a compression algorithm for SigComp, one should consider the desired compression ratio, processing and memory requirements, code size, implementation complexity, and intellectual property (IPR) issues [10]. Typical algorithms include the following:

- LZ77
- LZSS
- LZW
- DEFLATE
- LZJH

For example, LZ77 is a dictionary-based method. It tries to compress data by encoding phrases as small tokens that reference entries in the dictionary. LZ77 uses a sliding window to maintain the dictionary. The dictionary is a finite sliding window containing previously seen text. The compressor searches for the longest matching string up to a certain look-ahead buffer length. Based on the search results, the compressor sends the next character or phrase as a token (match_offset, match_length, next_char). The dictionary is updated by sliding the window to include the most-recently encoded phrases.

Another example is DEFLATE, a combination of LZ77 and Huffman coding. First, the data is encoded using LZ77, either as a literal character or as an (offset, length) pair. Then, the output elements (character, offset, length) from LZ77 are encoded using the Huffman coding.

The compression usually achieves its maximum efficiency after a few message exchanges have taken place. This is because the first message the compressor sends to the decompressor is only partially compressed, as there is no previous stored state to reference against. As one of the main goals of SigComp is to reduce the call setup time as much as possible, it is desirable to employ a mechanism to improve the compression efficiency from the first message.

RFC 3485 [11] defines a static dictionary for SIP and the session description protocol (SDP). The static dictionary is a collection of well-known strings that appear in most SIP and SDP messages. It constitutes a SigComp state that can be referenced in the first SIP message that the compressor sends out.

In addition to the static dictionary, a user-defined dictionary can also further improve compression efficiency. SigComp compressors can include the user-specific dictionary as a part of the initial messages to the decompressor, even before any time-critical signaling messages are generated from a particular application. This will improve the compression efficiency once the messages start to flow.

RFC 3221 [12] defines several mechanisms to improve the compression efficiency compared with per-message compression. These include dynamic and shared compression. Dynamic compression uses information from previously sent messages. To make sure these messages have been received by the decompressor, an explicit acknowledgement mechanism is used together with dynamic compression. For shared compression, the compressor makes use of received messages to improve compression efficiency.

5.2.3 Robustness Consideration

In addition to encoding the application messages using the chosen algorithm, the SigComp compressor is also responsible for ensuring that messages can be correctly decompressed even if packets are lost or misordered during transmission. The SigComp feedback mechanism can be used to acknowledge successful decompression at the remote endpoint. The following robustness techniques are usually employed in SigComp [10]:

- Acknowledgements using the SigComp feedback mechanism
- Static dictionary
- CRC checksum
- Announcing additional resources
- Shared compression

5.2.4 Universal Decompressor Virtual Machine

Decompression functionality for SigComp is provided by the UDVM. The UDVM is a virtual machine like the Java virtual machine, but with a key difference: it is designed solely for the purpose of running decompression algorithms.

The motivation for creating the UDVM is to provide flexibility in choosing a method to compress a given application message. Rather than choosing from a small number of pre-negotiated algorithms, compressor implementers have the freedom to select an algorithm of their choice. The compressed data is then combined with a set of UDVM instructions that allow the original data to be extracted, and the result is outputted as a SigComp message. As the UDVM is optimized specifically for running decompression algorithms, the code size of a typical algorithm is small (often less than 100 bytes). Moreover, the UDVM approach does not add significant extra processing or memory requirements compared with running a fixed preprogrammed decompression algorithm.

5.2.5 Summary

SigComp provides a compression mechanism for signaling protocols of VoIP applications, such as SIP. The compression efficiency depends on the compression algorithm used. With a typical SigComp implementation, SIP messages can often be compressed down to less than 50%, and sometimes even less than 10%, of the original message size. This can greatly reduce the message transmission delay over wireless channels and significantly improve critical VoIP performance metrics such as call setup delay.

References

1. C. Bormann, Burmeister, M. Degermark et al., "Robust header compression (ROHC): Framework and four profiles: RTP, UDP, ESP, and uncompressed," RFC 3095, July 2001.
2. L.-E. Jonsson, "Robust header compression (ROHC): Terminology and channel mapping examples," RFC 3759, April 2004.
3. L.-E. Jonsson, K. Sandlund, G. Pelletier et al., "Robust header compression (ROHC): Corrections and clarifications to RFC 3095," RFC 4815, February 2007.

4. S. Casner and V. Jacobson, "Compressing IP/UDP/RTP headers for low-speed serial links," RFC 2508, February 1999.
5. J. Rosenberg, H. Schulzrinne, G. Camarillo et al., "SIP: session initiation protocol," RFC 3261, June 2002.
6. R. Price, C. Bormann, J. Christoffersson et al., "Signaling compression (SigComp)," RFC 3320, January 2003.
7. A. Surtees, M. West, and A.B. Roach, "Signaling compression (SigComp) corrections and clarifications," RFC 4896, June 2007.
8. J. Ziv and A. Lempel, "A universal algorithm for sequential data compression," *IEEE*, vol. 23, pp. 1977, 337–343.
9. P. Deutsch, "DEFLATE compressed data format specification, version 1.3," RFC 1951, May 1996.
10. A. Surtees and M. West, "Signaling compression (SigComp) users' guide," RFC 4464, May 2006.
11. M. Garcia-Martin, C. Bormann, J. Ott et al., "The session initiation protocol (SIP) and session description protocol (SDP) static dictionary for signaling compression (SigComp)," RFC 3485, February 2003.
12. H. Hannu, J. Christoffersson, S. Forsgren et al., "Signaling compression (SigComp)—Extended operations," RFC 3321, January 2003.

6

QoS Monitoring of Voice-over-IP Services

Swapna S. Gokhale and Jijun Lu

CONTENTS

6.1 Introduction and Motivation

Voice over IP (VoIP) is an attractive choice for voice transmittal compared to traditional circuit-switching technology for the following reasons: lower equipment cost, integration of voice and data applications, lower bandwidth requirements, widespread availability of IP, and the promise of novel, value-added services [1]. In future, VoIP services will be expected to operate seamlessly over a converged network referred to as a Next Generation Network (NGN), comprising a combination of heterogeneous network infrastructures, including packet-switched, circuit-Switched, wireless, and wireline networks.

In order for VoIP services to be widely adopted, they must offer as good a quality of service (QoS) as the current Public Switched Telephone Network (PSTN) services. As a result, similarly to PSTN services, VoIP services must also be designed and optimized to offer superior QoS. Furthermore, during operation, QoS must comply with the service level agreements between the provider and the subscribers. In this chapter, we use the term QoS to refer to the actual quality that the end user of a service experiences. This notion of QoS is different from the traditional notions that define and measure QoS in terms of the performance parameters of the packet traffic flowing through the network. Examples of such parameters include delay, loss, and jitter. User-defined QoS, on the other hand, is concerned with metrics such as the response time, reliability, and availability of a service.

Despite the design and provisioning, a user may still experience poor QoS degradation caused by (i) higher than anticipated load and (ii) failure or weakening of the allocated resources. In addition to the these benign factors, intentional malicious attacks by adversaries may also threaten the QoS of VoIP services. Unlike the traditional PSTN, where the service logic resided in the trusted perimeters of the service provider, VoIP services reside in a more open environment and rely on network infrastructures involving standard protocols such as Transmission Control Protocol (TCP)/Internet Protocol (IP). As a result, these services are exposed to many more vulnerabilities that could potentially be exploited to threaten their quality. Thus, monitoring must also detect such anomalous or suspicious behavior in a timely manner to mitigate its impact.

In this chapter a QoS monitoring methodology for VoIP services is described. We consider the signaling performance of a VoIP session* as an indicator of its user-perceived QoS. The methodology relies on service-level data collected using open, standard Application Programming Interfaces (APIs) that were originally intended to facilitate rapid service creation across heterogeneous network infrastructures. We then discuss how the service-level data are aggregated and analyzed to monitor the signaling performance of VoIP sessions. We also describe how the data can be used to detect anomalous behavior. We discuss our experience in the experimental demonstration of the methodology using an example of an open, standard API environment, namely, Java APIs for Integrated Networks (JAIN) APIs [2] along with some insightful results. The architecture of a prototype monitoring system which encapsulates the methodology is also described. The methodology promotes *dual use* of open, standard APIs, one for rapid service creation and two for QoS monitoring. Also, by virtue of using open, standard APIs, the methodology can be used uniformly across heterogeneous network infrastructures and services provided by multiple vendors.

The layout of this chapter is as follows: Section 6.2 provides an overview of open, standard APIs. Section 6.3 describes the monitoring methodology. Section 6.4 discusses

---

* The terms session and call are used interchangeably in this chapter.

the experimental demonstration of the methodology and the prototype architecture. Section 6.5 summarizes related work. Section 6.6 offers conclusions and directions for future research.

6.2 Open, Standard APIs

Creating a VoIP service that operates seamlessly across heterogeneous networks often requires an understanding of not only the desired service capabilities but also of the different protocols, data formats, end-user devices, and capabilities of the different types of networks on which the service is to be offered. Ideally, the service developer should only focus on the essential characteristics of the service and not be distracted by the complexities of the underlying heterogeneous networks. An API can help service developers deal with some of these complexities by enabling the application provider to make requests of the underlying network. If an API is open, it can then enable the reuse of components and subsystems. If an API is adopted as a standard, it can enable interoperability between systems and applications.

The APIs for services offered over NGNs can be roughly grouped into three categories: resource APIs, network capabilities APIs, and external APIs [3]. The QoS monitoring methodology is based on resource APIs which are described in this section. The resource APIs provide access to low-level communication protocols and hide implementation details and message transmission formats. Applications developed using such APIs can be deployed across multiple networks and in multi-vendor environments. These APIs are termed as resource APIs as they control the network and special resources such as switches, media gateways, and messaging stores that adhere to a specific protocol. From the point of VoIP services, resource APIs would offer open and standardized interfaces to protocols used in VoIP including Session Initiation Protocol (SIP) [4], Megaco [5], and H.323 [6].

The APIs provide a layer of abstraction between network and application. They follow a provider/listener or an event-driven software architecture model [7]. An event model enables components to notify a state change to other components. Such a component is referred to as an event source, a producer, or a provider, whereas components that are interested in learning of the state change are referred to as event sinks, consumers, or listeners. The listeners can subscribe to the event sources to receive notification of state changes in either the synchronous or the asynchronous mode. In the synchronous mode, the event source or the provider suspends any further processing after notifying the listener, and waits for the listener to react to the event. On the other hand, in the asynchronous mode, the event source notifies the listener of the occurrence of an event, and continues with the normal processing. Typically, only one listener can subscribe to receive an event in the synchronous mode, whereas multiple listeners can subscribe in the asynchronous mode.

The platform implementing the API serves as the provider and informs the applications of the relevant events taking place in the network through the API. The applications interpret these events and take necessary actions to provide the services. In addition, an application may also initiate actions using the API that the platform will translate into appropriate protocol signaling messages delivered to the network. It is the job of the platform to interface to the underlying network and translate API methods and events to and from underlying signaling protocols as it sees fit.

6.3 QoS Monitoring Methodology

The QoS monitoring methodology comprises three activities: data collection, data aggregation, and data analysis. Each activity is described in separate subsections.

6.3.1 Data Collection

This methodology is based on the collection of service-level data, which consists of events raised by the platform implementing the open, standard APIs to inform the applications which provide services of the relevant happenings in the network. These events indicate important milestones that occur while providing a service and hence represent the progression of the service. Since service-level data are concerned with the experience of the user when the service is provided and not on how the service is provided, it should be possible to collect this data in an implementation-independent manner. The use of open, standard APIs to collect service-level data allows us to fulfill this objective easily.

To capture the events raised by the platform, Event Collection Applications (ECAs) are written. ECAs register with the platform to receive these events. The ECAs receive event notifications in an asynchronous mode and hence the platform will merely inform the ECAs of the event occurrences and continue processing. The ECAs will thus receive and record the stream of events raised by the platform through the APIs.

6.3.2 Data Aggregation

Service-level data may be aggregated in two ways before analyzing it for QoS monitoring. In the first aggregation, frequencies of event occurrences for different event types may be generated. These frequencies can provide an indication of the usage distribution of all the services across all the users, distribution of the use of services for a single user, and a group of users. In the second aggregation, events generated from a single session may be aggregated to infer the state machine representing the service offered in the session.

6.3.3 Data Analysis

The aggregated data are used to monitor the signaling performance and detect anomalous behavior.

6.3.3.1 Performance Monitoring

The signaling performance of a VoIP service is obtained using the second type of aggregation; by consolidating the service-level events raised from a single session. Thus, a performance estimate for specific operating conditions and service load may be obtained. Using this aggregation, performance may be monitored during operation. In addition, this technique may also be used to obtain a performance estimate prior to deployment of a service. Correlating the performance with the resource consumption can also provide guidance for provisioning prior to deployment.

6.3.3.2 Anomaly Detection

Statistical anomaly detection seeks to detect departures from the baseline or normal behavior of a service. It is therefore necessary to first establish the baseline or normal behavior.

On the basis of service-level data, baseline behavior may be defined at various levels of granularity, such as, the entire platform (for all services), for each service, for a collection of related services, a single user, and a group of users. For the baseline behavior of the overall platform, event counts for all services and users may be used. For the behavior of each service, event counts for that service may be used. Finally, for the behavior of a user or a group of users, events from individual sessions may be used. For example, aggregation of the events from sessions originating from user A may reveal important characteristics such as the average call duration, and most frequently called numbers which could be the attributes of user A's baseline behavior. Similarly, the average number of times a particular service is requested in an interval may form the baseline for that service.

A preliminary baseline behavior maybe established prior to the deployment of a service. This can be adapted continuously as additional event data becomes available after deployment. The event data collected and aggregated during operation will be used to detect anomalies from the baseline. Different techniques such as threshold and profile-based detection [8], machine learning [9], neural networks [10], and rule-based expert systems [11] may be used for detection.

6.4 Experimental Demonstration

In this section we discuss our experience in the experimental demonstration of the methodology and also provide some insightful results.

6.4.1 Experimental Test Bed

Many initiatives are concerned with the definition and the development of open, standard APIs to facilitate rapid service creation across heterogeneous networks. These include JAIN™ [12,13],* OSA/PARLAY [13], and 3GPP [14]. We chose JAIN APIs for the experimental demonstration for the following reasons: First, TSAS, 3GPP, and PARLAY define examples of external APIs and not resource APIs. JAIN on the other hand defines a rich set of integrated resource, and network capabilities APIs for both trusted and untrusted network access. Second, reference implementations of the JAIN APIs are available. Third, the use of Java™,† language enables us to exploit the most important benefit, namely, portability across different execution platforms. Fourth, since the JAIN architecture uses Java technology, in the future, this will also allow us to benefit from many existing and ongoing development efforts in the Java programming language space. A brief overview of JAIN APIs is presented in this section.

JAIN is a community of companies led by Sun Microsystems that develops standard, open, published Java APIs for NGNs consisting of integrated Internet Protocol (IP) or Asynchronous Transfer Mode (ATM), PSTN, and wireless networks [12]. The protocol layer in JAIN is based on Java standardization of specific protocols [15] (e.g., SIP [4], Media Gateway Control Protocol (MGCP) [16,17], H.323 [6], TCAP [18–20], ISUP [21], INAP/AIN [22], MAP [23], etc.). JAIN protocol APIs are examples of resource APIs, and these APIs

---

* JAIN is a trademark of Sun Microsystems.
† Java is a trademark of Sun Microsystems.

enable "service portability" by reshaping proprietary interfaces into uniform Java inter-faces. The JAIN layer translates the proprietary primitives into standard JAIN events and informs the application of the occurrence of these events through the use of the API. In addition, the application will be able to initiate actions using the API that the JAIN API layer will translate into vendor-specific proprietary primitives. Applications that are developed using resource APIs will have an underlying call model or a state machine, which will be specific to the service that is being provided by the application.

JAIN APIs have been defined for the different protocols used for VoIP including SIP [24] and H.323 [6]. Initially, we chose to use JAIN SIP APIs for the following reasons: Com-pared with H.323 [6], SIP is a more flexible solution, simpler and easier to implement, better suited to support intelligent user devices, as well as for the implementation of advanced features. Many believe that SIP, in conjunction with the MGCP [16], will be the dominant VoIP signaling architecture in the future [1].

The experimental infrastructure consists of public implementation of the JAIN SIP API available from NIST [25]. It contains Reference Implementation (RI), TCK, examples, some basic tools for JAIN-SIP-1.1 (JSR-32 maintenance release), and an SDP library that conforms to the public release of JSR 141 (JAIN-SDP) interfaces [25]. JAIN-SIP RI is a full implementa-tion of the SIP specification as given in RFC 3261 [4].

Session initiation protocol is a peer-to-peer protocol, with the peers called as user agents (UAs). The typical message flow between two UAs in SIP shown in Figure 6.1 was used in our experimentation. In Figure 6.1, UA1 initiates the session by sending an INVITE request to UA2. UA2 is alerted (i.e., the phone is ringing) and an interim response, "180 Ringing," is sent back to UA1. Subsequently, UA2 answers the phone, which generates an OK response back to UA1. UA1 acknowledges this response by sending an ACK message to UA2. After this INVITE/200/ACK three-way handshake [4], the session is established and the two UAs begin to exchange data. At the end of the data exchange, UA2 sends a BYE request to UA1. Upon receiving this request, UA1 sends a "200 OK" response back to UA2 and the session is closed bidirectionally. The signaling or the call set up delay for setting up a session among the two UAs was used as the performance metric.

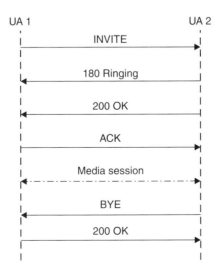

FIGURE 6.1 Typical SIP message flow.

The hardware platform hosting the infrastructure comprises two machines with the following configurations: The first one is a Dell OptiPlex GX260 (Intel Pentium 4 processor at 2.4 GHz, 1 GB of RAM, 40 GB hard driver and Intel PRO 1000 MT network adapter) and the other is an IBM ThinkPad T40 (Intel Pentium-M processor at 1.5 GHz, 512 MB of RAM, 40 GB hard driver and Intel PRO 100 VE network adaptor). Both computers are installed with Windows XP professional SP1 and are connected via a 100 M Ethernet connection across a LAN. Each machine hosts a UA.

6.4.2 Experimental Scenarios

The experimental scenarios require multiple simultaneous sessions to be sustained among the two UAs, which is achieved as follows: Our previous research, which studied the impact of intermediate proxies on the signaling delay [26], indicated that the signaling performance was not influenced by whether UA1 or UA2 initiated and terminated the call. As a result, we assume, without loss of generality, that UA1 initiates calls to UA2. UA1 accepts the following parameters as input: (i) distribution and the parameters of the interarrival time of the calls and (ii) distribution and the parameters of the call-holding time. Using the interarrival time distribution, UA1 generates a series of interarrival times $\{t_i\}$, $i = 1, 2, \ldots$ and computes the arrival time of each call S_i as $S_i = \sum_{j=1}^{i} t_j$. Using the information on call-holding time distribution, UA1 also generates the holding time of each call d_i, and computes the ending time of each call as $T_i = S_i + d_i$. At each arrival time S_i, UA1 automatically initiates a call to UA2 by sending an INVITE request and the call is established after a bi-directional three-way handshake. The transport protocol used is UDP. For messages that have a SDP body (e.g., INVITE request), the size of the different messages ranges from 450 bytes to 650 bytes. For messages without a SDP body (e.g., ACK request), the size ranges from 250 bytes to 400 bytes. This session is held for a duration d_i. At time T_i, which is the ending time of this session, UA1 sends a BYE request to UA2, which causes the call to be terminated. A test media is exchanged between the two UAs during the session. The test media is a QuickTime video clip (available from [27]), 3546 KB in size and 85 seconds in duration (audio property: 8000 sample rate, 8 bits per sample; video property: 160 * 120 frame size, 7.5 fps). It was sent repeatedly in the established session.

In our experiments, the interarrival time and the call-holding duration were assumed to follow an exponential distribution. Each experimental scenario is characterized by the number of simultaneous ongoing sessions among the two UAs, which is a function of the mean call-holding duration $\bar{d}_i$ and mean interarrival time $\bar{t}_i$. Initially, at the beginning of an experiment, there is a warm-up or a ramp-up period, when the number of ongoing calls between the two UAs continue to rise. Eventually, after a period of continuous increase, the rate of arrivals and departures of the sessions causes the system to reach a quasi-steady state where the number of simultaneous ongoing sessions is approximately constant with minor fluctuations. The signaling delay for the calls set up during ramp-up is likely to be lower than the delay of the calls set up after reaching the quasi-steady state. As a result, in order to avoid biasing the results, only the signaling delay measurements for the calls set up in the quasi-steady state are used to compute the signaling delay statistics.

6.4.3 Data Analysis

In this section we present the results of performance analysis and statistical anomaly detection based on service-level data.

TABLE 6.1

Average Number of Calls in Each Scenario

Interarrival Time ($\bar{T}_i$)	Mean Call Holding Duration ($\bar{d}_i$)						
	30	60	120	180	240	300	360
30	1	2	4	6	8	10	12
15	2	4	8	12	16	20	24
10	3	6	12	18	24	30	36
6	5	10	20	30	40	50	60
4	8	15	30	45	60	75	90

6.4.3.1 Performance Analysis

The objective of performance analysis is to estimate the signaling delay of a session between two UAs for different levels of load. Different levels of loads were emulated by changing the number of simultaneous ongoing sessions in the following manner. In the first set of experiments the mean interarrival time $\bar{T}_i$ was held constant, and the mean call-holding duration $\bar{d}_i$ was increased. In the second set $\bar{d}_i$ was held constant, and $\bar{T}_i$ was decreased. Increase in the mean call-holding duration or/and a decrease in the mean inter-arrival time increases the number of simultaneous ongoing sessions which results in an increase in the workload on the end-user device.

As indicated in Section 6.4.2, the system reaches a quasi-steady state after an initial ramp-up period. In Table 6.1 we summarize the average number of calls in the quasi-steady state for each combination of the mean interarrival time and call-holding duration, obtained as the ratio of the former to the latter. For each combination, once the system reaches a quasi-steady state, the signaling performance of more than 50 calls was measured and used to compute the average signaling delay.

Figures 6.2 and 6.3, respectively, illustrate the effect of the mean call-holding duration and interarrival time on the call set up delay. Figure 6.2 indicates that for a fixed mean interarrival time, the signaling delay increases as the mean call-holding duration increases because of the increase in the workload. On the other hand, when the mean call-holding duration is fixed, the call set up time decreases as the mean interarrival time increases because of a lighter workload as indicated in Figure 6.3.

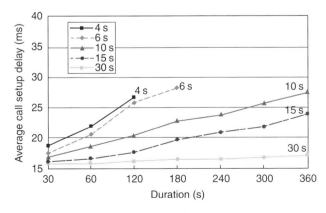

FIGURE 6.2 Signaling delay as a function of call-holding duration.

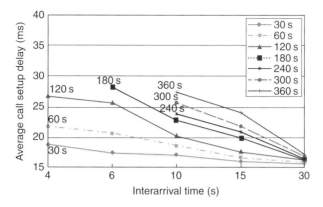

FIGURE 6.3 Signaling delay as a function of interarrival time.

Referring to Table 6.1, when the mean interarrival time is less than 6 seconds, some combinations of $\bar{T}_i$ and $\bar{d}_i$ did not yield any valuable results, because the machines were unable to sustain the number of simultaneous ongoing sessions that resulted in these low mean interarrival times. The hardware configuration used in our experiments was able to sustain less than 40 simultaneous calls without any problems. Hence, the signaling delay results were obtained only when the average number of calls in the quasi-steady state was less than 40. Figure 6.4 depicts the CPU utilization for the machine which hosts UA1 for each combination. From Figure 6.4, it can be observed that the CPU usage increases rapidly as the number of simultaneous calls increase, and reaches close to 100% at 18 calls in the quasi-steady state.

Although the general trend in the signaling delay as a function of the workload (and a function of the mean interarrival and call holding times) is intuitive and expected, such results obtained from the QoS monitoring methodology can be used to obtain a quantitative estimate of the signaling delay for a given level of load prior to deployment.

6.4.3.2 Anomaly Detection

To facilitate anomaly detection, we establish the baseline or the normal behavior in terms of the signaling performance of an incoming session request. The baseline signaling delay

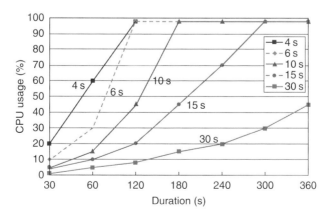

FIGURE 6.4 CPU usage of the machine with UA1.

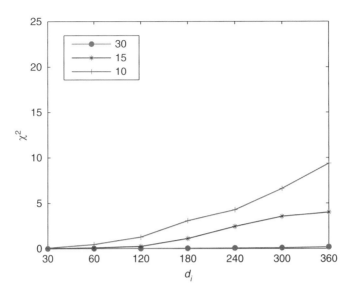

FIGURE 6.5   Anomaly detection using χ^2 test statistic.

was obtained by setting each of the mean interarrival time $\bar{t}_i$ and the mean call-holding duration $\bar{d}_i$ to 30 seconds. This combination refers to the first cell in the topmost row in Table 6.1, with only a single session in the quasi-steady state. For this combination the average signaling delay of 50 measurements was 15.82 ms.

The two UAs were then subject to abnormal operating conditions by increasing the number of simultaneous ongoing sessions between them. Abnormal operating conditions in this case thus constitute a Denial-of-Service (DoS) type attack. It is expected that the degree of abnormality will be determined by the level of increase in the number of simultaneous sessions. This number can be increased in two ways: (i) reducing the mean interarrival time $\bar{t}_i$, or/and (ii) increasing the mean call-holding duration $\bar{d}_i$. Abnormal conditions were emulated by considering different combinations of $\bar{t}_i$ and $\bar{d}_i$ and are depicted in all the cells other than the first one in the topmost row in Table 6.1. For each abnormal condition, signaling performance was measured for over 50 sessions after the system reached a quasi-steady state and the average signaling delay was computed.

The average signaling delay obtained for each combination was analyzed using the χ^2 test [28], with a baseline signaling delay of 15.82 ms to detect anomalous behavior. The values of the χ^2 statistic reported in Figure 6.5 indicate that for a given $\bar{d}_i$, the χ^2 test statistic increases as $\bar{t}_i$ decreases. Similarly, for a given $\bar{t}_i$, the χ^2 test statistic increases with $\bar{d}_i$. By placing an appropriate empirical threshold on the value of the χ^2 statistic, abnormal conditions can be flagged. Thus, the χ^2 statistic can be used effectively for anomaly detection based on signaling performance.

6.4.4 Prototype Architecture

The prototype monitoring system follows a loosely coupled, distributed system architecture [29] as shown in Figure 6.6. The system constitutes ECAs to collect the events raised by different resource APIs. These events are aggregated by aggregation applications and are then posted on a "whiteboard." The performance analysis and anomaly detection engines obtain these events from the whiteboard according to their requirements. Several

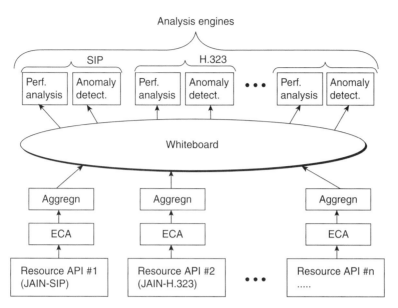

FIGURE 6.6   Architecture of the prototype QoS monitoring system.

technologies including JavaSpaces [30] and Java Messaging Service [31] may be used to implement the whiteboard. In the monitoring system, the ECA and aggregation application for each resource API are tightly coupled. This eliminates the need to transport huge volumes of raw data over the network. However, the loose coupling between aggregation applications and the analysis engines promotes incremental development, adding one analysis engine at a time. Further, it also promotes flexibility in that analysis engines to conduct different types of analysis in addition to performance monitoring and anomaly detection may be added at a later point. Also, collection, aggregation, and analysis of other resource APIs can be incorporated into the system readily in the future. In its current form, the system has the capability to collect, aggregate, and analyze the events generated by JAIN SIP APIs. In the future, similar capability for JAIN H.232 APIs will be added.

6.5 Related Research

In this section we summarize the related research and place our work in context.

6.5.1 VoIP Performance

Voice over Internet Protocol performance research can be roughly categorized into two groups: (i) performance of the voice packet traffic including delay, loss, and jitter and (ii) performance of the signaling or call set up.

Zheng et al. [32] study the delay and delay jitter at the IP packet level and its effect on VoIP by using a transient queuing solution. Mase et al. [33,34] propose enhanced end-to-end measurement-based admission control mechanisms for VoIP networks. Bilhaj et al. [35] present QoS control enhanced architecture for VoIP networks, which uses probe flow delay and average loss rate measurement systems. Awoniyi et al. [36] investigate the quality of voice communication over IEEE 802.11a WLANs taking realistic channel models into account. Muppala et al. [37] study the performance of VoIP traffic aggregates over DiffServ-enabled networks. Nguyen et al. [38] report performance results of laboratory experiments evaluating VoIP over satellite links under different link and traffic conditions. Furuya et al. [39] investigate the relationship between IP network performance criteria and voice quality for VoIP service experimentally. Wang et al. [40] propose and investigate a scheme that can improve the VoIP capacity in WLAN. Al-Najjar and Reddy [41] show that when VoIP calls are aggregated over a network link or path, the provider can employ a suitable linear-time encoding for the aggregated voice traffic, resulting in considerable quality improvement with little redundancy. Amir et al. [42] employ two protocols for localized packet loss recovery and rapid rerouting in the event of network failures in VoIP. Sulaiman et al. [43] evaluate the performance of VoIP in a 3G network. They also discuss voice quality in the presence of compression used to reduce the bandwidth requirements. Portoles-Comeras et al. [44] extend previous work on the impact of the number of hops on the quality of VoIP applications running over wireless multihop networks.

Compared to the performance of voice packet traffic, signaling performance has received relatively little attention. However, given that there are stringent requirements on the call set up delay of the PSTN, VoIP is expected to meet these requirements. E.721 [45] recommends an average delay of no more than 3.0, 5.0, or 8.0 seconds, respectively, for local, toll, and international calls. Practical set up delays in today's PSTN networks are much better than these standards suggest, as readers may attest from experience. AT&T [46], for example, claims that their average set up delay for the domestic network is less than 2.0 seconds.

Based on the ITU Q.725 [47] targets for the PSTN, Lin et al. [48] propose call set up delay targets for VoIP. Since the call set up time for PSTN has been reduced to a mere 1–2 seconds for toll calls, VoIP networks are expected to achieve the same level of delay. Furthermore, set up delays in the range of 2.5–5 seconds are acceptable for international calls. Eyers et al. [49] present a simulation study which targets the call set up delay based on UDP delay/loss traces for SIP [45] and H.323 [6] protocols. Kist et al. [50] investigate the call set up delay in 3GPP for possible sources and focus on the delays caused by the Domain Name System (DNS) and message propagation. Das et al. [51] analyze the performance of the H.323 call set up procedure over a wireless link using analytical techniques. De et al. [52] discuss the effects of different QoS mechanisms on the performance of two signaling protocols in VoIP (SIP and H.323). Wu et al. [53] study the signaling performance of SIP-T system using a queuing model. Fathi et al. [54] propose optimizing SIP session setup delay using an adaptive retransmission timer. They also evaluate SIP session set up performances with various underlying protocols such as TCP, UDP, and RLPs as a function of the FER.

Most of the given efforts address VoIP signaling performance in the context of long-distance calls routed over public wide area networks. Currently, however, VoIP is not really a viable option for such calls due to some outstanding issues with reliability and sound quality arising from the limitations in both Internet bandwidth and current compression technology [55]. As a result, most corporations today confine VoIP applications to their intranets. With more predictable bandwidth availability than the public Internet, intranets can support full-duplex, real-time voice communications [55].

When using VoIP across corporate intranets, issues such as the workload of the user devices (due to the number of simultaneous ongoing sessions and/or other applications that may be running), number of proxies that may be used to route the call and the message processing time at the user devices as well as at the proxies will have an impact on the call set up delay. When VoIP is used over the public Internet, the impact of these factors may be relatively insignificant due to the dominant issue of limited bandwidth. These factors, however, may become significant when VoIP is used across corporate LANs. The results obtained using the QoS monitoring methodology can allow an assessment of the impact of the workload on end-user devices on the signaling performance.

Corporate VoIP is also likely to be offered across heterogeneous network infrastructures. Corporations may also use implementations from different vendors simultaneously to provide VoIP services to retain bargaining power by avoiding tying down to a single vendor's solution. As a result, it is necessary to be able to monitor the signaling performance of VoIP sessions independent of the underlying networking technology as well as the implementation from a specific vendor. The use of open, standard APIs as advocated in the QoS monitoring methodology achieves the objective of platform independence and vendor neutrality.

6.5.2 VoIP Security

As the popularity of VoIP increases, the potential attacks targeted against VoIP applications will likely occur as attackers become more familiar with the technology through exposure and easy access [56]. Existing research efforts in VoIP security address one or more of the following three issues: (i) prevention and detection of the interception and modification of voice packets during transport, (ii) prevention of intrusions and attacks, and (iii) detection of intrusions and attacks.

A number of techniques such as IPSec can be used to ensure that the voice packets are not intercepted and modified during transit. For example, TLS [57] and IPsec [58] are two popular alternatives for providing security at the transport and network layer to guarantee message confidentiality and integrity for SIP. Megaco also uses IPsec as the security mechanism to prevent unauthorized protocol connections [5]. Typically, however, since these techniques incur additional processing, their impact on voice quality needs to be considered carefully. Guo et al. [59] propose a technique with a new hierarchical data security protection scheme, HDSP, which can maintain the voice quality degraded from packet loss and preserve high data security. Barbieri et al. [60] present the results of an experimental analysis of the transmission of voice over secure communication links implementing IPsec. Geneiatakis et al. [61] describe malformed message attacks against SIP elements and propose a new detection "framework" of prototyped attacks' signatures to detect and provide effective defense against this class. Martin and Hung [62] discuss a security policy from the perspective of communication and application security properties such as confidentiality, integrity, and availability for constructing a security framework for VoIP applications.

A number of techniques such as authentication, authorization, VPNs, and firewalls can be used to deter attacks on VoIP services. For example, SIP proxy servers, redirect servers, and registrars are required to support both mutual and one-way authentication and to implement Digest Authorization [4]. McBeth et al. [63] define an architecture for secure network voice along the lines of the DoD C4ISR architecture framework. Once again in this case, the impact of these approaches on the quality of a voice call needs to be considered. Aire et al. [64] regard VoIP security as an important factor in QoS and conduct a comparative analysis of the effects of firewall and VPN techniques on the quality of a call. Srinivasan et al. [65] propose a scheme for authenticating end users' identities with the outbound proxy server in that domain with the help of the registrar server. Cao and Jennings [66] demonstrate a mechanism for providing and authenticating response identify, based on SIP. Lakey [67] studies the security and disruption time (delay) of SIP and SIP signaling in wireless networks and services. Truong et al. [68] introduce a specification-based intrusion detection system to protect H.323 gatekeepers from both external and internal DoS attacks. Cao and Malik [69,70] outline the potential security issues faced by the critical infrastructure sectors as they transform their traditional phone systems into VoIP systems. A set of recommendations and best practices are offered to address the key issues of VoIP security as IP telephony is being introduced into critical infrastructure. Walsh and Kuhn [71] explain the challenges of VoIP security and outline steps for helping to secure an organization's VoIP network. Hung and Martin [72] discuss various VoIP security threats and possible approaches to tackle the threats in VoIP applications. Abdelnur et al. [73] present a security management framework for VoIP, which is capable of performing advanced security assessment tasks. Rippon [74] provides an analysis of the potential threats to the reliability and security of IP-based voice systems.

Prevention approaches provide the first line of defense, but cannot completely eliminate the occurrence of intrusions and attacks. To mitigate the damage caused by the attacks, it is necessary to detect the attacks in a timely manner. As attacks on network services have become prevalent, intrusion detection has emerged as a favored area of research in the past few years. Existing intrusion detection systems are based on two general approaches of detection: anomaly detection and misuse detection. Anomaly detection involves specifying the normal behavior of users or applications. Substantial deviations from this normal behavior can then be labeled as intrusive or at least suspicious. Misuse detection involves modeling of each known attack through the construction of a signature. Incoming activities that match a pattern in the library of attack signatures raise an alarm. Misuse detection is thus useful for detecting known attacks whereas anomaly detection is useful for detecting unknown attacks. Misuse detection has a low false positive rate, whereas anomaly detection suffers from a high false positive rate. Some IDSs employ both anomaly and misuse detection and are termed hybrid systems. Regardless of the method used for detection, an IDS can be alternatively classified as network-based or host-based depending on the input data upon which detection is based. A network-based IDS attempts to detect intrusions by analyzing the packet traffic that is transported over the network. A host-based IDS tries to identify intrusions by analyzing the activities at the hosts, mainly of users and programs. Host-based IDSs rely on many different data streams such as system call traces [75], resources consumed by the processes [76], file access events [76], network audit data collected at a host [77], call stack information [78], and commands issued by a user [79]. A few research efforts also use application data [81–83] as another form of host data. Some IDSs employ both network and host data for detection [84]. An excellent survey of intrusion detection techniques and systems is presented in [85].

While the general purpose intrusion detection systems can be used to detect attacks launched on VoIP services, Wu et al. [86] propose a specialized IDS for VoIP, which translates the incoming packet traffic into protocol-dependent information units. Niccolini et al. [87] analyze SIP intrusion detection and prevention requirements and propose an IDS/IPS architecture. Marshall et al. [88] describe security architecture for real-time services over IP (RSoIP) architecture against internal and external threats including denial-of-service (DoS) and distributed DoS (DDoS) attacks. Chen [89] proposes a method to detect DoS attacks that involve flooding SIP entities with illegitimate SIP messages. Sengar et al. [90] propose a VoIP intrusion detection system which utilizes not only the state machines of network protocols but also the interactions among them for intrusion detection.

Vuong et al. [91] discuss the drawbacks of host-based and network-based IDSs in the context of VoIP. Since VoIP-specific protocols rely on IPSec that renders the actual source, destination, and the content of the packets opaque while they are in transit, network-based IDSs may be ineffective. Although host-based IDSs can address this issue, they require significant processing power and memory capacity and hence cannot be employed in large-scale environments. These concerns are aggravated in lightweight devices such as mobile phones and handhelds over which VoIP services are likely to be offered. In addition, since VoIP services are time sensitive, additional processing required by any security solution must be kept to a minimum.

Intrusion detection based on application data can offer solutions to many of the challenges mentioned earlier. First, the data collected at the application level is generally not encrypted. Second, the processing performed by the application can be leveraged mitigating the need for additional processing for detection. Third, semantic information is available from application data which can aid in the understanding of the impact of the attack on the services provided by the application. Different types of application data including language library calls [80], log files generated by web servers [81–83], database servers [83], data access patterns generated by web portals [92], usage frequencies of the modules [93], internal states of software programs [94], and source code [95] have been used for detection. Signaling performance of a VoIP session can be considered to constitute application data. It is important to note that signaling performance data may be anyway collected during operation to monitor the QoS received by the customers. Thus, our approach may not require the collection of any additional data, which may also potentially mitigate the collection overhead.

6.6 Conclusions and Future Research

In this chapter we presented a QoS monitoring methodology for VoIP services. We considered the signaling performance of a VoIP session as an indicator of its user-perceived QoS. The monitoring methodology relied on the collection and analysis of service-level data. We used open, standard APIs that were originally intended to facilitate rapid service creation across heterogeneous network infrastructures to collect service-level data. We then described how the service-level data can be aggregated and analyzed to monitor the signaling performance of VoIP sessions. Furthermore, we also discussed how the data could be used to detect anomalous or suspicious behavior. We outlined our experience in the experimental demonstration of the methodology using an example of open, standard API environment, namely, JAIN (Java APIs for Integrated Networks) APIs along with some results. The architecture of a prototype monitoring system that encapsulates the monitoring methodology is also presented. The methodology promotes *dual use* of open, standard APIs, one for rapid service creation and two for QoS monitoring. Also, by the virtue of using open, standard APIs, the QoS monitoring methodology can be used uniformly across heterogeneous network infrastructures as well as across services provided by multiple vendors.

Our future research involves demonstrating the methodology using other resource APIs used for VoIP services such as the JAIN H.323 APIs. Similar to performance, reliability, and availability of VoIP services is also of paramount importance [96]. Extending the methodology to enable reliability and availability monitoring is also the topic of future research. Also, pattern-based misuse detection is a useful complement to anomaly detection for the purpose of detecting malicious attacks. Incorporating pattern-based misuse detection into the monitoring methodology is also the concern of future research.

References

1. D. Collins, *Carrier Grade Voice over IP,* McGraw-Hill, 2001.
2. D. Tait, J. Keizjer, and R. Goedman, JAIN: A new approach to services in communication networks, *IEEE Communications Magazine,* vol. 38, no. 1, January 2000, pp. 94–99.
3. VASA Services Workgroup, "Introduction to standardized communication APIs," April 2001 pp. 94–99.
4. J. Rosenberg, H. Schulzrinne, G. Camarillo, A. Johnston, J. Peterson, R. Sparks, M. Handley, and E. Schooler, "SIP: Session initiation protocol," RFC 3261 July 2002.
5. F. Cuervo, N. Greene, A. Rayhan, C. Huitema, B. Rosen, and J. Segers, "Megaco protocol version 1," RFC 3015 November 2000.
6. International Telecommunication Union (ITU), "Visual telephone systems and equipment for local area networks which provide a non-guaranteed quality of service," Telecommunication Standardization Sector of ITU, Recommendation H.323, Geneva, Switzerland, May 1996.
7. M. Shaw and D. Garlan, *Software Architecture: Perspective on an Emerging Discipline*, Prentice-Hall, Upper Saddle River, NJ, 1996.
8. A. Sundaram, "An introduction to intrusion detection," 1996. http://www.acm.org/crossroads/xrds2–4/xrds2–4.html.
9. T. Lane and C. E. Brodley, "An application of machine learning to anomaly detection," *Proceedings of 20th NIST-NCSC National Information Systems Security Conference*, 1997.
10. A. Ghosh and A. Schwartzbard, "A study in using neural networks for anomaly and misuse detection," *Proceedings USENIX Security Symposium*, 1999.
11. B. Mukherjee, L. T. Heberlein, and K. N. Levitt, "Network intrusion detection," *IEEE Network*, vol. 8, May 1994, pp. 26–41.
12. Sun Microsystems, "The JAIN APIs: Integrated network APIs for the Java platform," http://java.sun.com/products/jain.
13. Parlay Group. Parlay API. http://www.parlay.org.
14. 3GPP Homepage. http://www.3gpp.org/.
15. R. Bhat and R. Gupta, "JAIN protocol APIs," *IEEE Communications Magazine*, vol. 38, no. 8, January 2000, pp. 100–106.
16. M. Arango, A. Dugan, I. Elliott, C. Huitema, and S. Pickett, "Media Gateway Control Protocol (MGCP)," RFC 2705, October 1999.
17. Sun Microsystems, "JAIN MGCP API specification," http://jcp.org/jsr/detail/23.jsp.
18. American National Standards Institute (ANSI), "Signaling system number 7 (SS7)—Transaction capabilities application part (TCAP)," T1.114.
19. International Telecommunication Union (ITU), "Transaction capabilities application part (TCAP)," Recommendation Q.771–Q.775.
20. Sun Microsystems, "JAIN TCAP API Specification," http://jcp.org/jsr/detail/11. jsp.
21. Sun Microsystems, "JAIN ISUP API Specification," http://jcp.org/jsr/detail/17. jsp.
22. Sun Microsystems, "JAIN INAP API Specification," http://jcp.org/jsr/detail/35. jsp.
23. Sun Microsystems, "JAIN MAP API Specification," http://jcp.org/jsr/detail/29. jsp.
24. R. Jain, J. Bakker, and F. Anjum, "Java call control JCC and session initiation protocol," *IEICE Transaction on Communication*, vol. E84-B, no,12, December 2001, pp. 3096–3102.
25. National Institute of Standards and Technology (NIST), "JAIN-SIP project home," https://jain-sip.dev.java.net/.
26. S. S. Gokhale and J. Lu, "Signaling performance of SIP based VoIP," *International Symposium on Performance Evaluation of Computer and Telecommunication Systems (SPECTS 04)*, San Jose, California, July 2004, pp. 190–196.
27. National Institute of Standards and Technology (NIST), "IP Telephony Project," http://dns.antd.nist.gov/proj/iptel/.
28. R. A. Johnson and D. W. Wichern, *Applied Multivariate Statistical Analysis*, Prentice-Hall, Upper Saddle River, NJ, 1998.

29. S. S. Gokhale, "QoS assurance of next generation network (NGN) applications," *14th IEEE International Symposium on Software Reliability Engineering (ISSRE '03)*, Denver, Colorado, November 2003.

30. Sun Microsystems, "Jini specifications and API archive," http://java.sun.com/products/jini/.

31. Sun Microsystems, "Java Message Service (JMS)," http://java.sun.com/products/jms/.

32. L. Zheng, L. Zhang, and D. Xu, "Characteristics of network delay and delay jitter and its effect on voice over IP (VoIP)," *IEEE International Conference on Communications (ICC '01)*, 2001, pp. 122–126.

33. K. Mase, Y. Toyama, A. A. Bilhaj, and Y. Suda, "QoS management for VoIP networks with edge-to-edge admission control," *Global Telecommunications Conference (GLOBECOM '01)*, 2001, pp. 2556–2560.

34. K. Mase and Y. Toyama, "End-to-end measurement based admission control for VoIP networks," *IEEE International Conference on Communications (ICC '02)*, 2002, pp. 1194–1198.

35. A. A. Bilhaj and K. Mase, "Endpoint admission control enhanced systems for VoIP networks," *Proceedings of 2004 International Symposium on Applications and the Internet*, 2004, pp. 269–272.

36. O. Awoniyi and F. A. Tobagi, "Effect of fading on the performance of VoIP in IEEE 802.11a WLANs," *IEEE International Conference on Communications*, 2004, pp. 3712–3717.

37. J. K. Muppala, T. Bancherdvanich, and A. Tyagi, "VoIP performance on differentiated services enabled network," *Proceedings of IEEE International Conference on Networks (ICON '00)*, 2000, pp. 419–423.

38. T. Nguyen, F. Yegenoglu, A. Sciuto, and R. Subbarayan, "Voice over IP service and performance in satellite networks," *IEEE Communications Magazine*, vol. 39, no. 3, March 2001, pp. 164–171.

39. H. Furuya, S. Nomoto, H. Yamada, N. Fukumoto, and F. Sugaya, "Experimental investigation of the relationship between IP network performances and speech quality of VoIP," *10th Inernational Conference on Telecommunications (ICT '03)*, 2003, pp. 543–552.

40. W. Wang, S. C. Liew, and V. O. K. Li, "Solutions to performance problems in VoIP over a 802.11 wireless LAN," *IEEE Transactions on Vehicular Technology*, vol. 54, no. 1, January 2005, pp. 366–384.

41. C. Al-Najjar and A. L. Narasimha Reddy, "A service provider's approach for improving performance of aggregate Voice-over-IP traffic," *Proceedings of 14th IEEE International Workshop on Quality of Service (IWQoS '06)*, 2006, pp. 169–177.

42. Y. Amir, C. Danilov, S. Goose, D. Hedqvist, and A. Terzis, "An overlay architecture for high-quality VoIP streams," *IEEE Transactions on Multimedia*, vol. 8, no. 6, pp. 1250–1262, December 2006.

43. N. Sulaiman, R. Carrasco, and G. Chester, "Performance evaluation of voice call over an IP based network," *Proceedings of 41st Annual Conference on Information Sciences and Systems (CISS '07)*, 2007, pp. 392–395.

44. M. Portoles-Comeras, J. Mangues-Bafalluy, and M. Cardenete-Suriol, "Performance issues for VoIP call routing in a hybrid ad hoc office environment," *Proceedings of 16th IST Mobile and Wireless Communications Summit*, 2007, pp. 1–5.

45. International Telecommunication Union (ITU), "Network grade of service parameters and target values for circuit-switched services in the evolving ISDN," Telecommunication Standardization Sector of ITU, Recommendation E.721, Geneva, Switzerland, May 1999.

46. AT&T Webpage. http://www.att.com/attlabs/reputation/fellows/lawser.html.

47. International Telecommunication Union (ITU), "Signalling performance in the telephone application," Telecommunication Standardization Sector of ITU, Recommendation Q.725, Geneva, Switzerland, March 1993.

48. H. Lin, T. Seth, A. Broscius, and C. Huitema, "VoIP signaling performance requirements and expectations," Internet Engineering Task Force, June 1999. Internet Draft.

49. T. Eyers and H. Schulzrinne, "Predicting internet telephony call setup delay," *Proceedings of 1st IP-Telephony Workshop (IPTel '00)*, April 2000.

50. A. Kist and R. Harris, "SIP signaling delay in 3GPP," *6th International Symposium on Communications Interworking of IFIP (Interworking '02)*, Fremantle, WA, October 2002, pp. 211–222.

51. S. K. Das, E. Lee, K. Basu, N. Kakani, and S. K. Sen, "Performance optimization of VoIP calls over wireless links using H.323 protocol," *INFOCOM '02*, 2002.

52. B. S. De, P. P. Joshi, V. Sahdev, and D. Callahan, "End-to-end voice over IP testing and the effect of QoS on signaling," *Proceedings of 35th Southeastern Symposium on System Theory*, 2003.

53. J.-S. Wu and P.-Y. Wang, "The performance analysis of SIP-T signaling system in carrier class VoIP network," *17th International Conference on Advanced Information Networking and Applications (AINA '03)*, 2003.

54. H. Fathi, S. S. Chakraborty, and R. Prasad, "Optimization of SIP session setup delay for VoIP in 3G wireless networks," *IEEE Transactions on Mobile Computing*, vol. 5, no. 9, September 2006, pp. 1121–1132.

55. International Engineering Consortium (IEC), "Voice over Internet Protocol," http://www.iec.org/online/tutorials/.

56. Voice over IP Security Alliance. http://www.voipsa.org/.

57. IETF, "Transport layer security (tls)," http://www.ietf.org/html.charters/tls-charter.html.

58. IETF, "IP security protocol (ipsec)," http://www.ietf.org/html.charters/OLD/ipsec-charter.html.

59. J.-I. Guo, J.-C. Yen, and H.-F. Pai, "New voice over internet protocol technique with hierarchical data security protection," *IEE Proceedings of Vision, Image and Signal Processing*, vol. 149, no. 4, 2002, pp. 237–243.

60. R. Barbieri, D. Bruschi, and E. Rosti, "Voice over IPsec: Analysis and solutions," *Proceedings of 18th Annual Computer Security Applications Conference*, 2002, pp. 261–270.

61. D. Geneiatakis, G. Kambourakis, T. Dagiuklas, C. Lambrinoudakis, and S. Gritzalis, "A framework for detecting malformed messages in SIP networks," (CD-ROM), *Proceedings of 14th IEEE Workshop on Local and Metropolitan Area Networks (LANMAN '05)*, 2005.

62. M.V. Martin and P.C.K. Hung, "Towards a security policy for VoIP applications," *Canadian Conference on Electrical and Computer Engineering*, 2005, pp. 65–68.

63. M. S. McBeth, R. Cole Jr., and R. B. Adamson, "Architecture for secure network voice," *Proceedings of Military Communications Conference (MILCOM '99)*, 1999, pp. 1454–1457.

64. E. T. Aire, B. T. Maharaj, and L. P. Linde, "Implementation considerations in a SIP based secure voice over IP network," *7th AFRICON Conference in Africa (AFRICON '04)*, 2004, pp. 167–172.

65. R. Srinivasan, V. Vaidehi, K. Harish, K. L. Narasimhan, S. L. Babu, and V. Srikanth, "Authentication of signaling in VoIP applications," *Asia-Pacific Conference on Communications*, 2005, pp. 530–533.

66. F. Cao and C. Jennings, "Providing response identity and authentication in IP telephony," *1st International Conference on Availability, Reliability and Security (ARES '06)*, 2006.

67. E. T. Lakay and J. I. Agbinya, "Security issues in SIP signaling in wireless networks and services," *International Conference on Mobile Business (ICMB '05)*, 2005, pp. 639–642.

68. P. Truong, D. Nieh, and M. Moh, "Specification-based intrusion detection for H.323-based voice over IP," *Proceedings of 5th IEEE International Symposium on Signal Processing and Information Technology*, 2005, pp. 387–392.

69. F. Cao and S. Malik, "Security analysis and solutions for deploying IP telephony in the critical infrastructure," *Workshop of the 1st International Conference on Security and Privacy for Emerging Areas in Communication Networks*, 2005, pp. 171–180.

70. F. Cao and S. Malik, "Vulnerability analysis and best practices for adopting IP telephony in critical infrastructure sectors," *IEEE Communications Magazine*, vol. 44, no. 4, April 2006, pp. 138–145.

71. T. J. Walsh and D. R. Kuhn, "Challenges in securing voice over IP," *IEEE Security & Privacy Magazine*, vol. 3, no. 3, May–June 2005, pp. 44–49.

72. P. C. K. Hung and M. V. Martin, "Security issues in VoIP applications," *Canadian Conference on Electrical and Computer Engineering*, 2006, pp. 2361–2364.

73. H. Abdelnur, V. Cridlig, R. State, and O. Festor, "VoIP security assessment: methods and tools," *1st IEEE Workshop on VoIP Management and Security*, 2006, pp. 29–34.

74. W. J. Rippon, "Threat assessment of IP based voice systems," *1st IEEE Workshop on VoIP Management and Security*, 2006, pp. 19–28.

75. R. Chinchani, S. Upadhyaya, and K. Kwait, "Towards the scalable implementation of a user level anomaly detection system," *Proceedings of Military Communications Conference (MILCOM '02)*, 2002, pp. 1503–1508.

76. S. Han and S. Cho, "Rule-based integration of multiple measure-models for effective intrusion detection," *IEEE International Conference on Systems, Man and Cybernetics*, 2003, pp. 120–125.

77. T. E. Daniels and E. H. Spafford, "A network audit system for host-based intrusion detection system," *Proceedings of Annual Computer Security Application Conference (ACSAC '00)*, 2000, pp. 178–187.

78. H. H. Feng, O. M. Koesnikov, P. Folgla, W. Lee, and W. Gong, "Anomaly detection using call stack information," *Proceedings of Symposium on Security and Privacy (S&P '03)*, 2003, pp. 62–75.

79. J. Marin, D. Ragsdale, and J. Surdu, "A hybrid approach to the profile creation and intrusion detection," *Proceedings of DARPA Information Survivability Conference and Exposition II (DISCEX '01)*, 2001, pp. 12–14.

80. A. K. Jones and Y. Liu, "Application intrusion detection using language library calls," *17th Annual Computer Security Applications Conference (ACSAC '01)*, 2001, pp. 442–449.

81. G. Vigna, W. Robertson, V. Kher, and R. A. Kemmerer, "A stateful intrusion detection system for world-wide servers," *19th Annual Computer Security Applications Conference (ACSAC '03)*, 2003, pp. 34–43.

82. L. Wang, G. Yu, G. Wang, and D. Wang, "Method of evolutionary neural network based intrusion detection," *2001 International Conference on Info-tech and Info-net (ICII '01)*, 2001, pp. 13–18.

83. W. Shu and T. D. H. Tan, "A novel intrusion detection system model for securing Web-based database systems," *25th Annual International Computer Software and Applications Conference (COMPSAC '01)*, 2001, pp. 249–256.

84. A. Abimbola, Q. Shi, and M. Merabti, "NetHost-Sensor: A novel concept in intrusion detection systems," *Proceedings of 8th IEEE International Symposium on Computers and Communication (ISCC '03)*, 2003, pp. 232–240.

85. Y. Bai and H. Kobayashi, "Intrusion detection systems: technology and development," *Proceedings of 17th International Conference on Advanced Information Networking and Applications (AINA '03)*, 2003, pp. 710–715.

86. K. Wu and D. S. Reeves, "Capacity planning of DiffServ networks with best-effort and expedited forwarding traffic," *Telecommunication Systems*, vol. 25, no. 3–4, 2004, pp. 193–207.

87. S. Niccolini, R. G. Garroppo, S. Giordano, G. Risi, and S. Ventura, "SIP intrusion detection and prevention: recommendations and prototype implementation," *1st IEEE Workshop on VoIP Management and Security*, 2006, pp. 47–52.

88. W. Marshall, A. F. Faryar, K. Kealy, G. de los Reyes, I. Rosencrantz, R. Rosencrantz, and C. Spielman, "Carrier VoIP security architecture," *12th International Telecommunications Network Strategy and Planning Symposium (NETWORKS '06)*, 2006, pp. 1–6.

89. E. Y. Chen, "Detecting DoS attacks on SIP systems," *1st IEEE Workshop on VoIP Management and Security*, 2006, pp. 53–58.

90. H. Sengar, D. Wijesekera, H. Wang, and S. Jajodia, "VoIP intrusion detection through interacting protocol state machines," *International Conference on Dependable Systems and Networks (DSN '06)*, 2006, pp. 393–402.

91. S. Vuong and Y. Bai, "A survey of voip intrusions and intrusion detection systems," *6th International Conference on Advanced Communication Technology*, 2004, pp. 317–322.

92. R. Sion, M. Atallah, and S. Prabhakar, "On-the-fly intrusion detection for Web portals," *Proceedings of International Conference on Information Technology: Computers and Communications (ITCC '03)*, 2003, pp. 325–330.

93. J. C. Munson and S. Wimer, "Watcher: The missing piece of the security puzzle," *Proceedings of 17th Annual Computer Security Application Conference (ACSAC '01)*, 2001, pp. 230–239.

94. A. K. Ghosh, J. Wanken, and F. Charron, "Detecting anomalous and unknown intrusions against programs," *Proceedings of 1998 Annual Computer Security Application Conference*, 1998, pp. 259–267.

95. D. Wagner and D. Dean, "Intrusion detection via static analysis," *Proceedings of IEEE Symposium on Security and Privacy (S&P '01)*, 2001, pp. 156–168.

96. C. R. Johnson, Y. Kogan, Y. Levy, F. Saheban, and P. Tarapore, "VoIP reliability: a service provider's perspective," *IEEE Communications Magazine*, vol. 42, no. 7, July 2004, pp. 48–54.

7

Current and Future VoIP Quality of Service Techniques

Barry Sweeney and Duminda Wijesekera

CONTENTS

7.1 What is Quality of Service?

Quality of service (QoS) is a context-dependent term that expresses qualitative measures for service consumers. In this chapter, QoS is defined as the capability to provide resource assurance and service differentiation in an Internet Protocol (IP) network. Resource assurance is a commitment expressed by the network to provide appropriate service to suit the requirements of the application, such as bandwidth, packet loss, jitter, and latency. Service differentiation is the ability of the network to treat packets of different applications in different ways. QoS for Voice over Internet Protocol (VoIP) is usually described in terms of service and voice quality. Service quality is associated with availability, post-dial delay, and call-completion rates [1]. Voice quality is the user's experience. This chapter focuses on the QoS resource assurance and service differentiation approaches and mechanisms that affect voice quality.

Quality of service is typically measured in VoIP using two metrics: intelligibility and latency. Intelligibility is a numerical indication of the perceived quality of voice after compression and/or transmission and is expressed with one of two methods for end-to-end intelligibility (handset to handset).* The first is based on the International Telecommunications Union Standardization Sector (ITU-T) Recommendations P.800[†] [2] and P.862[‡] standards, which expresses the intelligibility as the mean opinion score (MOS) with a number between 1 and 5, with 5 being optimal (i.e., not achievable). P.862 uses a comparison of the output-coded signal to the received-coded signal. This means that the test devices establish a sample of voice before the test begins. The sender test device transmits the digital representation of the voice signal and the received test device compares it to the original sample. In this model, delay is a separate measurement than MOS since it is factored out using synchronization processes. Since it is an active approach, it makes it difficult to evaluate voice performance in real-time.

A better model for measuring VoIP call quality is based on the E-Model described in ITU-T Recommendation G.107[¶] [3] and G.108[§] and is described in detail in the Telecommunications Industry Association (TIA) TSB-116A, "Telecommunications—IP Telephony Equipment—Voice Quality Recommendations for IP Telephony," which expresses the QoS as an R factor with a range from 1 to 100 based on a variety or network and equipment impairments that are captured through passive techniques [4]. The E-Model was originally targeted in the VoIP environment for VoIP transmission planning. However, many vendors are adopting the E-Model for providing voice quality measurements in the Call Detail Records (CDRs) to capture call quality and conduct trend analysis. MOS is the more widely used QoS measurement due to the familiarity of the telecommunications community with the significance of the MOS numbers and is the measurement used in this chapter although conversions between R-Factor and MOS are possible.** The legacy time division multiplexing (TDM) wireline (as compared to wireless) MOS target for commercial voice service providers is 4.0 (R-Factor of 80) and is referred to as "Toll Quality." Due to the proliferation of wireless voice systems, users have come to accept a lower quality of voice services as a tradeoff for mobility referred to as "Conversational Quality," which has a MOS of 3.8. Figure 7.1 compares user level of satisfaction to the MOS score as defined in ITU-T P.800 and G.109 standards [5].[††]

In addition to the measurement of MOS, another factor that affects the user experience is the latency of the bearer stream. In this chapter, latency is defined as the one-way packet transfer delay for all successful and erroneous packets (includes packets discarded within the jitter buffer) averaged over the measurement interval with the measurements points being the handsets of the two end instruments (discussed in detail later in the chapter).

The ITU-T recommendation Y.1291[‡‡] [6] designates three planes that are used to provide QoS: control plane, data plane, and management plane. The control plane is associated

---

* P.564 is not considered since it is not handset to handset and only addresses IP network impairments.
† http://www.itu.int/rec/T-REC-P.800/en
‡ http://www.itu.int/rec/T-REC-P.862/en
¶ http://www.itu.int/rec/T-REC-G.107/en
§ http://www.itu.int/rec/T-REC-G.108/en
** The conversion formula for converting R Factor to MOS is $MOS = 1 + R * .035 + R * (R - 60) * (100 - R) * 7 * 10^{-6}$ for the ranges discussed in this chapter.
†† http://www.itu.int/rec/T-REC-G.109/en
‡‡ http://www.itu.int/rec/T-REC-Y.1291-200405-I/en

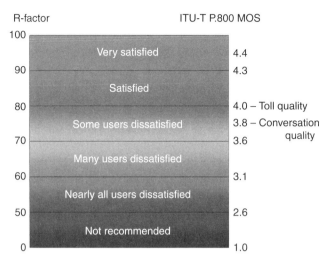

FIGURE 7.1 Comparison of user satisfaction with MOS.

with mechanisms for controlling the traffic flows in the system, using call admission control (CAC), routing, or bandwidth reservation mechanisms. The data plane is focused on affecting individual packets with the aid of buffer management, packet marking, queuing and scheduling, traffic classification, and traffic conditioning. Finally, the management plane and is focused on the operation, administration, and management (OA&M) aspects using service level agreements (SLAs) and traffic restoration. This chapter focuses on the current and emerging approaches that could be used to provide QoS for VoIP in the control and data planes.

7.2 Why is QoS Needed?

For those who remember the days of dial-up modems, consumers did not always have access to high speed Internet connections. Even in today's environment of high speed optical core networks, IP edge connections to the core network are sometimes bandwidth constrained due to cost or infrastructure limitations (my parents can only get dial-up modem service in Maine). Compounding the problem is that carrier networks are designed for data applications that run over the transmission control protocol (TCP) and whose performance are not as sensitive to latency, packet loss, and jitter as VoIP. Even with proper traffic engineering, unexpected surges can cause congestion on normally uncongested networks, such as during the 9/11 attack or when the Starr Report was uploaded. In times of congestion, QoS techniques ensure that acceptable resource assurance and service differentiation is provided for VoIP subscribers.

7.3 Factors that Affect QoS

Mean opinion score is affected by a number of impairments. The sources of the impairments differ depending on the technology. In VoIP, the major factors that affect MOS are jitter, echo, speech compression, and packet loss. The first source of impairment is jitter, which is defined in this chapter to be the latency variation during the measurement interval. All VoIP end instruments are designed to compensate for jitter by using a de-jitter buffer. Most VoIP end instruments employ an adaptive rate de-jitter buffer that expands and contracts in size, dependent on the measured jitter. Another factor that affects the size of the de-jitter buffer is the CODEC. For example, if the CODEC used is G.711 with 20 ms samples, the de-jitter buffer will occur in increments of 20 ms (i.e., 40, 60, or 80 ms). If the CODEC used is G.711 with 10 ms samples, the de-jitter buffer will occur in increments of 10 ms (i.e., 20, 30, or 40 ms). The primary effect of jitter control is its impact on latency and packet loss. If the jitter exceeds the size of the de-jitter buffer, the VoIP packet is discarded resulting in increased packet loss. If the jitter buffer size is increased to minimize the packet loss, then the end-to-end latency is increased due to the time spent in the de-jitter buffer resulting in a degradation of the user experience caused by high latency. Therefore, a trade-off between packet loss and latency is required when sizing the de-jitter buffer and it is usually sized to be one to two times the sample interval if a static de-jitter (vs. adaptive) buffer is used.

The second source of impairment is echo. Echo is defined in this chapter as the audible leak-through of a speaker's voice into her own receive path and is manifested in VoIP in two ways. The first manifestation is when the speaker hears her own words a short period of time after she says something. However, as long as the delay between talking and hearing is less than 25 ms, the human brain compensates for the echo so that it is not noticeable. This type of echo is often called sidetone. The second manifestation of echo is the perceived echo and is caused when a speaker's voice is returned to the speaker from the "leak-through" at the distant end and is typically caused by the remote end instrument. It should be noted that "leak-through" in digital portions of the connection does not occur. Because the delay associated with long-distance calls is typically above 25 ms, perceived echo is noticeable and must be eliminated.

Most of the "leak-through" occurs at the end instrument or in an analog tail circuit to the end instrument. Often the "leak-through" occurs in the analog portions of the end instrument and is caused by poor insulation between transmit and receive wires. Another source is acoustic echo, which is when the air and parts of the telephone provide a coupling between the speaker and the microphone. This effect is more significant when a speaker phone is used. Echo becomes more distracting as the loudness increases and is typically expressed as the talker echo loudness rating (TELR) and for VoIP phones is typically from 62 dB to 65 dB for low delay sessions. As the delay increases, the effect of the echo on the MOS increases with the MOS measuring approximately 4.0 with a delay of 250 ms when the TELR is 65 dB [7].* Therefore, echo cancellation focuses on attenuating the echo part of the signal so that it is not as loud and disappears when compared to the conversation (echo 6 dB quieter than the conversation). One issue that is unique to VoIP is that the bearer path typically exists only between the two VoIP end instruments. This makes it difficult to centrally locate an echo canceller and vendors have designed echo cancellers into the media gateways and some VoIP end instruments in order to compensate for echo.

The third impairment affecting MOS is caused by speech compression. Speech compression is defined in this chapter to be the use of predictive coding to reduce the bit rate needed for transmission of the voice bearer stream. The impairment caused by speech compression is usually referred to as the equipment impairment factor (Ie) and results from the distortion of the voice during the compression process. The Ie value for G.711,

---

* TIA/TSB-116-A Figure 7.4.

CODEC	Bit rate in kbps	Ie	MOS
G.711	64	0	4.0
G.726	32	7	3.85
G.728	16	7	3.6
G.729A	8	11	3.7
G.723.1	6.3	15	3.6
G.723.1	5.3	19	3.9

FIGURE 7.2 Comparison of CODECs attributes.

which does not use compression, is 0 and the Ie value generally increases proportional to the level of compression (although new CODECs compensate to minimize the Ie value). Figure 7.2 shows a comparison of some attributes of several CODECs.*

The effect of distortion is more pronounced as the latency increases. For example, a MOS of 4.0 is achievable for G.711 when latency is less than 250 ms, whereas it is only achievable for G.729A when the latency is less than 125 ms.

The last factor that affects MOS is packet loss. It is defined as the ratio of the packets transmitted successfully to the number of packets transmitted during the measurement interval expressed as a percentage. Compared to TDM, VoIP is tolerant of packet loss. In TDM, packet loss was manifested in bit error rates that resulted in noise that degraded the MOS. In VoIP, packet loss is the result of lost, corrupted, or discarded voice samples (packets). Most conversations involve a large percentage of time either listening to silence or with one talking. During those times, it is difficult to detect lost packets. Typically, lost packets only become noticeable when listening to the other person talk. To compensate for a single lost packet, a VoIP waveform CODECs (i.e., G.711) end instrument typically plays the previous packet. A VoIP non-waveform CODECs (e.g., G.723.1) end instrument typically uses forward error correction (FEC) to compensate for lost packets.† Both techniques are referred to generically as packet loss concealment (PLC) techniques. The result is that the listener does not hear the dropout unless multiple consecutive packets are lost because of the small sample size (i.e., 10–20 ms).

Another reason is that the probability of multiple lost packets occurring sequentially is relatively small. For instance, a 2% packet loss will result in two samples being lost during a second of speech (assumes a 10 ms sample size). The probability of the lost packets occurring consecutively is relatively small. Since the end instrument corrects for the loss of the first packet and the dropout is only for 10 ms, most listeners cannot detect the loss unless they are in a quiet room with limited background noise. With the use of PLC and the natural distribution of packet loss, testing has shown that the listener typically does not detect packet loss until it is around 3–5%. The sources of packet loss in VoIP systems are consistent with those found in IP networks with the addition of packet loss due to the de-jitter buffer, which is caused by a packet being discarded when the end-to-end jitter exceeds the de-jitter buffer size.

---

* The MOS values are representative and actual measurements vary.
† Speech packet n contains speech sample m and m 1 1, packet n 1 1 contains speech sample m 1 1 and m 1 2.

As mentioned previously, latency is the second of two metrics that are used to measure QoS. Most VoIP systems are designed to minimize latency to avoid "talk over." "Talk over" is a condition that occurs in a conversation when one person starts to talk at the same time as the other is talking due to a delay in hearing the start of the remote party's words and is caused by high latency. "Talk over" is most noticeable in international calls and calls that transit satellite networks. In a typical conversation, after one person speaks, there is a slight pause (approximately 200 ms) before the other person speaks. This phenomenon is called "turn-taking" and when the latency approaches the "turn-taking" pause a loss in synchronicity in the conversation occurs and results in "talk over." In VoIP, "talk over" typically occurs at around 220 ms, depending on the person.

In addition to the "talk over" issue, long pauses caused by latency often change the meaning of a statement (i.e., "Will you marry me (delay) (delay) Yes" is interpreted differently than "Will you marry me (immediate) Yes"). The sources that affect the end-to-end VoIP latency are shown in Figure 7.3 and the three sources of latency that telecom engineers influence are CODEC, de-jitter buffer, and the switching delays (discussed in detail in the QoS mechanisms discussion).

Most of the Voice QoS standards available today are based on the impact of jitter, speech compression, echo, packet loss, and latency of end-to-end TDM systems. These standards are being adapted for VoIP slowly, and the result of the delay is that most SLAs for VoIP are conservative due to the use of TDM measurement values instead of VoIP measurement values. In addition, most SLAs provided by IP service providers are focused on data applications and measurement intervals are often 30-day averages, measured at the network layer (vs. handsets), and only use a small subset of measurement points. Because a typical voice call is considered dropped if packet loss is 100% for a continuous 6–10 s interval, a 30-day average can hide a very large number of periods with unacceptable performance. Measuring at the network layer ignores the latency and packet loss that occurs above the network layer, such as CODEC and de-jitter buffer delays. Finally, limiting the measurements to a limited number of paths will result in unobserved degradations occurring between nodes that are not measurement points.

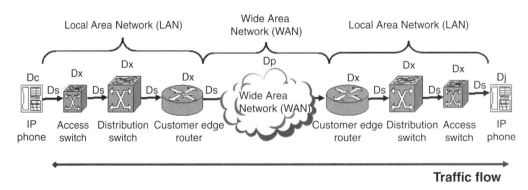

Dc – CODEC delay (includes packetization & algorithmic delays)
Ds – Serialization delay
Dj – De-jitter buffer delay
Dp – Propagation delay
Dx – Switching delay (includes processing delays)
Dt – Total 1-way delay
Dt = Dc + Ds + Dp + Dx + Dj

FIGURE 7.3   End-to-end latency sources.

7.4 QoS Mechanisms

As mentioned earlier, there are three distinct approaches for achieving QoS for VoIP called the control plane, data plane, and management plane approaches. Since the focus of this chapter is on the voice quality aspect of QoS, both the control plane and data plane QoS mechanisms will be discussed and management plane mechanisms are deferred. In discussing the mechanisms, it is important to discriminate between mechanisms that are currently in use and ones that may become viable in the future [8]. The following section focuses on currently available mechanisms and then discusses future candidate mechanisms.

7.4.1 Current Control Plane QoS Mechanisms

The first discussion focuses on the control plane and deals with controlling the flows (versus individual packets) in the network. The mechanisms used today by telecom engineers to provide control plane QoS are traffic engineering, CAC, multiprotocol label switching, and QoS routing. The first requirement of all commercial approaches using control plane QoS is to conduct accurate traffic engineering. Typically, traffic engineering involves converting legacy voice DS0s to IP throughput needs. The conversion factor varies depending on the CODEC, IP version, and the sample interval. The IP throughput calculation should include IP overhead and user signaling packets. A simple conversion method is to multiply the existing DS0s from an enclave by the conversion factor (G.711 with IPv4 and 20 ms samples has a conversion factor of 89.2 kbps including signaling).* A more thorough calculation uses the Busy Hour Erlang B. For example, if the following assumptions are made with a Busy Hour Erlang B of 25, the access circuit should be designed with an allocation of 2.5 Mbps for VoIP traffic to include signaling as shown in the following:

```
Assumptions:
Call arrival distribution      =    Poisson
CODEC type                     =    G.711 (coding rate: 64,000 bits/sec)
Frame size                     =    20 ms interval time (0.020 sec)
Samples/packet                 =    80 samples per packet
Frames/packet                  =    1
Frames/second                  =    50
Frame size/packet              =    160 bytes
Frames/Erlang                  =    50
Packets/second/Erlang          =    50
Packet size (for Ethernet)     =    246 bytes (assumes IPv6)

Access bandwidth formula       =    Busy Hour Erlang B * Packet Size *
                                    Packets/Second/Erlang * 8 bits/byte
Access bandwidth = 25 * 246 * 50 * 8 = 2,460,000 bps or 2.5 Mbps.
```

The next requirement for the control plane approach is to use CAC to ensure that the offered call load stays within the traffic engineered load. CAC is a function found within the call control agent (CCA) of a Softswitch or local session controller (LSC), which are the IP equivalent of today's end offices. CAC is typically stated in terms of a call count or

---

* RTCP adds 5% additional overhead.

budget. For example, a CAC call count of 10 means that the CAC will block all call requests after it has 10 active and/or pending calls. A budget-based CAC uses the bin-packing algorithm using actual bandwidth requirements of each call dependent on the used CODEC. For example, if the CAC budget is 1 Mbps, the CAC can allow 11 G.711 (20 ms sample) calls to be active or 30 G.729 calls (assumes 33.2 kbps per call) or a combination of both CODECs that add up to less than 1 Mbps.

The second control plane QoS mechanism discussed in this chapter is QoS routing to control the paths over which the flows travel. For example, premium customers may be routed over the direct path between two points, whereas routine customers are routed over suboptimal paths to avoid congestion on the preferred paths. The path may be suboptimal because it involves more propagation delay or more switching delays even though it meets the QoS requirements.

The final control plane mechanism is the use of MPLS* [9] to provide QoS. In current VoIP networks, MPLS is primarily used to provide logical separation of traffic, fast failure recovery (FFR) around local failures, and to minimize switching delays within the core network. Sometimes service providers are required to logically separate user traffic and the MPLS virtual private network (VPN) capability allows the service provider to achieve that goal. In this case, the use of IPsec to encrypt the VPN is not always implemented due to encryption performance degradations. Another use of MPLS is for recovery of end-to-end connectivity around local failures. Most routing vendors who support MPLS on their platforms claim a FFR around a local failure to be under 50 ms. Although the recovery from the failure is noticeable to the VoIP subscriber, the call will remain active. The final benefit of MPLS is that it reduces the end-to-end latency due to the label switching feature. It is important to note that MPLS-Traffic Engineering (MPLS-TE) is a feature of MPLS, but the feature is typically not associated with VoIP QoS on a per flow basis, but may be applied on an aggregate basis to improve performance.

7.4.2 Current Data Plane QoS Mechanisms

The second approach is the data plane approach and is focused on providing QoS on a per packet basis. The mechanisms used by the data plane approach include traffic conditioning, differentiated services (DiffServ), and per hop behaviors (PHBs). In combination with CAC, traffic conditioning is conducted at the IP layer to ensure that the offered VoIP load remains within the engineered traffic limits. Traffic conditioning is defined in RFC 2475† [10], but for our purpose is focused on metering, policing, and shaping. Metering is the process of measuring the traffic in order to perform shaping and policing. Policing is the discarding of packets based on specified rules. Finally, shaping is the process of delaying packets to cause a traffic stream to conform to a desired traffic profile. Within VoIP, traffic conditioning is normally associated with customer edge routers (CER) found at the enclave boundary and the CER access circuit to the service provider's provider edge routers (PER). Normally, traffic conditioning is applied to both ends of the access circuit if congestion is likely to occur.

In congested networks, it is necessary to discriminate between VoIP packets and other packets in order to ensure that the VoIP packets are given preferential treatment. The method used by commercial VoIP vendors is to mark a field in the IP header of the VoIP packet called the differentiated services field (DS Field). The DS Field has a six-bit field

---

* http://www.ietf.org/rfc/rfc3031.txt
† http://www.ietf.org/rfc/rfc2475.txt

(allowing for 64 possible settings) within the IP header of the bearer and signaling packets and is described in RFC 2474.* Once a packet is marked, the marking is called the differentiated services code point (DSCP) of the packet and is used for the application of PHBs. Typically, the DSCP is marked by the end instrument or by the first LAN switch encountered by the VoIP packet. Normally, the signaling, bearer, and network management packets are marked with different DSCPs.

As mentioned previously, DSCPs are set so that PHBs can be applied. PHBs are defined in RFC 2475,† but for our purpose are externally observable forwarding behaviors applied at a DS-compliant router or LAN switch to a DS behavior aggregate. PHBs may be specified in terms of their resource (e.g., buffer, bandwidth), priority relative to other PHBs, or in terms of their observable traffic characteristics (e.g., latency, packet loss, and jitter). In VoIP, the behavior aggregates are usually segmented into bearer, signaling, and network management. Typically, PHBs are associated with the access link and other WAN connections. However, if the LAN is not properly designed and congestion occurs, PHBs can be implemented within the LAN switches to mitigate congestion.

Because the VoIP bearer packets are sensitive to jitter, packet loss, and latency, they receive the expedited forwarding (EF) PHB as described in RFC 3246‡ [11]. The EF PHB is a forwarding treatment for a particular DS aggregate that ensures that the departure rate of the aggregate's packets equal or exceed a configurable rate independent on the intensity of any other traffic attempting to transit the node. The effect of applying the EF PHB to the VoIP bearer packets is that the latency, packet loss, and jitter experienced by the VoIP packets is minimized.

In contrast to the bearer packets, the VoIP signaling packets use the TCP, which tolerates losses. Even though signaling is more tolerant of network impairments, it must receive preferred PHBs to minimize call setup delays and to terminate calls during overload conditions. Depending on the vendor preferences, the VoIP signaling packets may receive the EF PHB or may receive the assured forwarding (AF) PHB as defined in RFC 2597¶ [12]. The performance of the AF PHB depends on the resources allocated to the AF class of the packet, the current load of the AF class, and, in case of congestion within the class, the drop precedence of the packet. Each AF PHB has three drop precedence and user signaling is normally given the lowest drop probability. Sometimes, VoIP vendors place both the network control and VoIP signaling packets in an AF queue separate from all other traffic.

The final aggregate behavior is the network management behavior aggregate. This aggregate normally consists of two types of traffic: the bulk traffic associated with call detail records and Syslog files; time sensitive traffic associated with SNMP traps and alerts. The network management traffic receives the AF PHB and sometimes the bulk traffic is queued in a different AF queue than the time sensitive network management traffic. Figure 7.4 shows the points where QoS mechanisms can be applied.

7.4.3 Future Control Plane QoS Mechanisms

The mechanisms discussed earlier are referred to as "Open Loop" QoS mechanisms because the data plane is de-coupled from the control plane and there is no feedback from

---

* http://www.ietf.org/rfc/rfc2474.txt
† http://www.ietf.org/rfc/rfc2475.txt
‡ http://www.ietf.org/rfc/rfc3246.txt
¶ http://www.ietf.org/rfc/rfc2597.txt

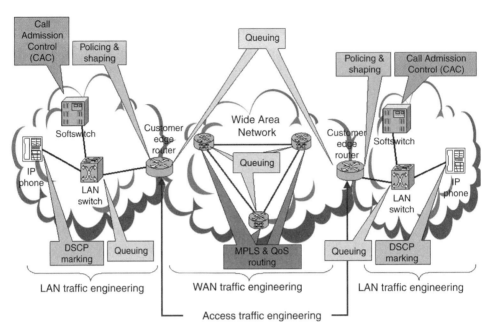

FIGURE 7.4   Current QoS mechanisms.

the data plane to the control plane. The second approach is referred to as "Closed Loop" and is focused on providing QoS through the signaling plane and involves coupling the control plane to the data plane by providing a feedback mechanism from the network to the CAC. The "Closed Loop" tools vary in their maturity and majority of them are still in the research and development stage at the time of this writing (early 2008). The following section provides an overview of the different approaches to include their strengths and weaknesses.

7.4.3.1 Resource Reservation Protocol

The most mature "Closed Loop" approach at this point is the resource reservation protocol (RSVP). It is defined in RFC 2205* [13] as a protocol that makes resource reservations for both unicast and many-to-many multicast applications, adapting dynamically to changing group membership as well as to changing routes. In VoIP, RSVP is initiated by the LSC or Softswitch, on behalf of the end instrument, to the router. Upon receiving a call request, the LSC or Softswitch sends a request to the router for a reservation to be established to the distant end. This step in the call setup process is called a precondition and is described for the session initiation protocol (SIP) in RFC 3312† [14]. The router will convert the LSC or Softwitch request into a RSVP request. Only if the request is satisfied does the LSC or Softswitch continue with the call setup signaling. RSVP is used in the WAN where bandwidth limitations require some mechanism to provide QoS. An example of a call-signaling flow using preconditions with SIP is shown in Figure 7.5.

---

* http://www.ietf.org/rfc/rfc2205.txt
† http://www.ietf.org/rfc/rfc3312.txt

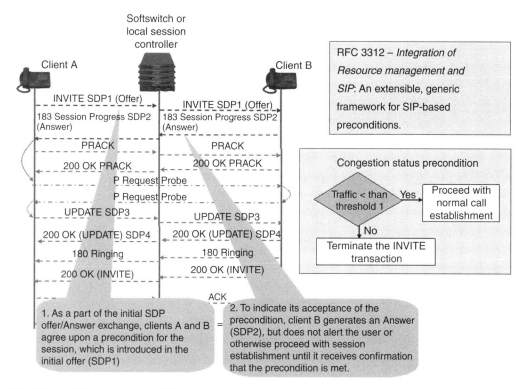

FIGURE 7.5 SIP precondition call flow.

The RSVP requests usually result in resources being reserved in each RSVP aware node along the data path. RSVP does not require that every node in the data path support RSVP. Typically, it is only enabled on nodes where congestion occurs. A single RSVP request results in resources being reserved in one direction. Therefore, both ends of the VoIP session must initiate a RSVP request to support the bi-directional bearer stream. RSVP does not have a routing function and is designed to operate with existing routing protocols. A benefit of this approach is that RSVP maintains a "soft" state in routers and adapts automatically to routing changes.

The limitation usually associated with RSVP is scalability and is caused by the need of RSVP to maintain the state on every node of the path for every VoIP call, which incurs processing and memory resources. The scalability issue has been addressed by the emergence of an approach that aggregates RSVP sessions in the network core routers and is described in RFC 3175.* Because the reservations of the core routers are aggregated to decrease the number of reservations, the largest numbers of reservations are maintained at the enclave level where aggregation is not feasible. Simulations have shown that midlevel routers have the capacity to easily support the typical traffic engineered load. The primary limitation today with RSVP is that most router vendors have not embraced it for their products and it is primarily limited to Cisco routers. As a result, the VoIP community has been hesitant to embrace RSVP and therefore commercial deployments are limited. Figure 7.6 shows a typical VoIP call setup flow when RSVP is enabled.

---

* http://www.ietf.org/rfc/rfc3175.txt

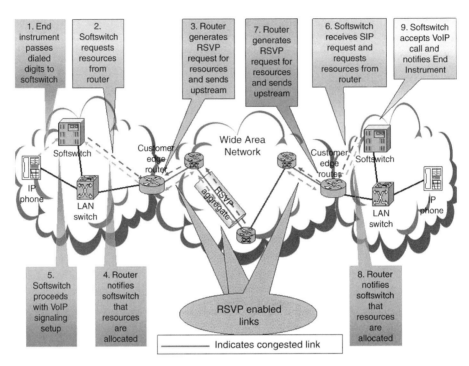

FIGURE 7.6   Typical VoIP RSVP call flow.

7.4.3.2 Explicit Congestion Notification

The next approach is still in the research and development phase and is called explicit congestion notification (ECN). ECN was originally described in RFC 3168* for TCP traffic to provide an early warning mechanism to the TCP stack that congestion was occurring in the network. The ECN bits are the last two bits in the TOS field of the IP header and the ECN bits combined with the DSCP bits are the eight bits found in that field as shown in Figure 7.7.

The IETF and commercial industries are investigating the feasibility of using the ECN bits as a mechanism for reporting network congestion to the CAC function in the Softswitches and LSC via the end instruments [15].† The basic principle is that routers have the ability to mark the ECN field of VoIP bearer packets (or probe packets sent by the end instrument) when the router experiences congestion. When a VoIP call is being established, the end instrument (or Softswitch or LSC on its behalf) transmits a probe to the distant end. The ECN-enabled routers in the path will mark the packets using the ECN bits to indicate the levels of congestion. The distant end will return the packet and upon receipt, the originating node will know the level of congestion in both directions. This approach recognizes that asynchronous routing may occur. Upon receipt of packets that indicate congestion, the Softswitch or LSC may decide to block the call from entering the network until the congestion is alleviated.

---

*   http://www.ietf.org/rfc/rfc3168.txt
†   J. Babiarz, K. Chan, and V. Firoiu, "Congestion notification process for real-time traffic," 2005, draft-babiarz-tsvwg-rtecn-05.

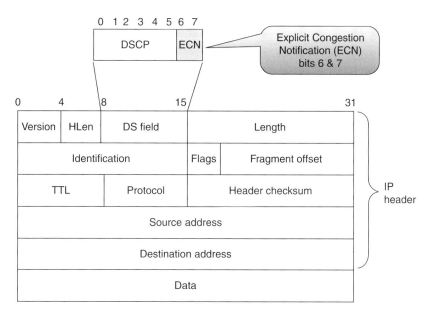

FIGURE 7.7 Explicit network congestion bits.

Like RSVP, the ECN approach only requires that it be enabled on routers where congestion occurs. In addition, it is also designed to work with the call-signaling precondition step discussed in the RSVP section discussed earlier and described in RFC 3312. The 2 bits within the ECN field allow the router to indicate up to four states. The current thoughts on the markings of the ECN bits are shown in Figure 7.8.

After a session is established, the marking of the ECN bits allow network providers to monitor the network for congestion and determine whether their packets are being marked inappropriately. The use of the packets for monitoring congestion is straightforward, but the detection of inappropriately marked packets is complex and is beyond the scope of this chapter. The primary issue with this approach is that commercial routing

Traffic load status in the network	Bit 6	Bit 7	Action taken by communication server
Unknown	0	0	End system is not ECN-capable
Not congested	1	0	Admit all calls
1st traffic level (1st level of congestion)	1	1	Block admission of regular calls (Admit emergency '911' calls and public service calls)
2nd traffic level (2nd level of congestion)	0	1	Future use

FIGURE 7.8 ECN markings.

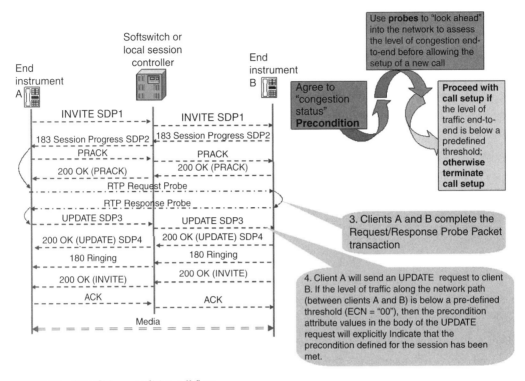

FIGURE 7.9   SIP ECN precondition call flow.

vendors have not implemented this approach in their products due to a lack of perceived customer demand. Another weakness of the approach is that it is reactive in nature in that it cannot predict congestion. Finally, a third weakness is that it delays the call setup until the probes are received and processed. Figure 7.9 shows a SIP with precondition call flow using the ECN approach.

7.4.3.3 Bandwidth Broker

The final "Closed Loop" approach discussed in this chapter is the bandwidth manager approach. Like the ECN approach, the bandwidth manager approach is still in the research and development phase. However, several VoIP vendors are investing in the bandwidth manager approach and it is likely that a commercial version will be available within the next five years. There are several bandwidth manager options that are being investigated. The first option is whether the bandwidth manager should be link or path-based. The path-based approach calculates an end-to-end path for every call, whereas the link-based approach only uses the routing topology information to calculate the load on each link and does not maintain an end-to-end perspective. Independent of the approach taken, the bandwidth manager has several responsibilities. The first responsibility is to maintain a list of paths or links in the IP network and the allocated voice capacity for each. The bandwidth manager then periodically polls the routers, Softswitches, and LSC to determine their utilization and routing behavior. Based on the inputs, the bandwidth manager calculates new budgets for every Softswitch and LSC and pushes the information down to the appliance.

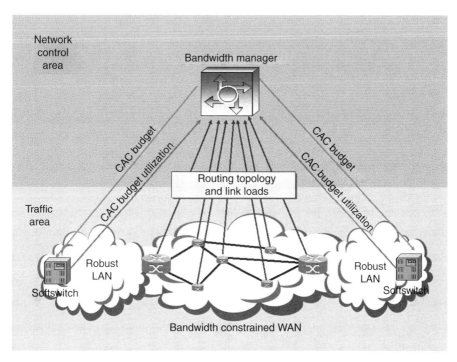

FIGURE 7.10 Basic elements of a bandwidth manager.

One impetus for the bandwidth manager approach is the large number of standards that are under development for the approach. In particular, the 3rd Generation Partnership Project (3GPP) has defined an end-to-end QoS that incorporates a policy decision function that maps in part to the bandwidth broker functions described earlier. In addition, the European Telecommunications Standards Institute (ETSI), Telecommunications and Internet Protocol Harmonization over Networks (TIPHON), and Services and Protocols for Advanced Networks (SPAN) have continued that work and published several architecture documents with the bandwidth manager. Figure 7.10 shows the basic elements of a bandwidth manager.

7.5 Summary

Voice over Internet Protocol is the future for voice services and the race for voice vendors to bring VoIP and converged services has begun. QoS is essential to achieve an acceptable MOS in IP networks. As discussed in this chapter, service providers typically combine mechanisms in the control and data plane to achieve their QoS objectives. These QoS mechanisms are sufficient to meet the constraints of today's service offerings; however, it is likely that they will be insufficient to meet the future needs. In response, the VoIP vendor community is working with the standards community and service providers to develop "Closed Loop" QoS mechanisms to provide closed loop mechanisms that couple the data and control planes to overcome the obstacles of the future.

References

1. ITU-T, "Y.1530—Call processing performance for voice service in hybrid IP networks pre-published," November 2007, http://www.itu.int/rec/T-REC-Y.1530/en (accessed January 26, 2008).
2. ITU-T, "P.800—Methods for subjective determination of transmission quality," August 1996, http://www.itu.int/rec/T-REC-P.800/en (accessed January 26, 2008).
3. ITU-T, "G.107—The E-model, a computational model for use in transmission planning," March 2005, http://www.itu.int/rec/T-REC-G.107-200503-I/en (accessed January 26, 2008).
4. ITU-T, "Y.1541—Network performance objectives for IP based services," February 2006, http://www.itu.int/rec/T-REC-Y.1541-200602-I/en (accessed January 26, 2008).
5. ITU-T, "G.109—Definition of categories of speech transmission quality," September 1999, http://www.itu.int/rec/T-REC-G.109-199909-I/en (accessed January 26, 2008).
6. ITU-T, "Y.1291—An architectural framework for support of quality of service in packet networks," May 2006, http://www.itu.int/rec/T-REC-Y.1291-200405-I/en (accessed January 26, 2008).
7. Telecommunications Industry Association (TIA), "Technical service bulletin (TSB)-116A—Telecommunications—IP telephony equipment—Voice quality recommendations for IP telephony," March 2006.
8. G. Huston, "Next steps for the IP QoS architecture," RFC 2990, November 2000, http://www.ietf.org/rfc/rfc2990.txt (accessed January 26, 2008).
9. A. Viswanathan and R. Callon, "Multiprotocol label switching architecture," RFC 3031, January 2001, http://www.ietf.org/rfc/rfc3031.txt (accessed January 26, 2008).
10. S. Blake, D. Black, M. Carlson, E. Davies, Z. Wang, and W. Weiss. "An architecture for differentiated services," RFC 2475, December 1998, http://www.ietf.org/rfc/rfc2475.txt (accessed January 26, 2008).
11. A. Charny, J. Bennet, K. Benson, J. Le Boudec, W. Courtney, S. Davari, V. Firoiu, and D. Stiliadis, "An expedited forwarding PHB (Per-Hop Behavior)," RFC 3246, March 2002, http://www.ietf.org/rfc/rfc3246.txt (accessed January 26, 2008).
12. J. Heinanen, F. Baker, W. Weiss, and J. Wroclawski, "Assured forwarding PHB group," RFC 2597, June 1999, http://www.ietf.org/rfc/rfc2597.txt (accessed January 26, 2008).
13. R. Branden, L. Zhang, S. Berson, S. Herzog, and S. Jamin, "Resource reservation protocol (RSVP)—Version 1: Functional specification," RFC 2205, September 1997, http://www.ietf.org/rfc/rfc2205.txt (accessed January 26, 2008).
14. G. Camarillo, W. Marshall, and J. Rosenberg, "Integration of resource management and session initiation protocol (SIP)," RFC 3312, October 2002, http://www.ietf.org/rfc/rfc3312.txt (accessed January 26, 2008).
15. J. Babiarz, K. Chan, and V. Firoiu, "Congestion notification process for real-time traffic," Internet, Draft 2005, http://www.potaroo.net/ietf/all-ids/draft-babiarz-tsvwg-rtecn-05.txt (accessed January 26, 2008).

8

Measurement and Analysis on the Quality of Skype VoIP

Zhen Ren and Haining Wang

CONTENTS

Skype is currently the most popular Voice over Internet Protocol (VoIP) software used by Internet users. Focusing on the quality of voice of this popular VoIP application, we survey the existing performance measurement studies on Skype. Skype has been developing attractive features for its IP telephony services. We investigate the impact of network-base factors, such as packet loss, delay, etc., on the quality of voice with respect to different kinds of services: Skype client-to-client audio conversation, Skype client Public Switched Telephone Networks (PSTN) to telephone call, group conference, and Skype calls in a mobile environment.

8.1 Introduction

The technology of VoIP brings easy-to-use and cost-saving communication. Different from traditional telephony services, the packet infrastructure of IP network achieves the flexibility in network transmission. Regular data files and multimedia streams are carried in packets, and the network routers between the source and destination are unaware of the upper application level details. The benefit of network resource sharing, together with the network convergence, motivates the wide development and deployment of VoIP. For the past few years, the rapid growth of Skype has evidenced the advantages of IP telephony service over the traditional PSTN.

Skype was created by the entrepreneurs Niklas Zennström and Janus Friis. Since August 2003, when the Skype's first public beta version was released, the number of Skype users has experienced tremendous growth. In April 2006, Skype registered user accounts reached 100 million, and soon increased to 200 million in the second quarter of 2007. Up to September 30, 2007, the number of unique Skype user accounts is 246 million. It is reported that 10,140,836 concurrent users were online in October 30, 2007 [1]. Skype is well regarded for offering stable and free VoIP service to its users, and is also inexpensive for calling from a PC to a phone, which can either be a landline or a cell phone [2]. Its software integrates functions of audio conference, instance message, file transfer, voice mail, call forwarding, and maintenance of buddy list, providing a variety of attractive services for Skype users.

The recent investigations on Skype focus on different aspects of IP telephony services: performance, security, network topology, and so on. In VoIP, performance issue is important, because of the real-time nature of VoIP and the best-effort IP service model. It is very challenging to increase the service quality to the PSTN level, but Skype is currently making outstanding progress. In this study, Skype is measured in different models and analyzed in different scenarios. Skype supports multi-party conversation, but in the context of group conference there are no standard metrics to characterize the performance of a conferencing application. A new subjective metric is proposed and applied on Skype conferencing.

In this chapter, we survey the measurement studies on Skype service quality under different network environments. We analyze the metrics used for evaluating Skype service quality and classify the advantages and disadvantages. The basic measurements for voice quality include subjective methods such as MOS and objective models such as E-Model and PSEQ. The measurements mainly focus on quantifying network factors, such as network delay, jitter, and packet loss, which directly affect the voice quality.

The remainder of the chapter is organized as follows: Section 8.2 introduces the functionalities of Skype. Section 8.3 presents some of the frequently used models for assessing the perceptual quality of IP telephony. Section 8.4 surveys on these models applied to specific working scenarios of Skype, in group conference and mobile environments. Section 8.5 concludes the chapter.

8.2 Skype Overview

Skype is unique in its network structure, as shown in Figure 8.1. The same developing team of Skype formerly produced the well-known peer-to-peer file-sharing application KaZaa, and the Skype network is organized as a peer-to-peer overlay structure using a

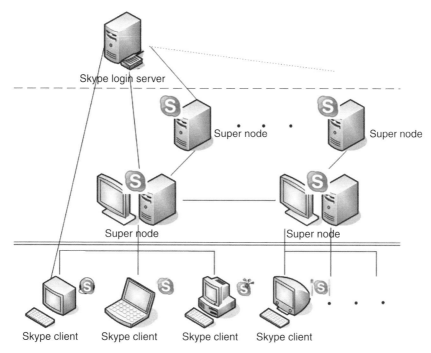

FIGURE 8.1   Skype decentralized topology: super nodes and normal clients.

specific protocol. Different from other VoIP applications that are based on the more conventional centralized server model, Skype does not include central servers except for a login server, with which the Skype clients would communicate when logging in.

The function of Skype begins with a login process, when the Skype clients first register to the login server. They are then connected to a super node. Clients join the network after this process.

Skype network consists of normal Skype clients and super nodes [3]. Multiple clients can connect to one super node. The strategy of selecting a Skype super node is not very clear. According to previous studies [3], if a client has a public IP, sufficient CPU power, memory, and network bandwidth, and stays online for a long time, it is very likely to be selected as a super node. The super nodes still run normal routines as clients, but at the same time, they are connected to each other forming the backbone of the Skype network.

In the conventional VoIP process, the establishment of an audio session consists of two parts: signaling and multimedia transition. In Skype, none of these two stages involve the login server. Hence, when considering the Skype telephony functions, we can exclude the login server from the Skype network topology. Skype clients and super nodes form an overlapping, peer-to-peer topology. A client delivers its voice through the super node it connects to.

The peer-to-peer network infrastructure is one of the main reasons for Skype's success. On one hand, the decentralized overlay structure ensures high performance for the demanding real-time telephony service while on the other, the network is cheap to maintain and easy to scale. The Skype stores its user directory among the nodes across the network, rather than in a complex and costly centralized global server. In this manner, the network can scale very easily to a large size without too much degradation on the overall performance. The Skype application works out the efficient network path, and takes as much resource as it can access to ensure quality of service. The use of efficient audio coding with iLBC, iSAC, or a third unknown codec, is another reason that the quality of Skype VoIP services is guaranteed.

Skype develops the following key feature functions, which make it very competitive among its peer VoIP software.

Connection to PSTN Phones   Skype supports call between a client on the Internet and the traditional telephone network, including mobile network, with SkypeOut and SkypeIn. The Internet and PSTN are connected by two kinds of specialized servers, which convert the VoIP packets to the traditional telephony signals and vice versa.

Group Conference   The latest version of Skype allows users to establish audio conference containing up to nine people online at the same time. In the group conference case, Skype does not use a specialized central server either. The audio data are collected from group members, mixed at one of the end points, and then distributed to all participants.

NAT and Firewall Traversal   Skype's NAT and firewall traversal ability ensures that it can work behind almost all kinds of NATs and firewalls. It is conjectured that the Skype client uses a variation of STUN and TURN protocols. Moreover, as Skype can use both UDP and TCP as its transport layer protocol, it can easily switch to TCP transition if the UDP flow is blocked by some firewalls.

Security   The website of Skype claims that it uses 256-bit Advanced Encryption Standard (AES) for encryption, and yet a 1536 to 2048 bit RSA for negotiating symmetric AES keys to secure the conversations carried over Skype.

8.3 VoIP Quality of Service

Unlike the circuit-switched PSTN, which is dedicated to deliver reliable telephony service, IP network is constructed with the "best effort" service model. The packets could be delayed or lost during transmission. A great deal of research work has been done to improve the quality of VoIP services, such as load balancing and routing backup. But first, the quality of service has to be measured, and the network factors have to be quantified.

One way to determine the quality of VoIP service is to measure it from a user perceptive view. Models have been developed to evaluate the perceptive quality of voice. As shown in Figure 8.2, the measurement methods can be categorized into subjective and objective.

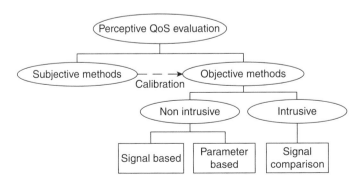

FIGURE 8.2   User-perceived QoS measurements.

Subject methods simply depend on user's opinions to measure the quality of voice, whereas objective methods are divided into intrusive methods and non-intrusive methods, according to whether both the original signal and the transmitted signal are monitored to compare the degradation. Non-intrusive methods can be further divided into signal- and parameter-based methods.

8.3.1 Mean Opinion Score (MOS)

Mean Opinion Score [4] is the most commonly applied subjective method. It is a set of standard and subjective tests, in which test sentences are read aloud by both male and female speakers over the communications medium. A number of listeners are then asked to actually rate the quality of voice. All individual scores from the tests are collected to calculate an arithmetic mean, resulting in a final MOS score. The score is normally between 1 and 5, with 5 being the best and 1 the worst, as listed in Table 8.1.

On one hand, the MOS scoring process requires the participation of human beings, hence it is often time-consuming and expensive to operate. On the other hand, MOS gives a prime criterion for the perceptual quality assessments, and provides the base for many objective measurements.

8.3.2 Quality Factors

In VoIP, the following factors have direct impact on the quality of voice

Audio Codec: Audio codec is used to digitize normal voice, so that the voice can be packetized and transmitted across the IP network. It is usually chosen with respect to bandwidth limits, and different codecs have different impact on the coded voice signal. G.711 μ/A, G.723.1, and G.729A are popular codecs in IP telephony. Skype claims that it also uses iLBC and iSAC.

Delay: Delay can be induced from several links along the communication path: audio codec, data packetization, serialization, network transmission, jitter buffer, etc. VoIP service has the real-time requirement in its nature. Usually if the delay accumulated at the receiver is less than 150 ms, it does not much affect the perceptual quality. Otherwise, the impairment of delay must be considered. A simple and end-to-end metric would be mouth-to-ear (M2E) delay [5] measured between the very end of the speaker and listener.

Packet Loss: During network transmission, packet loss is inevitable. Especially in the VoIP context, a packet that arrives late (i.e., beyond a certain threshold) is equivalent to a packet loss.

In the IP packet networks, the transportation impairment cannot be totally removed. Compared to traditional dedicated telephone connections, queuing in the IP routers induces much more unsteady delay during packet transmission. Besides, it is one important source

TABLE 8.1

Mean Opinion Score

MOS	Quality of the Speech	Impairment
5	Excellent	Imperceptible
4	Good	Perceptible, but not annoying
3	Fair	Slightly annoying
2	Poor	Annoying
1	Bad	Very annoying

of delay jitter, causing uneven arrival and packet disorder, and a full queue will result in packet losses.

Moreover, different implementations of silence suppression, jitter buffer, voice synchronization, and design of network, may further complicate the measurements.

In IP telephony technology, buffer is commonly used to smooth the network jitter, but the process transforms jitter into delay and packet loss. In the following, we introduce a few objective models that are developed to measure these factors.

8.3.3 Perceptual Evaluation of Speech Quality (PESQ)

The Perceptual Evaluation of Speech Quality (PESQ) [6] is one of the most widely used objective methods for measuring VoIP quality of voice. PESQ is the result of an integration of the Perceptual Analysis Measurement System (PAMS) and PSQM99, an enhanced version of Perceptual Speech Quality Measure (PSQM). It is based on signal comparison, but different from its predecessor PSQM, which is designed to assess speech codecs. PESQ takes more elements into consideration, such as coding distortions, errors, packet loss, delay and variable delay, and filtering in analog network components. The PESQ model collects signals from both the sender and receiver side for comparison, and the outcome is the estimation of perceived quality of received voice, which can be mapped to scores from subjective listening tests, for example, MOS scores.

First, PESQ uses time alignment algorithm to estimate the delay impact of the network. Based on the set of delays, PESQ uses a perceptual model to compare the signal on the sender side with the aligned degraded signal on the receiver side. This model transforms signals from both sides into one internal representation, which maintains the voice feature of perceptual frequency and loudness. The entire process can be structured in the following stages [7], as illustrated in Figure 8.3:

Step 1. Signaling Preprocessing The preprocess takes signal frequency as input, and applies time alignment to the signals. The time alignment algorithm computes the delay between the sender and receiver, and compares the confidence of having two delays in a certain time interval with the confidence of having a single delay for that interval.

Step 2. Perceptual Modeling Signals from the sender and receiver sides are transformed separately into one internal representation. The transformation applies mapping in both time and frequency domains (32 ms or 256 samples for 8 kHz, similar to the PSQM), and it also utilizes a signal filter for typical telephone network bandwidth (in order to not affect the PESQ measurement).

Step 3. Cognitive Modeling This phase is associated with noise computation. The model first calculates a difference between reference signal and distorted signal for each time-frequency cell. A positive difference indicates the presence of noise, whereas a negative difference indicates a minimum noise presence such as codec distortion. These values are considered to produce a MOS score prediction. The model makes it easy to discover the time jitter, identify which frames are involved, and which frames are affected by the delay and should be erased to prevent a bad score.

The final PESQ score is in the range of -0.5 to 4.5. For most cases, the results are in the MOS-like listening quality score range between 1.0 and 4.5.

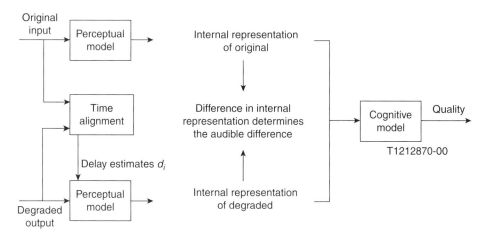

FIGURE 8.3 Overview of PESQ process.

8.3.4 E-Model

E-Model [8,9] was developed by the European Telecommunications Standards Institute (ETSI), and standardized by the ITU as G.107. Currently it is among the most popular objective measurement methods. E-Model belongs to non-intrusive parameter-based method category, and therefore it does not need signals from both source and destination to measure the quality of voice. The model quantifies the network factors, and can be used to improve network environment for better IP telephony services.

E-Model characterizes damaging factors to estimate the quality degradation. It calculates a base value, then considering multiple network factors, the degradation is expressed as damage factors, and subtracted from this base. The output of E-Model is the "Rating Factor" R, which can be mapped to MOS scores with the range of 0–100, as shown in Table 8.2 and the typical range of R is between 50 and 94. The values below 50 are unacceptable, and the maximum rating for a typical telephone connection is 94. Thus, 94 is often taken as the base value for E-Model.

The R factor could be different under different codecs, and for wideband codecs the R factor may increase to above 100. For example, the R factor for an unimpaired connection may be equal to 110.

The R factor is calculated as

$$R = R_0 - I_s - I_d - I_e + A + W.$$

TABLE 8.2

R Factor to MOS Value Mapping

R Value Range	MOS Score	User Satisfaction
90–100	4.3–5.0	Very satisfied
80–90	4.0–4.3	Satisfied
70–80	3.6–4.0	Some dissatisfaction
60–70	3.1–3.6	More dissatisfaction
50–60	2.6–3.1	Most dissatisfaction
0–50	1.0–2.6	Not recommended

TABLE 8.3

E-Model Parameter Description

Parameters	Description
R_0	The basic signal-to-noise ratio based on sender and receiver loudness ratings and the circuit and room noise.
I_s	The sum of real-time, or simultaneous speech transmission impairments, e.g., loudness levels, sidetone and PCM quantizing distortion.
I_d	The sum of delay impairments relative to the speech signal, e.g., talker echo, listener echo, and absolute delay.
I_e	The equipment impairment factor for special equipment, e.g., low bit-rate coding (determined subjectively for each codec and for each percentage of packet loss and documented in ITU-T Recommendation G.113).
A	The advantage factor adds to the total and improves the R-value for new services.
W	The wideband correction factor.

I_d and I_e indicate the network impact on speech signals. The parameters of E-Model are explained in Table 8.3.

E-Model can be simplified; for example, Cole and Rosenbluth's model [10] takes only two parameters: loss and delay. The loss rate and delay time are translated into delay impairment and equipment impairment factor by a mapping derived from experimental MOS score curve. The simplified E-model is presented as

$$R = 94.2 - I_e - I_d;$$

$$MOS = 1 + 0.035 * R + 7 * 10^{-6} * R * (R - 60) * (100 - R).$$

8.4 Quantifying Skype QoS

Skype is highly recommended by the users, because of its cost-saving advantage over traditional telephones. It also overmatches its peer softphones on the quality of service.

8.4.1 Skype Quality of Service

The performance of Skype was compared with another commonly used VoIP client software MSN messager by Gao and Luo [11]. The set up delay and M2E delay for the audio conversation functions were compared. The experiments were performed under different network environments including LAN, WLAN, DSL, GPRS, and CDMA2000x, with the clients located in same or different domains.

Results

Connection Setup Delay: The connection setup delay is measured for the two softwares. In most cases, Skype outperformed MSN messager with shorter delays. Because Skype can traverse almost all types of NAT and firewalls, it can work under every testing condition, whereas MSN can only set up communication for clients in the same domain or no NAT/firewall is involved. The setup delay for Skype varies due to the traversal process.

M2E Delay: The M2E delay is the the mouth-to-ear delay measured in the conversation. The experiments show that when the network condition is good, that is, when in the same domain or no NAT/firewall exists, MSN has shorter M2E delays in many such cases. But both MSN and Skype has acceptable M2E delays, which does not much effect the quality of voice.

8.4.2 Skype Connection to PSTN

The design of Skype also considers the compatibility with the traditional telephone network. In order to connect to PSTN, common VoIP solutions set a conjunction point between the two different networks, and signals are converted at this point. In Skype, the connecting servers are SkypeIn and SkypeOut, respectively.

SkypeIn represents the IP end point as a normal phone number to the PSTN side. When a PSTN number is dialed by Skype client software, the call is directed to SkypeOut, which translates the PSTN number and setups the connection. During the conversation, SkypeIn/ SkypeOut are responsible to convert the analog signal of voice data to digital format and packetize them at the IP side. At the same time, SkypeIn and SkypeOut servers are connected to the Skype network as normal clients. Even functionally speaking, they run completely different tasks from the ordinary client.

The quantification of quality of voice in Skype-to-PSTN scenario is difficult, because the delay can be generated in both network transition and signal converting stages, and there is less information about the SkypeIn/SkypeOut path. An alternative is to consider in an end-to-end way, and measure the E2M delay during the conversation.

SkypeIn/SkypeOut provide high performance service for connecting PC to landline, but compared to Skype-to-Skype telephony, the delay increase is obvious, caused by the signal converting process and capacity limitation of SkypeIn/SkypeOut.

8.4.3 Skype Group Conversation

Skype also supports group conversation. According to previous analysis on Skype traffic, Skype uses end-mixing strategy in group conversation, that is, one of the members in the group is chosen as the leader. The leader is responsible for accepting audio from all other members, mixing the speeches, and forwarding them to all other participants.

As illustrated in Figure 8.4, in the end-mixing topology network, normal participants transmit their own speech data to the mixer, or leader, where the voices are mixed and sent back to them. Normal participants get a mixed copy of the voices from all other members in the conference, including the mixer. The mixer collects all the individual voices, and also generates a copy for its own use.

Although many measurement methods have been developed for two-party conversation, few studies have been done on the multi-party scenario. In a multi-party conversation, the additional audio-mixing process should be considered, but the impact of this process has not yet been well studied.

Fu et al. [12] proposed a new metric Group Mean Opinion Score (GMOS) based on two-party MOS. The key idea of GMOS was to calculate a subjective score based on MOS from each pair of speaker and listener in the conference. Among the users in the voice conference, the quality of voice perceived by each participant can be very different. Because of the heterogeneous network conditions, even for the same listener, the voice from different speaker may not be of the same quality. In GMOS model, assuming that the session has N participants, $\mathcal{P}_1, \mathcal{P}_2, \ldots, \mathcal{P}_N$, each participant provides MOS scores for the rest $N - 1$ others.

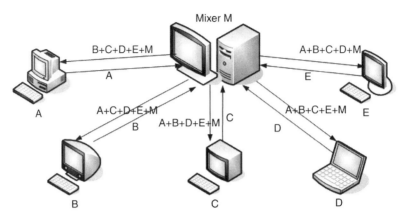

FIGURE 8.4   Topology of multi-party Skype conversation.

Hence from one participant's perspective, say the ith participant $\mathcal{P}_i$, his/her score to the overall quality of the conference session is his/her GMOS towards this conference. And it can be computed as

$$GMOS_i(MOS_i(1),\ldots, MOS_i(N), \alpha) = AVE + \alpha(AVE - MIN)U(-\alpha)$$
$$+ \alpha(MAX - AVE)U(\alpha). \qquad (8.1)$$

where α is used to adjust the listener's subjective opinion towards the quality of voice and $MOS_i(k)$ is the MOS score set by participant i for participant k.

$$AVE = \frac{\sum_{k=1}^{N-1} MOS_i(k)}{N-1}, \qquad (8.2)$$

$$MAX = \max\{MOS_i(1),\ldots, MOS_i(N)\}, \qquad (8.3)$$

$$MIN = \min\{MOS_i(1),\ldots, MOS_i(N)\}, \qquad (8.4)$$

$$\alpha \in [-1, 1], \qquad (8.5)$$

$$U(x) = \begin{cases} 1 & x > 0, \\ 0 & x \le 0. \end{cases} \qquad (8.6)$$

Experimental Setup
Using the new metric, Fu et al. [12] performed experiments of three-person, four-person, and five-person conferences, in which the participants were in different locations, as shown in Figure 8.5. MOS and GMOS scores were collected from pairs of listeners and speakers.

Results
The experimental results of α are close to 0. The listeners score with $\alpha \in (0, 1]$ are positive towards the voice quality, whereas the listeners score with $\alpha \in [-1, 0)$ are negative towards the voice quality. About 10% of the GMOSes are beyond the MIN and MAX range, that is,

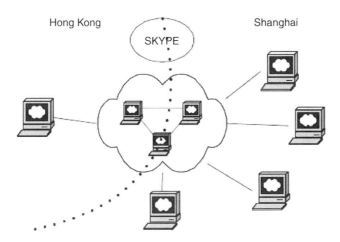

FIGURE 8.5 Network setting of Skype conference. (*Source:* Fu et al., "Performance metrics and configuration strategies for group network communication." 07INQOS, June 2007. With Permission.)

about 50% of the values of $\alpha \in [-0.2, 0.2]$. For the 392 αs, the average is 0.093 with the standard deviation of 0.464. Thus statistically, these subjects are positive on average for Skype.

Using the existing objective quality assessment model for two-party conversation, the MOS for each pair can be calculated. For example, Fu et al. [12] developed a Two Step Mapping Method (TSMM) [12], which uses E-Model to estimate MOS scores in one conference, and then compute GMOS based on these MOSes. As a result, the calculated GMOS is very close to the subjectively collected GMOS of the conference.

8.4.4 Skype Over Wireless Environment

With the wide deployment of wireless networks, people would question if the mobile environment can provide sufficient data transfer rates to support the VoIP applications. Hoßfeld et al. [13] have analyzed the achievable and actual quality of IP-based telephony calls on Skype under wireless environments.

The study contains two parts: performance measurements in a real Universal Mobile Telecommunications System (UMTS) network and in a test environment which emulate rate control mechanisms and changing system condition of UMTS networks. In the experiments, PESQ is used to evaluate quality of voice, packet loss, inter-packet delay, and throughput, and capture the influence of network-based factors. Also, Network Utility Function (NUF) is applied to describe the impact of the network on the quality of voice eventually perceived by end-users.

As the PESQ model requires signals from both the sender and receiver, the voice data were recorded in wav files. Therefore, the degradation of the quality of voice because of the Skype iLBC codec should be considered. In the experiments, the PESQ value of 3.93 was used as reference for the result of audio codec degradation.

The PESQ reduction caused by the network connectivity between sender and receiver is described by the Network Utility Function (NUF) U_{Netw}: $PESQ_{rcvd} \simeq U_{Netw} * PESQ_{sent}$.

The UNF takes the impact of several common network factors into consideration:

$$U_{Netw} = \prod_i U_i = U_m * U_s. \quad U \in \{0,1\}. \tag{8.7}$$

Here the m-Utility Function (m-UF) U_m captures the variation of the mean throughput during a certain observation window ΔW. m_{sent} is the average throughput of sender, and m_{rcvd} is the average throughput of receiver. Assuming a linear dependence on the loss ratio: $1 + \max\{1 - (m_{rcvd}/m_{sent}), 0\}$, then U_m can be calculated as $U_m = \max\{1 - k_m l, 0\}$, where k_m denotes the degree of utility reduction.

U_s denotes the s-Utility Function (s-UF), which captures the change of the standard deviation of the throughput from s_{sent} to s_{rcvd} during an observation window ΔW. To calculate the standard deviation, the throughput values are averages during a short interval of ΔT. The relative change of the standard deviation is denoted as $\sigma = (s_{rcvd} - s_{sent})/s_{sent}$.

Then, U_s can be calculated as

$$U_s = \begin{cases} \max\{1 - k_s^+ \sigma, 0\} & \text{for } s_{sent} < s_{rcvd} \\ 1 & \text{for } s_{sent} < s_{rcvd} \\ \max\{1 - k_s^- \sigma, 0\} & \text{for } s_{sent} < s_{rcvd} \end{cases} \tag{8.8}$$

Here the parameter k_s^+ reflects the decrease of U_s when the standard deviation doubles, while k_s^- is used when the standard deviation is vanishing.

Experimental Setup

First, a bottleneck scenario is set up in a LAN, which includes a traffic-shaping router to measure the effect of bandwidth variation on VoIP services. Skype clients are then connected to the Internet via a public UMTS operator, and the uplink and downlink of VoIP traffic are monitored.

Figure 8.6 shows the experimental setup in the UMTS scenario, as well as the service degradation caused by Skype codec.

Results

In the bottleneck LAN experiment, the dynamically changing condition in UMTS is emulated by generating different packet loss rates. Skype is able to detect packet losses and treat them as an indication of congestion in the network. It then increases the throughput of the sender by sending packets with larger payload. The result of Skype throughput and PESQ score are presented in Figure 8.7.

In the real UMTS experiment, the packet interarrival time and PESQ scores are measured for uplink and downlink, and listed in Tables 8.4–8.7.

In this study, as expected, the packet losses degrade the PESQ value in the bottleneck LAN experiment. Moreover, due to network jitter and the use of a different codec, the PESQ values in the public UMTS environment are worse than those in the bottleneck LAN emulation. However, the quality of voice is still acceptable, indicating that the capacity offered by UMTS is sufficient for supporting mobile VoIP calls.

8.4.5 Skype User Satisfactions

In addition to the measurement of the perceptual quality of voice for Skype, Chen et al. [14] presented a model to quantify a new objective index of user satisfaction.

The concept of user satisfaction is based on the assumption that if users are not satisfied with the service, they will terminate the conversation sooner, resulting in a shorter session. Hence a new metric, User Satisfaction Index (USI), is built upon call durations.

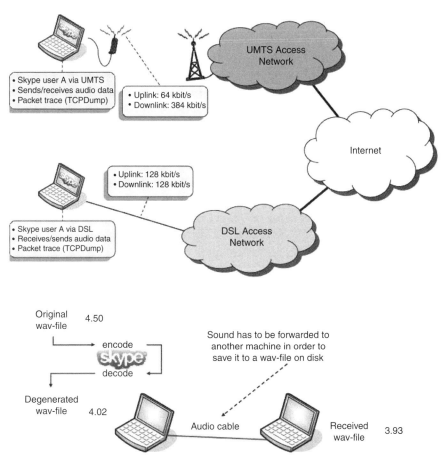

FIGURE 8.6 Skype conversation on UMTS. (*Source:* T. Hoßfeld et al., "Measurement and analysis of Skype VoIP traffic in d 3G UMTS systems." University of Wurzberg, December 2005. With permission.)

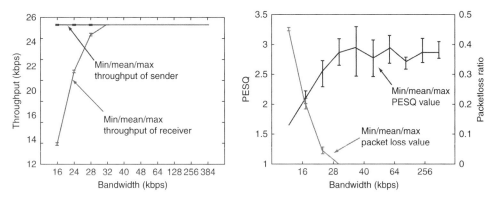

FIGURE 8.7 Bottleneck LAN scenario. (*Source:* T. Hoßfeld et al., "Measurement and analysis of Skype VoIP traffic in d 3G UMTS systems," University of Wurzberg, December 2005. With permission.)

TABLE 8.4

Key Performance Measures for UMTS Uplink Scenario

	Throughput m_{sent} (bps)	Deviation s_{sent} (bps)	Goodput m_{rcvd} (bps)	Deviation s_{rcvd} (bps)	PESQ Q
μ	18071.58	2300.95	18055.23	3497.57	2.24
σ	8.84	568.87	21.20	858.38	0.16

Source: T. Hoßfeld et al., "Measurement and Analysis of Skype VoIP Traffic in d 3G UMTS systems," University of Wurzberg, December 2005. With permission.

TABLE 8.5

Received Packets in the UMTS Uplink Scenario

Payload	Number	Mean PIT	Std. PIT
3 byte	3	20.02 s	10.73 ms
108 byte	847	61.97 ms	35.00 ms
112 byte	28	1.92 s	27.05 ms

Source: T. Hoßfeld et al., "Measurement and Analysis of Skype VoIP Traffic in d 3G UMTS systems," University of Wurzberg, December 2005. With permission.

TABLE 8.6

Key Performance Measures for UMTS Downlink Scenario

	Throughput m_{sent} (bps)	Deviation s_{sent} (bps)	Goodput m_{rcvd} (bps)	Deviation s_{rcvd} (bps)	PESQ Q
μ	18023.77	1848.15	18007.08	2172.39	2.49
σ	48.16	282.70	51.64	284.97	0.085

Source: T. Hoßfeld et al., "Measurement and analysis of Skype VoIP traffic in d 3G UMTS systems," University of Wurzberg, December 2005. With permission.

TABLE 8.7

Received Packets in the UMTS Downlink Scenario

Payload	Number	Mean PIT	Std. PIT
3 byte	6	9.46 s	4.49 s
21 byte	14	1.73 s	3.58 s
108 byte	817	61.32 ms	16.00 ms
112 byte	16	3.20 s	45.02 ms

Source: T. Hoßfeld et al., "Measurement and analysis of Skype VoIP traffic in d 3G UMTS systems," University of Wurzberg, December 2005. With permission

Similar to the measurement of the perceptual quality of voice, in the new model, common network factors are analyzed against call durations. The statistical study on experimental traces indicates that the bitrate and jitter are closely related to the session time. As a result, the USI of a session is defined as follows:

$$\text{USI} = -\beta^t \mathbf{Z} \tag{8.9}$$

$$= 2.15 * \log(\text{bitrate}) - 1.55 * \log(\text{jitter}) - 0.36 * \text{RTT}, \tag{8.10}$$

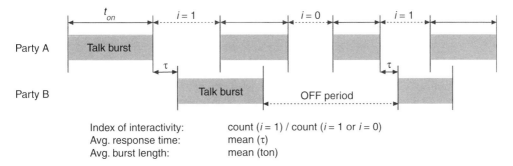

FIGURE 8.8 Voice interactivity metrics. (*Source:* K. T. Chen et al., "Quantifying Skype user satisfaction," *SIGCOMM '06*, Pisa, Italy, September 2006. With permission.)

The new model only includes simple parameters, which are easy to measure. Furthermore, USI is able to assess the voice interactivity and smoothness of the conversation. The authors proposed three metrics as illustrated in Figure 8.8.

The index of interactivity denotes the responsiveness, which is the degree of alternation between statement and response. The response time denotes the delay between two talk bursts. Including the burst lengths, the three metrics are related to the quality of conversation, and thus they can directly affect USI. The correlation test verifies the relationship between USI and the interactivity metrics. Overall, USI provides another effective quantification method for measuring VoIP service quality.

8.5 Summary

Many studies have been conducted on the measurement and analysis of Skype's quality of service. Utilizing existing methodology on Skype, these studies quantified the factors that affect the quality in different Skype working scenarios: common Skype client-to-client audio conversation, group conference, and Skype calls in mobile environments.

As a widely used subjective quality of voice measurement, MOS connects user's perceptual feeling with the objective quality of voice. Most objective measurements quantify influencing factors in different aspects, and then map the results to MOS scores. For Skype group conferencing, GMOS is a subjective metric based on two-party MOS. MOS scores are collected from each pair of speaker and listener in the conference. Then an overall opinion of the conference service quality is derived. This metric is aimed to help VoIP software optimize their group conversation qualities by adjusting the network topology.

A study of Skype in a controlled LAN scenario shows that Skype changes its behavior when it detects a congestion, thus concludes that network bandwidth would effect the throughput of Skype. Furthermore, in the realistic UMTS mobile environment, Skype quality of voice is measured using PESQ, with NUF quantifying relative network factors. The result indicates that the mobile environment is less reliable and the performance of Skype is degraded, but the network capacity is still sufficient to meet the basic quality requirements.

Finally, a metric based on Skype session time is developed, indicating the degree of user satisfaction. The quantified result is a user satisfaction index, which reveals the responsiveness and talk burst length of the speech.

References

1. "Skype—Usage and traffic," http://en.wikipedia.org/wiki/Skype.
2. "Editor's review of Skype," http://www.download.com/Skype/3000-2349 4-10225260.html.
3. S. A. Baset and H. Schulzrinne, "An analysis of the Skype peer-to-peer internet telephony protocol," *Proceedings of the INFOCOM '06*, Barcelona, Spain, April 2006.
4. ITU-T Recommendation P.800, "Methods for subjective determination of transmission quality," 1996.
5. W. Jiang, K. Koguchi, and H. Schulzrinne, "QoS evaluation of VoIP end-points," *IEEE International Conference on Communications, ICC '03*, 2003.
6. ITU-T Recommendation P.862, "Perceptual evaluation of speech quality (PESQ), an objective method for end-to-end speech quality assessment of narrowband telephone networks and speech codecs," February 2001.
7. F. De Rango, M. Tropea, P. Fazio, and S. Marano, "Overview on VoIP: Subjective and objective measurement methods," *IJCSNS International Journal of Computer Science and Network Security*, vol. 6, no. 1B, January 2006.
8. ITU-T Recommendation G.107, "The E-model, a computational model for use in transmission planning," March 2005.
9. "The E-model, R factor and MOS, overview," *Psytechnics*, December 2003.
10. R. G. Cole and J. H. Rosenbluth, "Voice over IP performance monitoring," *ACM SIG-COMM Computer Communication Review*, vol. 31, no. 2, April 2001, pp. 9–24.
11. L. Gao and J. Luo, "Performance analysis of a p2p-based VoIP software," *Proceedings of the Advanced International Conference on Telecommunications and International Conference on Internet and Web Applications and Service (AICTICIW '06)*, 2006.
12. T. Z. Fu, J. Chiu, D. M. Lui, and C. S. Liu, "Performance metrics and configuration strategies for group network communication," *Fifteenth IEEE International Workshop on Quality of Service, IWQoS '07*, June 2007.
13. T. Hoßfeld, A. Binzenhöfer, M. Fiedler, and K. Tutschku, "Measurement and analysis of Skype VoIP traffic in 3G UMTS systems," *University of Würzburg Institute of Computer Science Research Report Series,* Report No. 377, December 2005.
14. K. T. Chen, C. Y. Huang, P. Huang and C. L. Lei, "Quantifying Skype user satisfaction," *ACM SIGCOMM Conference on Applications, Technologies, Architectures, and Protocols for Computer Communicate (SIGCOMM '06)*, Pisa, Italy, September 2006.

9

QoE Assessment and Management of VoIP Services

Akira Takahashi

CONTENTS

9.1 Introduction

The quality of voice communication services such as conventional Public Switched Telephone Networks (PSTN) and Voice over Internet Protocol (VoIP) should be discussed in subjective terms. For example, if some noise is added to the speech signal transmitted over networks, this causes no problem if humans cannot hear it. In contrast, even if the network transmits the speech signal without errors or noise, the quality becomes bad when the terminal device causes a serious degradation such as acoustic echo.

Traditionally, the quality of telephone services has often been called quality of service (QoS). Although the definition of QoS given by the International Telecommunication Union—Telecommunication Standardization Sector (ITU-T) also includes the subjective quality perceived by users, the term QoS has been used to represent network performance, excluding terminal and human factors, in end-to-end services. Therefore, ITU-T defines the new term "quality of experience" (QoE) as follows [1]:

> The overall acceptability of an application or service, as perceived subjectively by the end-user.

Notes

1. Quality of experience includes the complete end-to-end system effects (client, terminal, network, services infrastructure, etc.).

2. Overall acceptability may be influenced by user expectations and context (ITU-T Recommendation P.10/G.100).

Quality of experience includes the speech quality of VoIP services, the usability of telephone terminals, the functionality of terminals and network services, and the propriety of price, for example. However, this section only deals with speech quality because this is the most fundamental aspect of "telephone quality" and has been well studied for decades.

First, we definitely need a means of quantifying the QoE level. There are basically two approaches: One is to directly measure the quality of speech by a psycho-acoustic method, which is called "subjective quality assessment." The other is to estimate the subjective quality, which is evaluated by a subjective quality assessment, solely from physical characteristics of the speech transmission system/network and/or speech signal itself. This is called "objective quality assessment."

To achieve a sufficiently high QoE, we need to maximize the subjective quality evaluated as indicated previously, under some realistic conditions, for example, with limited network resources and cost. We can divide this process into two steps: quality design and quality management.

Quality design is usually performed prior to providing services. In this process, we determine the requirements for various quality-design parameters, such as end-to-end delay, packet-loss rate, and echo return loss, for example. However, we can design services more efficiently with an objective quality assessment such as the E-model, which was standardized as ITU-T Recommendation G.107 [2]. It is introduced in a later section.

Quality should be managed in an in-service environment because the QoE of the service must be monitored constantly or periodically to find system failures and transmission degradation due to congestion, for example. To do this, objective quality-assessment methodologies are indispensable. In particular, algorithms with a very light computational load are desirable because, in some scenarios, one needs to estimate the QoE on a residential

VoIP gateway. In addition, the algorithm only utilizes limited information, such as packet-header information, and is required to be very efficient in quality estimation.

In this chapter, we first give an overview of the subjective quality-assessment methodologies that are standardized in ITU. Then, taking into account the importance of objective quality assessment in the practical use of VoIP services, we introduce various objective quality-assessment approaches and their state-of-the-art conditions.

9.2 Subjective Assessment Technologies

Subjective quality assessment is the most fundamental method to quantify speech quality. Before conducting a subjective quality experiment, one has to be very clear about what to evaluate. In addition, the testing methodology as well as other conditions such as source speech materials, subjects, and listening devices must be determined. This section summarizes the elements that should be taken into account in the design of a subjective quality assessment.

9.2.1 Quality Factors

In a subjective quality-assessment experiment, the critical question is: "What should be evaluated in the test?" The various factors that determine the quality of VoIP services include, but are not limited to, the following:

- Coding distortion
- Packet-loss degradation
- Talker/listener echo
- Delay
- Loudness
- Sidetone
- Speech bandwidth
- Inductive noise

In determining the quality-assessment method, the method that can best evaluate the quality factor(s) intended to be tested under given conditions, in a sufficiently sensitive manner, must be chosen.

9.2.2 Quality Attributes

The quality of VoIP services has three attribute: listening, speaking, and conversational qualities. For example, coding distortion and packet-loss degradation can be characterized by listening-quality assessment, in which subjects are asked to listen to speech samples and rate their quality. However, talker echo, by which speakers hear their own voice, electrically reflected at a hybrid circuit, can only be characterized in terms of talking quality. Similarly, the effect of delay cannot be evaluated in a one-way communication, and therefore we should use a conversational method for evaluating it. Although the final evaluation of the quality attribute of VoIP services is conversational, listening and talking qualities should not be neglected because they provide useful diagnostic information.

9.2.3 Subjects

An experimenter often uses only a few subjects in a subjective test. However, we should use as many subjects as possible because the subjective score should be an average of individual preference to guarantee its reproducibility. Due to limitations of time and expense, a good compromise can be found between 24 and 40 subjects in ordinary subjective tests carried out within ITU-T.

ITU-T Recommendation P.800 [3] gives the following guidelines for the selection of subjects in a conversational test. Subjects taking part in the conversation test are chosen at random from the normal telephone-using population, with the provisos that

a. they have not been directly involved in work connected with the assessment of performance of telephone circuits, or related work such as speech coding; and

b. they have not participated in any subjective test whatever for at least the previous six months, and not in a conversation test for at least one year.

If the available population is unduly restricted, then allowance must be made for this fact in drawing conclusions from the results. No steps are taken to balance the numbers of male and female subjects unless the design of the experiment requires it. Subjects are arbitrarily paired in the experimental design prior to the test and remain thus paired for its duration.

9.2.4 Sound-Proof Booth

The subjective experiment to study VoIP services is a psycho-acoustic test, so it must be conducted in an acoustically shielded chamber(s). A sound-proof booth specifically designed to be used in the subjective quality assessment of telephone communications is shown in Figure 9.1. There are more simplified booths that are less expensive. For testing normal handset telephone services, P.800 requires that the volume of the chamber is less than 20 m^3, with a reverberation time less than 500 ms, and the noise floor level is NC25 [4] or NR25 [5]. Apparently, we need a pair of such booths when conducting a conversational test.

9.2.5 Listening/Terminal Devices

In a listening-only test, using the modified intermediate reference system (IRS)-receiving characteristics defined in ITU-T Recommendation P.830 [6] is recommended. This can be achieved by simply using handset/headset devices with such characteristics, or by prefiltering the speech samples with these characteristics. In a conversational or talking test, handset terminals with the modified IRS-sending/receiving characteristics are used. For wideband VoIP services, which have a speech bandwidth of 7 kHz and are recently becoming popular, one should use the sending/receiving characteristics specified in ITU-T Recommendation P.341 [7].

9.2.6 Environmental Noise

We need to take into account the environmental noise at the sending and receiving ends. These should be carefully considered, particularly in the listening-only test, where one

FIGURE 9.1 Sound-proof booth.

needs to add the environmental noise at the sending side prior to the playback of speech samples. That is, we should use source speech with the desirable environmental noise when processing the speech through a system under test. When simulating a typical room environment, Hoth noise [8] is often used. Other noise sources [9] are also used, particularly in evaluating mobile systems.

9.2.7 Rating Method

There are several rating methods used in subjective quality assessment of speech communication services although the most well-known and widely used is absolute category rating (ACR), which is introduced in the following subsection.

9.2.7.1 Absolute Category Rating

The ACR test has been used to characterize telephone services for a long time. In the ACR test, subjects are asked to evaluate the quality of speech/conversation on the five-point scale shown in Table 9.1. A procedure of ACR tests for evaluating the listening quality is illustrated in Figure 9.2. The average value of opinion votes over all the subjects is called the mean opinion score (MOS). The cumulative distribution of opinion score is demonstrated in Figure 9.3. This figure implies that we can suppress the probability of

TABLE 9.1

ACR Scores

Score	Category
5	Excellent
4	Good
3	Fair
2	Poor
1	Bad

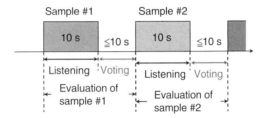

FIGURE 9.2   ACR procedure.

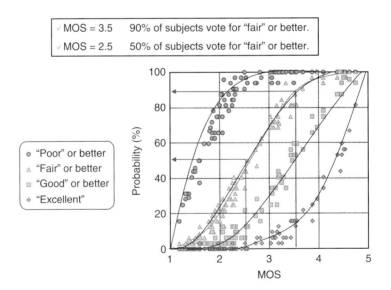

FIGURE 9.3   Distribution of opinion votes.

"poor" or "bad" to less than 10% if we design the service such that it achieves a MOS of 3.5 or better.

Although the MOS is a simple and convenient indicator of telephone services, we must be very careful about its reproducibility. That is, subjective experiments are perceptual and cognitive test, so the resultant score is affected by various internal/external experimental factors, one of which is the quality balance in a test. Imagine that you conduct a

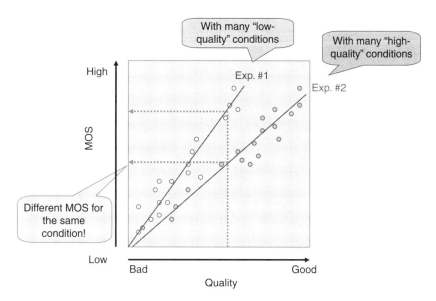

FIGURE 9.4 Experimental dependence of MOS.

subjective listening experiment in which most speech samples have very high quality. Subjects tend to give a low score even if the quality degradation is subtle. Now, you evaluate the same speech sample again but in a different test in which there are many low-quality speech samples. In this case, subjects are more or less used to the quality degradation and give a relatively high score for that speech sample. In this way, you will obtain a different MOS for the same speech sample in a different experimental context (see Figure 9.4). This is often called the "experimental dependence of MOS."

To avoid this problem, one needs to design a well-balanced subjective experiment. However, that is sometimes not easy for inexperienced experimenters. One effective method is the use of the reference system called modulated noise reference unit (MNRU), which is standardized as ITU-T Recommendation P.810 [10]. MNRU adds amplitude-correlated Gaussian noise to original speech. The resultant signal sounds like pulse coded modulation (PCM)-coded speech. The quality of MNRU is controlled by the signal-to-noise ratio (SNR), which is called the Q-value. By using multiple MNRU signals with various Q-values, we design the quality balance of speech samples used in a subjective test. A typical combination of Q-values is, for example, 0, 5, 10, 15, 20, 25, 30, 35, and 40 dB.

The use of MNRU further removes the experimental dependence of MOS. The definition of an "equivalent-Q" value, for which the quality of target speech is equivalent to that of MNRU, is illustrated in Figure 9.5. While the absolute value of MOS may differ from experiment to experiment, the relative quality between MNRU and target speech is expected to be preserved over different subjective experiments. Thus, the equivalent-Q value is not affected by the quality balance of a subjective test.

Care should also be taken when we compare MOS from different laboratories because MOS is also dependent on language and culture. Japanese MOS and Western MOS obtained under exactly the same coding conditions are plotted in a graph in Figure 9.6. This indicates that Japanese MOS tends to be less than those of other countries. This may be due to the interpretation of evaluation categories in different languages (excellent, good, fair, poor, and bad), for example, source speech characteristics, and cultural effects in voting for the maximum score.

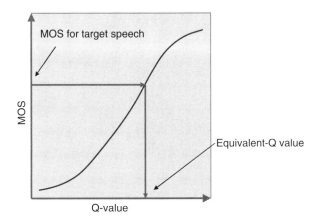

FIGURE 9.5   Definition of equivalent-Q value.

9.2.7.2 Degradation Category Rating

In evaluating the fidelity of a system, the ACR method is not necessarily the best one. If we evaluate the coding distortion of speech with background noise by the ACR method, for example, the MOS becomes low even for uncoded conditions simply due to the existence of background noise. We would not like to obtain this result. In such a case, the DCR method is appropriate.

In the DCR method, subjects are asked to evaluate the quality of target speech compared to reference speech, which is usually without degradation. The five categories used in the DCR test are shown in Table 9.2, and the flow of the DCR test procedure is shown in Figure 9.7. The average value over all subjects is called degradation MOS (DMOS). (Note that the term "DMOS" is also used for differential MOS, which is defined as a difference between the reference and target MOS. This is completely different from degradation MOS.)

9.2.7.3 Comparison Category Rating

In evaluating noise-reduction systems, for example, we do not know whether the quality of processed speech is always worse than the original because we can expect a quality improvement by removing the noise. In this sense, the five categories in the DCR method

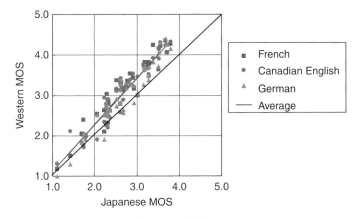

FIGURE 9.6   Relationship between Japanese and Western MOS.

TABLE 9.2

DCR Scores

Score	Category
5	Degradation is inaudible
4	Degradation is audible but not annoying
3	Degradation is slightly annoying
2	Degradation is annoying
1	Degradation is very annoying

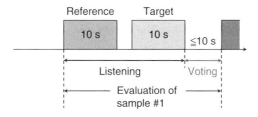

FIGURE 9.7 DCR procedure.

TABLE 9.3

CCR Scores

Score	Category
3	Much better
2	Better
1	Slightly better
0	About the same
−1	Slightly worse
−2	Worse
−3	Much worse

are not appropriate. Thus, we sometimes use the comparison category rating (CCR) method for such a purpose.

In the CCR test, we ask subjects to compare two speech samples and rate the quality of the second speech sample in comparison with the first according to the categories shown in Table 9.3. In the CCR procedure, the order of the reference and target samples is chosen at random for each trial. In half of the trials, the reference sample is followed by the target sample. In the remaining trials, the order is reversed and collected scores must be multiplied by −1. Then, the resultant average score is called the comparison MOS (CMOS).

9.2.8 Analysis of Results

In interpreting the results of subjective quality assessment, one needs to conduct a statistical analysis in addition to the calculation of a MOS. This includes the standard deviation and confidence interval of the average score. These numbers should be reported with the MOS data.

9.3 Objective-Assessment Technologies

Subjective quality assessment quantifies users' experience without estimation; thus, it is the most reliable method to evaluate the QoE of VoIP services. Subjective quality assessment is a psycho-acoustic experiment in which the actual users are involved, however, it is time consuming and expensive. In addition, special facilities such as booths shown in Figure 9.1 are required. Moreover, it is not applicable to in-service and real-time quality management in principle. Therefore, an objective means for estimating subjective quality solely from objective characteristics of VoIP services is desired. This is called "objective quality assessment." Objective quality assessment is not a method for deriving a simple indicator of objective performance, but a method for deriving a good indicator of subjective quality in an objective manner.

Objective quality-assessment models can be categorized into several approaches from the viewpoints of application scenarios and information used in a model (see Table 9.4).

9.3.1 Media-Layer Model

A media-layer model uses a speech waveform as input for the model. This can be further distinguished into two categories (see Figure 9.8); one is a full-reference model, which uses source speech as well as degraded speech, and the other is a no-reference model, which uses only degraded speech. By definition, a full-reference model is superior to a no-reference model in terms of quality-estimation accuracy. However, using source speech is often impossible in some application scenarios such as in-service quality monitoring at a user's premises. For this purpose, only no-reference models are applicable.

The study of media-layer objective models of speech was started with SNR as a means for evaluating PCM-coded speech. In the latter half of the 1980s, several objective models that exploited spectral distortion rather than waveform distortion were proposed as objective quality-assessment methods that are more applicable to the evaluation of low-bitrate codecs. However, due to their lack of accuracy in estimation, none was standardized as an ITU-T Recommendation. Later, a model based on the Bark spectral distortion provided adequate accuracy and was the basis for Recommendation P.861 "Perceptual Speech Quality Measure (PSQM)" in 1998.

TABLE 9.4

Categories of Objective Quality-Assessment Models

	Media-Layer Model	Parametric Packet-Layer Model	Parametric Planning Model	Bitstream Model	Hybrid Model
Input information	Speech waveform	Packet header information	Quality design/ management parameters	Payload information (not decoded)	(Combination of any information)
Scenario	Codec optimization Service bench-marking	In-service quality management	Quality planning In-service quality management	In-service quality management	In-service quality management
Standard	P.862 P.563	P.564	G.107	—	P.CQO-L

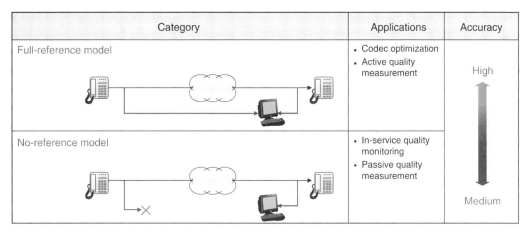

Category	Applications	Accuracy
Full-reference model	• Codec optimization • Active quality measurement	High
No-reference model	• In-service quality monitoring • Passive quality measurement	Medium

FIGURE 9.8 Two types of media-layer models.

This approach exhibits sufficient performance only under error-free coding conditions, and hence, in general, it is inapplicable to the evaluation of VoIP speech, which often suffers from packet loss. Therefore, the next target for standardization was the development of a method that is applicable to the evaluation of the effects of discontinuous degradation such as packet loss in VoIP. Consequently, a compromise between the algorithms of perceptual analysis measurement system (PAMS), which utilizes different perceptual modeling than PSQM and has quite a sophisticated time-alignment scheme, and PSQM+, which is an extension of PSQM, was standardized as Recommendation P.862 "Perceptual Evaluation of Speech Quality (PESQ)" in 2001. This recommendation describes a "full-reference" objective model, that is, it requires input test speech as a reference, so its main application is active measurement in which test speech samples are fed into the system under test, and the original speech is compared with the post-transmission speech.

A very rough block diagram of the PESQ algorithm is shown in Figure 9.9. First, the model synchronizes the source and degraded signals in the time domain. This is extremely important because the algorithm quantifies the distortion by subtracting the spectrum of degraded speech from that of original speech. If two signals are not aligned correctly, the resultant distortion does not make sense. Then, the algorithm transfers both signals into the Bark spectrum domain and represents the intensity by loudness, so that the estimates obtained by the algorithm better fits the human perceptions. The distortion is defined as the difference between source and degraded speech. Finally, the algorithm aggregates the sequential distortion data based on its cognitive model. For more details, readers are recommended to read Recommendations [11,12].

Some problems regarding the implementation of Recommendation P.862 have recently been reported to ITU-T. An implementers' guide is now available as Recommendation P.862.3 [13], so equipment vendors and network operators and providers will be aware of the problems and therefore be able to use the Recommendation appropriately. ITU-T also standardized a no-reference model as Recommendation P.563 [14]. On the other hand, Alliance for Telecommunications Industry Solutions (ATIS) adopted a different model, which is called auditory non-intrusive quality estimation plus (ANIQUE+) [15]. These algorithms can be powerful tools when one needs to monitor the quality of VoIP services in a single-ended way, that is, without access to source speech data.

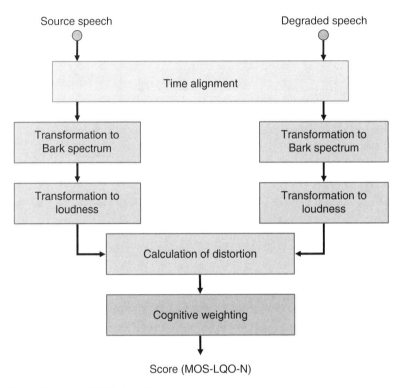

FIGURE 9.9   Block diagram of PESQ algorithm.

9.3.2 Parametric Packet-Layer Model

ITU-T has also standardized the framework and performance criteria for an objective quality-assessment methodology based solely on IP-packet information (not speech in the payload) for use in real-time quality monitoring. This is Recommendation P.564 [16]. P.564 does not recommend any specific algorithms. If an algorithm satisfies the functionality and performance requirements, then it is called "P.564-compliant." P.564 has three operation modes: dynamic operation mode, static operation mode, and embedded operation mode.

The dynamic operation mode works with RTCP-XR [17], which is an extended report of the RTCP protocol, and conveys quality characteristics of a terminal and network. RTCP-XR includes information such as packet-loss rate in networks, packet-discard rate at the terminal jitter buffer, and burstiness of packet loss/discard. A P.564 algorithm estimates the resultant listening quality of speech using this information. The resultant quality estimates reflect actual end-to-end speech quality because the packet discard at the terminal jitter buffer, which is not usually available in networks but a very crucial quality factor in VoIP services, can be taken into account.

The static operation mode requires a predetermined calibration file, in which the characteristics of the terminal such as jitter buffer behavior are included. The static operation mode can be used at a network midpoint for monitoring VoIP quality based on packet capturing. If one knows which terminal is used by the end user, he/she can estimate the packet discard rate at the terminal jitter buffer by taking into account both the terminal jitter-buffer characteristics and delay jitter in networks.

The last mode, which is the embedded operation mode, assumes that a P.564-compliant algorithm is implemented in the terminal so that the algorithm has direct access to the jitter buffer. The model reports, for example, the quality estimates to a quality-management center.

9.3.3 Parametric-Planning Model

Parametric-planning models have long been studied in the ITU-T, and several models were proposed in the early 1980s as candidates for an ITU-T standard. However, the working group was unable to settle on a single algorithm as the international standard and consequently created an informative document in which four different models were introduced.

A new model called the "E-model" was proposed for ITU-T standardization in the 1990s. The E-model is based on the OPINE model from NTT [18], in which all the factors responsible for quality degradation are summed on a psychological scale. The E-model and the TR model from AT&T [19] produce similar output: a score on a psychological scale is produced as an index of overall quality. In the E-model, the quality degradation introduced by speech coding, bit error, and packet loss is treated collectively as an "equipment-impairment factor." ITU-T standardized the E-model as Recommendation G.107 in 1998. It was also adopted by the European Telecommunications Standards Institute (ETSI) and Telecommunications Industry Association (TIA) as a network planning tool [20,21] and has become the most widely used quality-estimation model in the world.

The E-model has 21 input parameters that represent terminal, network, and environmental quality factors. Its output is called the "R-value," which is a function of the five quality factors calculated from the 21 input parameters (Figure 9.10). First, the degrees of quality degradation due to individual quality factors such as loudness, echo, delay, and distortion are calculated on the same psychological scale. These values are then subtracted from the reference value.

Recommendation G.107 provides a set of "default values," which can be used when network planners assume that terminals and the usage environment are "normal." An index of the overall quality should thus represent perceptions of a user using a normal terminal under normal circumstances.

Although R-values produced by the E-model have some correlation with the subjective conversational MOS and are useful in network planning, they are not necessarily accurate as estimators of subjective quality. In particular, the validity of the additive property assumed in the E-model is sometimes questionable [22,23]. The accuracy of E-model prediction is being thoroughly studied by ITU-T.

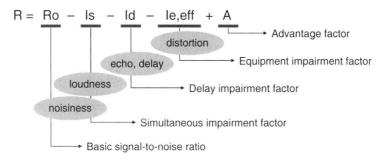

$$R = Ro - Is - Id - Ie,eff + A$$

- Advantage factor
- distortion
- Equipment impairment factor
- echo, delay
- Delay impairment factor
- loudness
- noisiness
- Simultaneous impairment factor
- Basic signal-to-noise ratio

FIGURE 9.10 R-value calculated by E-model.

9.3.4 Hybrid Model

In practical quality measurement of VoIP services, using as much information as possible to obtain reliable quality estimates is desired. From this viewpoint, combining the earlier-mentioned technologies into "hybrid models" often works. In particular, utilizing packet-transmission characteristics in a no-reference media layer model may improve the estimation accuracy.

9.4 Conclusion

In this chapter, we overviewed technologies that quantify human perceptions of speech quality. First, we reviewed various subjective quality-assessment methodologies, which try to directly measure subjective quality by psycho-acoustics testing. Then, we introduced objective quality-assessment methodologies, which estimate subjective quality solely from physical characteristics of VoIP systems. These are very effective, particularly for in-service quality management that requires real-time quality evaluation without human subjects.

References

1. ITU-T Recommendation P.10/G.100, "Vocabulary for performance and quality of service," January 2007.
2. ITU-T Recommendation G.107, "The E-model, a computational model for use in transmission planning," June 2006.
3. ITU-T Recommendation P.800, "Methods for subjective determination of transmission quality," July 2006.
4. L. L. Beranek, "Noise and vibration control," McGraw-Hill, 1971, pp. 564–566.
5. ISO 1996.
6. ITU-T Recommendation P.830, "Subjective performance assessment of telephone-band and wideband codecs," February 1996.
7. ITU-T Recommendation P.341, "Transmission characteristics for wideband (150–7000 Hz) digital hands-free telephony terminals," February 1998.
8. D. F. Hoth, "Room noise spectra at subscribers' telephone locations," *J.A.S.A.*, vol. 12, April 1941, pp. 99–504.
9. NTT-AT, "Ambient noise database," CD-ROM.
10. ITU-T Recommendation P.810, "Modulated noise reference unit (MNRU)," February 1996.
11. ITU-T Recommendation P.862, "Perceptual evaluation of speech quality (PESQ): An objective method for end-to-end speech quality assessment of narrow-band telephone networks and speech codecs," February 2001.
12. ITU-T Recommendation P.862.1, "Mapping function for transforming P.862 raw result scores to MOS-LQO," November 2003.
13. ITU-T Recommendation P.862.3, "Application guide for objective quality measurement based on Recommendations P.862, P.862.1, and P. 862.2," November 2005.
14. ITU-T Recommendation P.563, "Single ended method for objective speech quality assessment in narrow-band telephony applications," May 2004.
15. ANSI ATIS 0100005-2006.

16. ITU-T Recommendation P.564, "Conformance testing for narrowband voice over IP transmission quality assessment models," July 2006.
17. IETF RFC3611, "RTP control protocol extended reports (RTCP XR)," April 2003.
18. N. Osaka, K. Kakehi, S. Iai, and N. Kitawaki, "A model for evaluating talker echo and sidetone in a telephone transmission network," *IEEE Transactions on Communications*, vol. 40, no. 11, 1992, pp. 1684–1692.
19. J. R. Cavanaugh, R. W. Hatch, and J. L. Sullivan, "Models for the subjective effects of loss, noise and talker echo on telephone connections," *B.S.T.J.*, vol. 55, no. 9, November 1976, pp. 1319–1371.
20. ETSI ETR250, "Speech communication quality from mouth to ear for 3.1 kHz handset telephony across networks," July 1996.
21. TIA/EIA TSB116, "Voice quality recommendations for IP telephony," March 2001.
22. A. Raake, "Speech quality of heterogeneous networks involving VoIP: Are time-varying impairments additive to classical stationary ones?," *Proceedings of the 1st ISCA Tutorial and Research Workshop on Auditory Quality of Systems*, April 2003, pp. 63–70.
23. A. Takahashi, A. Kurashima, and H. Yoshino, "Objective assessment methodology for estimating conversational quality in VoIP," *IEEE Transactions on Audio Speech and Language Processing*, vol. 14, no. 6, November 2006, pp. 1984–1993.

10

Delay Performance and Management of VoIP System

Zhibin Mai, Shengquan Wang, Dong Xuan, and Wei Zhao

CONTENTS

10.1 Introduction

Voice-over-Internet Protocol (VoIP) has been identified as a critical real-time application in the network quality of service (QoS) research community. Transmission of voice traffic must meet stringent requirements on packet delay as it is an important factor affecting quality of call. The International Telecommunication Union (ITU) recommends that a one-way delay between 0 and 150 ms is acceptable in Recommendation G.114 [1].

In traditional telephony, there is a call admission control (CAC) mechanism. That is, when the number of call attempts exceeds the capacity of links, the request for setting up new calls will be rejected while all calls in progress continue to be unaffected. Most current IP networks have no CAC and, therefore, only offer best-effort services. New traffic may continue entering the network even beyond the network capacity limit, consequently causing both the existing and new flows to suffer packet loss and/or significant delay. To prevent these occurrences and provide delay guarantees, a CAC and other delay mechanisms have to be introduced in IP networks in order to ensure that sufficient resources are available to provide delay guarantees for admitted new and existing calls.

Several CAC mechanisms, such as site-utilization-based CAC (SU-CAC) and link-utilization-based CAC (LU-CAC) have been used in existing VoIP systems. However, none of the current VoIP systems can really provide delay guarantees to VoIP networks as they are not able to effectively apply and support the CAC mechanisms. For example, the SU-CAC mechanism performs admission control based on the pre-allocated resource to the sites, which represents a host or a network with different sizes and demands an approach to perform resource pre-allocation to sites at the configuration time. Unfortunately, current VoIP systems, such as the Cisco's VoIP system [2], have not been able to define such an approach. Resource pre-allocation in these systems is performed in an ad hoc fashion. The main approach of resource pre-allocation is over-provisioning. The overprovision mechanism provides abundant network resources so that applications behave as if the network is unloaded. It is the simplest but the most expensive method. Hence, delay-guarantee service cannot be effectively provided. Another case is the LU-CAC mechanism. With this mechanism, admission control is based on the utilization of the individual link bandwidth. This mechanism needs resource reservation on individual links in a network. Current VoIP systems rely on resource reservation protocols, such as RSVP, to perform explicit resource reservation on all routers along the path of traffic in the network. Such a resource reservation approach will introduce significant overhead to the core-routers and, hence, greatly comprise the overall network performance.

In [3–6] we developed a series of delay-analysis theories and utilization-based admission control mechanisms to provide delay-guaranteed services for real-time applications in static-priority scheduling networks. Given the traffic model, the network topology, and the traffic deadline requirement, for any input of link utilization, we compute the worst-case delay (deterministic case) or delay distribution (statistical case) with our delay-analysis methods. Then, we can verify whether or not the utilization is safe to make the end-to-end delay meet the deadline requirement using the utilization-based admission control. In [7], we designed and developed a delay-provisioning system for VoIP, which has been realized in the Internet2 VoIP Testbed at Texas A&M University. We systematically evaluate our proposed delay-provisioning system in terms of admission delay and admission probability. Our data show that, if a VoIP system is enhanced by our delay-provisioning system, the overall system can achieve high resource utilization while invoking relatively low overhead.

In this chapter, we illustrate our work [2–7] on delay management for the VoIP system. In Section 10.2, we review the current delay-management mechanisms for VoIP system. In Section 10.3, we introduce our work on delay analysis for utilization-based admission control. In Section 10.4 and Section 10.5, we introduce the integration of the delay-provisioning system with existing VoIP system and illustrate the performance of the utilization-based admission control of the VoIP system. In Section 10.6, we provide the conclusion.

10.2 Overview on Delay Management

Most current IP networks can only offer best-effort services. New traffic may keep entering the network even beyond the network capacity limit, consequently causing both existing and new flows to suffer packet loss and/or significant delay. To prevent these occurrences and provide delay guarantees for VoIP traffic, delay-management architecture must be introduced in IP networks in order to ensure that sufficient resources are available to provide delay guarantees for both admitted new and existing calls.

The delay-management architecture of VoIP System can be partitioned into two planes: data plane and control plane [8]. Mechanisms in data plane include packet classification, shaping policing, buffer management, scheduling, loss recovery, and error concealment. They implement the actions that the network needs to take on user packets, in order to enforce different class services. Mechanisms in control plane consist of resource provisioning, traffic engineering, admission control, resource reservation, connection management, etc. They allow the users and the network to reach a service agreement, and let the network appropriately allocate resources to ensure delay guarantees to the calls that have been admitted.

Delay-management mechanisms in both data plane and control plane work together to provide delay guarantee to voice in IP networks. At the configuration time, resource provisioning of control plane determines the portion of resources to be allocated for voice traffic. At run time, upon a new call arrival, admission control of control plane will decide the acceptance of the new call based on the amount of provisioned resources and the usage of the resources by the admitted calls. Once the call is admitted, the end-hosts start sending voice packets to the network. Delay mechanisms of data plane will be invoked to perform traffic-enforced functions such as traffic classification, shaping, buffer management, scheduling, error control etc, directly on the voice packets. As such, resource provisioning and CAC mechanisms have to be in place to provide end-to-end delay guarantees in VoIP systems.

Call admission control algorithms can be roughly grouped in two broad categories: (1) measurement-based CAC algorithm and (2) parameter-based CAC algorithm. The measurement-based CAC algorithm uses network measurement to estimate current load of existing traffic. It has no prior knowledge of the traffic statistics and makes admission decisions based on the current network state only [9–13]. Measurement-based admission control can only provide soft (rather than deterministic) delay guarantees for voice traffic. It relies on the measurement period to reflect the dynamics of network status. Shorter measurement period, which means measurement is conducted more frequently, can reflect the networks better, but consumes more network resources; and that longer measurement period cannot reflect the network dynamics well, but cost less network resources. Also, in some cases, the probe messages are not payload messages themselves, which may not be able to reflect how the network treats the real voice packets. The parameter-based CAC algorithm uses the parameters of resource and service to decide whether the network can accommodate the new connection while providing the end-to-end delay guarantees. The parameters are used to compute a deterministic bound imposing that, in whichever traffic situation, the end-to-end delay guarantees are provided for all flows [3–6,14–16]. Parameter-based admission control can provide delay-guaranteed services to applications, which can be accurately described, such as VoIP. However, when the traffic is very bursty, it is very difficult to describe the traffic characteristics. Then parameter-based admission control may result in overbooking the network resource and hence lowering network utilization.

Utilization-based CAC (such as SU-CAC and LU-CAC used in this chapter) belongs to parameter-based CAC category, where the parameters are the requested bandwidth utilization for each new connection and the available bandwidth utilization in the resource. Through effective resource provisioning and appropriate system (re)configuration steps, the delay guarantee test at run time is reduced to a simple utilization-based test. As long as the utilization of links along the path of a flow is not beyond a given bound, the performance guarantee of the end-to-end delay can be met. Utilization-based CAC renders the system scalable.

In [3–6], we developed a series of delay-analysis theories for the resource provisioning and utilization-based admission control mechanisms to provide delay guarantees for real-time applications, such as VoIP. Based on the research results in [3–6], we designed and developed a delay-provisioning system for VoIP in [7] to support SU-CAC and LU-CAC to provide delay guarantees for existing VoIP systems. In the following section, a delay-analysis technique is illustrated to provide resource provisioning for utilization-based admission control.

10.3 Delay Analysis for Utilization-based Admission Control

In utilization-based admission control, utilization is defined as the portion of resource on average. This mechanism [16] makes an admission control decision based on a predefined safe resource utilization bound; that is, for each task admission request, as long as the used resource utilization plus the requested resource utilization are not beyond the predefined safe resource utilization bound that is computed offline, the service guarantee can be provided. Utilization-based admission control involves only a simple utilization test and eliminates the delay computation at the admission time. Utilization-based admission control renders the admission control very efficient. However, it tends not to be effective, since it does not take into account the dynamics of tasks in the admission process.

We assume that the network topology is known in advance, which includes the potential end-to-end path information and link bandwidth information. Each voice traffic flow will be regulated by a leaky bucket with burst size σ and average rate ρ at the entrance of the network. Link k is of capacity C_k. The link bandwidth utilization allocated to voice traffic is assumed to be u_k at link k. Since the deadline requirement can be either deterministic or statistical, resulting in deterministic services and statistical services, we classify the delay analysis as utilization-based deterministic delay analysis and utilization-based statistical delay analysis, respectively.

Utilization-based Deterministic Delay Analysis: If the deadline requirement is deterministic, we can bound the worst-case queuing delay by the following theorem:

Theorem 1. The worst-case queuing delay d_k suffered by any voice packet with the highest priority at the buffer of output link k is bounded by

$$d_k \leq \frac{c_k - 1}{c_k - u_k} u_k \left(\frac{\sigma}{\rho} + Y_k \right), \tag{10.1}$$

where $c_k = \sum_{j \in L_k} C_j / C_k$, $Y_k = \max_{R \in S_k} \sum_{s \in R} d_s$, L_k is the set of the input links of output link k with capacity C_k, and S_k is the set of all subroutes used by voice packets with the highest priority upstream from output link k.

The proof can be found in Appendix A. In (10.1), the value of Y_k, in turn, depends on the delays d_s's experienced at output link s other than k. Then, we have a circular dependency. Therefore, d_k can be determined iteratively. Furthermore, the end-to-end worst-case delay can be obtained, which only depends on the link utilization u_k, the parameters for voice traffic (burst size σ and average rate ρ), and the network topology.

Utilization-Based Statistical Delay Analysis: If the deadline requirement is probabilistic, we can bound delay-violation probabilities as follows:

Theorem 2. In this case, d_k is a random variable and D_k is denoted as its deadline. The violation probability of delay for any voice packet with the highest priority suffered at the buffer of output link k is bounded by

$$P\{d_k > D_k\} \leq \begin{cases} \dfrac{1}{\sqrt{2\pi}} \exp\left(-24\dfrac{1-u_k}{u_k^2}\dfrac{D_k}{\sigma/\rho}\right), & u_k \geq \dfrac{D_k}{\sigma/\rho} \\[4mm] \dfrac{1}{\sqrt{2\pi}} \exp\left(-6\dfrac{1-u_k}{u_k^3}\left(u_k + \dfrac{D_k}{\sigma/\rho}\right)^2\right), & u_k < \dfrac{D_k}{\sigma/\rho} \end{cases} \tag{10.2}$$

The proof can be found in Appendix B. The end-to-end deadline violation probability can be bounded as

$$P\left\{d^{e2e} > \sum_{k \in R} D_k\right\} \leq 1 - \prod_{k \in R}(1 - P\{d_k > D_k\}). \tag{10.3}$$

10.4 Integration of the Delay-Provisioning System with Existing VoIP System

In this section, we would like to use the Cisco VoIP system to illustrate how our delay-provisioning system can be integrated with the existing VoIP system. Figure 10.1 shows our delay-provisioning system integrating with the commercial VoIP system.

The delay-provisioning system consists of three kinds of components (see Figure 10.2):

- QoS manager (QoSM): The QoSM implements three basic functions: (1) provides user interface to control and monitor the components, which are in the same QoS domain; (2) provides registration to the distributed agents and coordination among the distributed agents in the same QoS domain; and (3) cooperates with the peer QoSMs that belong to other QoS domains.

- Call admission control agent (CACA): CACA does delay analysis and makes admission control to provide QoS guarantees for VoIP. It consists of two modules: utilization computation module and admission decision-making module. The utilization computation module performs delay analysis and computes the maximum bandwidth utilization. It supports both utilization-based deterministic delay analysis and utilization-based statistical delay analysis. It usually runs at the configuration time. The computed utilization will be allocated to either links in the LU-CAC mechanism or to sites in the SU-CAC mechanism. At run time, the

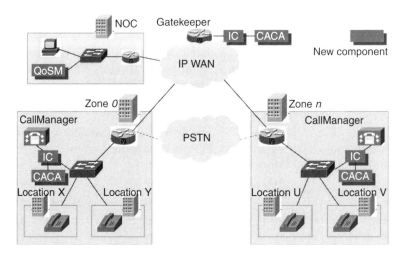

FIGURE 10.1   Architecture of the delay-provisioning system.

admission decision-making module will make an admission decision for each incoming call request, based on the allocated bandwidth utilization (by the utilization computation module) and the currently consumed bandwidth.

- Integration component (IC): IC integrates CACA into existing VoIP systems and provides call signaling processing modules to monitor and intercept call setup signaling from Gatekeeper or Call Manager, withdraws the useful message and passes it to CACA, and executes call admission decision made by CACA.

Generally speaking, there are two approaches for the delay-provisioning system to integrate with the existing VoIP systems to execute the call admission decision:

- Front-end approach: In this approach, the call setup requests must pass through the delay-provisioning system before reaching the existing call admission decision unit (e.g., CallManager). The call setup responses must also pass through the delay-provisioning system before coming back to the call request endpoint. The delay-provisioning system can directly enforce its call admission decision to the call setup request by adding, modifying, or dropping signaling between the endpoint and the existing admission decision unit.

- Back-end approach: In this approach, the call setup requests and responses will be forwarded to the delay-provisioning system by the existing call admission decision unit (e.g., Gatekeeper). The delay-provisioning system will indirectly execute its call admission decision to the call setup request by negotiating with the existing call admission decision unit.

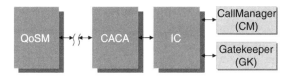

FIGURE 10.2   Components of the delay-provisioning system.

In general, the back-end approach has the following advantages over the front-end approach: (1) The implementation of the IC will not be very complicated since the existing system normally allows the IC to selectively receive and process the signaling; (2) The consistency of the signaling can be easily achieved since the IC directly interacts with existing call admission decision unit; (3) The integration overhead is little because only the existing admission control unit is aware of the agent. However, the back-end approach requires the existing system to have the ability that the call setup requests and responses can be redirected to the external application, for example, our CACA, whereas the front-end does not. In the following, we will apply the above two integration approaches to integrate our delay-provisioning system with the Cisco VoIP systems.

10.4.1 Integration with Cisco VoIP Gatekeeper

Cisco Gatekeeper is a built-in feature of Cisco IOS in some Cisco Router series and is a light-weight H.323 gatekeeper. The Registration Admission Status (RAS) signaling that the Cisco Gatekeeper handles is H.323-compatible. Cisco Gatekeeper provides interface for external applications to offload and supplement its features. The interaction between the Cisco Gatekeeper and the external application is completely transparent to the H.323 endpoint.

As shown in Figure 10.3, the back-end approach is adopted for the IC to intercept the call signaling between the CACA and the Gatekeeper. The IC handles the H.323 RAS signaling and communicates with the Cisco IOS Gatekeeper. The communication between the Cisco IOS Gatekeeper and the IC is based on Cisco's propriety protocol, Gatekeeper Transaction Message Protocol (GKTMP) [17]. GKTMP provides a set of ASCII RAS request/response messages between Cisco Gatekeeper and the external application over a Transmission Control Protocol (TCP) connection. There are two types of GKTMP messages: (1) GKTMP RAS messages that are used to exchange the contents of RAS messages between the Cisco IOS Gatekeeper and the external application and (2) trigger registration messages used by the external application to indicate to the Cisco Gatekeeper which RAS message should be forwarded. If an external application is interested in receiving certain RAS messages, it must register the requests for the messages with the Cisco Gatekeeper. In our implementation, the IC is interested in receiving the following four RAS messages from the Gatekeeper: admission request (ARQ), location confirm (LCF), location reject (LRJ), and disengage request (DRQ). All of the four messages will be automatically registered to Cisco Gatekeeper once the CACA is functional. Due to space limitations, we do not list out all the possibilities of how the IC processes the RAS message. Figure 10.3a illustrates a successful call-request procedure. Figure 10.3b illustrates a simple tearing down procedure, where the IC will update the status of network resource once the message DRQ is received.

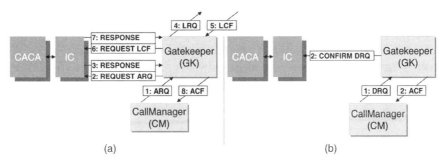

(a)   (b)

FIGURE 10.3 An Illustration of a successful call request procedure in IC for the Gatekeeper.

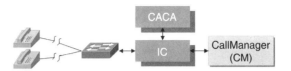

FIGURE 10.4   An illustration of the communication protocol in IC for CallManager.

10.4.2 Integration with Cisco VoIP CallManager

Cisco CallManager is a comprehensive and heavyweight VoIP processing application. It can interact with endpoints using multiple protocols, for example, Skinny Call Control Protocol (SCCP), H.323, Session Initiation Protocol (SIP), etc. In the delay-provisioning system, we implement SCCP, a popular signaling protocol in Cisco VoIP system, in the IC for CallManager. To the best of our knowledge, Cisco CallManager does not provide an interface for external applications to supplement its CAC mechanism as Cisco Gatekeeper does. In this case, only the front-end approach can be adopted to intercept the call signaling of a CallManager. Figure 10.4 shows the basic idea of this method.

Since the basic idea and the procedure of the signaling process in both the IC for CallManager and the IC for Gatekeeper are similar, we would like to highlight the difference in intercepting the SCCP using the IC for CallManager. The CallManager is unaware of the IC. It directly sends the implicit grant permission message (i.e., "StartMediaTransmission") to the endpoints of the admitted call. However, in case the CACA makes a decision to deny the call because of the lack of available bandwidth, it should not let the message "StartMediaTransmission" be received by the endpoints. One of the approaches is to continue dropping the message from the CallManager until the CallManager terminates the TCP connection after a finite timeout. There are two problems for this approach: (1) caller does not get any indication whether the call is accepted or not and (2) waiting time (about 60 s) is too long for the caller. To compensate for the two problems, the IC can explicitly indicate the caller by sending the busy tone message "StartToneMessage" to the endpoint and prevent the CallManager from sending a message to the endpoint by sending the call terminating message "OnHookMessage" to the CallManager. However, additional messages from the IC would interfere with the synchronization of the TCP connection between the endpoint and the CallManager. To speedup the resynchronization of the TCP connection, and limit the impact on the CallManager, the IC will send a TCP RESET packet to the endpoint.

10.5 Performance Evaluation

To measure the overhead introduced to admission and the overall bandwidth utilization of the system, we choose two measurement metrics: admission latency and admission probability. Admission latency is used to measure the overhead of admission and admission probability is a well-known metric used to measure the overall bandwidth utilization. The higher the admission probability, the higher the overall bandwidth utilization achieved.

10.5.1 Admission Latency

In this section, we run a suite of experiments to evaluate the admission latency in two VoIP systems: (1) the one with our CACA and (2) the one without our CACA. Due to the different design and implementation methodology of CACA for CallManager and Gatekeeper, we run experiments for both cases. The experiments are run in the Internet2 VoIP Testbed in Texas A&M University.

10.5.1.1 Call Admission Control Agent for Cisco Gatekeeper

In this experiment, we tried 300 calls for each CAC mechanism. The call signaling crosses two Cisco Call Mangers and two Cisco Gatekeepers from a Cisco IP phone in Texas A&M University to another IP phone in Indiana University. To show the introduced overhead by our delay-provisioning system, we have two sets of data: local admission latency and round-trip admission latency.

Figure 10.5a shows the distribution of local admission latency between receiving ARQ and sending out LRQ by Gatekeeper. Figure 10.5b shows the distribution of round-trip admission latency between receiving ARQ and sending ACF out by Gatekeeper. Table 10.1 gives us the summary of the distribution of admission latency for each case in terms of the mean value and standard deviation. The local admission latency excludes the network latency and the processing latency on the other side. It shows a more accurate latency introduced by our delay-provisioning system, which is shown by the standard deviation of the latency distribution in Table 10.1. The round-trip admission latency gives us the view of the overall admission latency.

With CACA, the introduced latency is about 4.4 ms. The overall latency is very acceptable and the introduced latency is pretty small. To measure the introduced latency, we measured the admission latency between receiving ARQ and sending out LRQ from the Gatekeeper; not the additional latency from the CACA directly. Here, the additional admission latency includes not only the admission latency introduced in CACA, but also the additional latency in Gatekeeper caused by interaction between Gatekeeper and CACA, which cannot be measured directly.

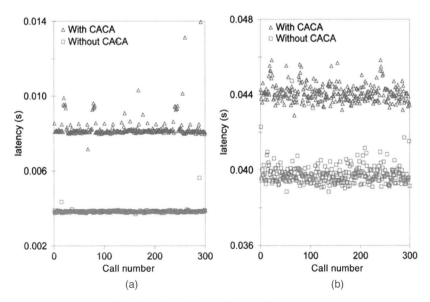

(a)

(b)

FIGURE 10.5 The distribution of local and round-trip admission latency.

TABLE 10.1

The Mean Value and Standard Deviation of Latency Distribution

	Local Admission Latency (ms)		Round-trip Admission Latency (ms)	
	Mean Value	Standard Deviation	Mean Value	Standard Deviation
With CACA	8.286	0.863	44.302	2.665
Without CACA	3.850	0.277	39.870	1.530

10.5.1.2 Call Admission Control Agent for Cisco CallManager

In this experiment, we also initiated 300 calls for each CAC mechanism. The call signaling crosses one Cisco Call Manger between two Cisco IP phones in Texas A&M University. To show the introduced overhead by our delay-provisioning system, we have two sets of data: local admission latency and round-trip admission latency. Figure 10.6a shows the distribution of local additional admission latency which is introduced by CACA in processing one call-signaling message. Figure 10.6b shows the distribution of round-trip admission latency. Table 10.2 gives us the summary of the distribution of admission latency for each case in terms of the mean value and standard deviation. With CACA, the introduced latency is about 1.2 ms (i.e., additional latency). The overall latency is acceptable and the introduced latency is quite small.

10.5.2 Admission Probability

To make the data convincing, the measure of admission probability requires a high volume of calls in the VoIP system. However, it is not feasible or realistic to produce a high volume of calls in the VoIP system: First, our delay-provisioning system is only deployed in

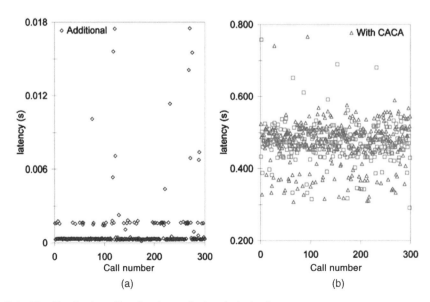

FIGURE 10.6 The distribution of local and round-trip admission latency.

TABLE 10.2

The Mean Value and Standard Deviation of Latency Distribution

	Latency (ms)	
	Mean Value	**Standard Deviation**
With CACA	476.002	94.796
Without CACA	479.367	92.114
Additional	1.202	3.080

Internet2 VoIP Testbed in Texas A&M University, where simultaneous calls from many sites are not available. Second, even in the fully deployed VoIP system, a high volume of calls for the experiment will affect the operation of VoIP heavily. Admission probability can only be measured by simulation. In this section, we run a suite of simulation to evaluate the admission probability for the LU-CAC mechanism and the SU-CAC mechanism, respectively.

Traditionally, call arrivals follow a Poisson distribution and call lifetimes are exponentially distributed. This call mode can approximate the realistic call mode very well. In our simulation, we use this call mode to simulate calls by Mesquite CSIM 19 toolkits for simulation and modeling. In the simulation, overall requests for call establishment in the network form a Poisson process with rate λ, while call lifetimes are exponentially distributed with an average lifetime of $\mu = 180$ s for each call. All calls are duplex (bi-directional) and use G.711 codec, which has a fixed packet length of (160 + 40) bytes (RTP, UDP, IP headers, and two voice frames) and a call flow rate of 80 kbps (including 64 kbps payload and other header).

Two different network topologies are chosen for the simulation: an Internet2 backbone network and a campus network. The Gatekeeper and CallManager are configured to perform CAC in the Internet2 environment and in the campus environment, respectively.

10.5.2.1 Internet2 Backbone Network

Abilene is an advanced backbone network that supports the development and deployment of the new applications being developed within the Internet2 community. Figure 10.7 shows the core map of the Abilene network used in our simulation. There are 11 core node routers, each located in a different geographical area. All backbone links are either OC48 or OC192. The call route will be chosen uniformly randomly from the set of all pairs of core node routers. Suppose that the end-to-end deadline for queuing is 10 ms, then the maximum utilization is 0.195 for deterministic service, that is, under the condition that about 19.5% link bandwidth is used for voice traffic, the end-to-end delay for any voice packet can meet the deadline requirement. The maximum utilization is 0.307 for statistical service with deadline violation probability 10^{-6}, that is, under the condition that about 30.7% link bandwidth is used for voice traffic, any voice packets may miss the deadline with small probability 10^{-6} at most. λ ranges from 100.0 to 1000.0.

Figure 10.8 shows the admission probabilities for the voice call in the two CAC mechanisms as a function of arrival rates. We find that the LU-CAC mechanism can achieve a much higher admission probability than SU-CAC for both deterministic service and statistical service. Statistical service can achieve a much higher admission probability than deterministic service, as expected.

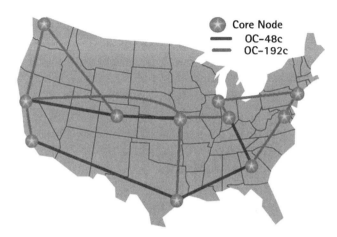

FIGURE 10.7   The core map of Abilene network (September 2003). (*Source:* "Abilene network backbone," http://abilene.internet2.edu, 2003.)

10.5.2.2 Campus Network

Figure 10.9 shows the campus network topology used in our simulation. The link bandwidth is either 100 Mbps or 155 Mbps. The call route will be chosen uniformly randomly from the set of all pairs of sites (0–18). Suppose that the end-to-end deadline is 10 ms for queuing, then the maximum utilization is 0.208 for deterministic service, that is, under the condition that about 20.8% link bandwidth is used for voice traffic, the end-to-end delay for any voice packet can meet the deadline requirement. The maximum utilization is 0.332 for statistical service with a deadline violation probability 10^{-6}, that is, under the condition that about 33.2% link bandwidth is used for voice traffic, any voice packets may miss the deadline with small probability 10^{-6} at most. λ ranges from 1.0 to 10.0.

Figure 10.10 shows the admission probabilities for the voice call in the two CAC mechanisms as a function of arrival rates. The admission probabilities in the two CAC mechanisms are different. Similar to the observation made in Internet2 Backbone network, the

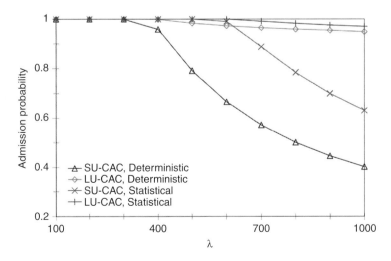

FIGURE 10.8   The admission probability in Abilene network.

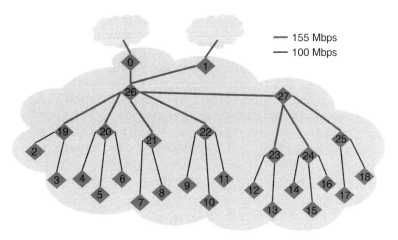

FIGURE 10.9 A campus network topology.

LU-CAC mechanism can achieve much higher admission probability than SU-CAC for both deterministic service and statistical service. Statistical service can achieve a much higher admission probability than deterministic service, as expected.

10.6 Conclusion

In this chapter, we reviewed the general delay-management architecture for the VoIP system and illustrated our work on delay analysis for utilization-based admission control that can effectively provide delay guarantee for VoIP traffic. We also presented our work on design and implementation of a delay-provisioning system that can be seamlessly integrated into the current Cisco VoIP system.

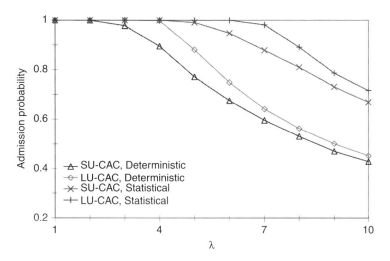

FIGURE 10.10 The admission probability in the campus network.

The integration of our proposed delay-provisioning system with existing VoIP systems has practical applicability for different types of networks: (1) Closed networks where all traffic is under control. Our system can directly work in such networks. (2) Semi-open networks, such as Internet2, where all highest priority traffic can be known, although it cannot be controlled. In our delay-provisioning system, voice traffic has the highest priority and low priority traffic has no impact on voice. Although VoIP is assigned the highest priority (the single real-time priority) in this chapter, the theoretical results in our previous work [3–6] support systems with other applications which have a higher priority than VoIP applications. Our approach can still work by considering other higher priority traffic in the utilization computation. (3) Open networks, such as the Internet, where traffic cannot be controlled and is difficult to predict. Our approach can provide statistical guarantees as long as the traffic on the Internet can be modeled. To our knowledge, modeling Internet traffic is an open issue and is beyond the scope of this study. At the current stage, we can measure the Internet delay at the egress and ingress points of custom networks and dynamically change the bandwidth allocation at custom networks to compensate for the fluctuation of the delay in the Internet. Certainly, such an approach cannot provide guaranteed services. However, if the Internet delay (backbone delay) is small (a general case), the end-to-end delay is predictable.

Appendix A

The worst-case delay d_k suffered by voice packet at link k can then easily be formulated in terms of the aggregated traffic constraint function $F_k(I)$ and the service capacity C_k of the server according to [18]:

$$d_k = \frac{1}{C_k} \max_{I < \beta_k}(F_k(I + d_k) - C_k I), \tag{A-1}$$

where β_k is the maximum busy interval.

We now show that the aggregated traffic function $F_k(I)$ can be bounded by replacing the individual traffic constraint functions $F_{j,k}(I)$ with a common upper bound, which is independent of input link j.

The delay at each server can now be formulated without relying on traffic constraint functions of individual flows. The following theorem in fact states that the delay for each flow on each server can be computed by using the constraint traffic functions at the entrance to the network only. Lemma B.1. The aggregated traffic of the voice flows at link k coming from input link j is constrained by

$$F_{j,k}(I) = \begin{cases} C_{j,k}I, & I \le \tau_{j,k} \\ n_{j,k} \cdot (\eta_k + \rho I), & I > \tau_{j,k} \end{cases} \tag{A-2}$$

where

$$\tau_{j,k} = \frac{n_{j,k} \cdot \eta_k}{C_{j,k} - n_{j,k} \cdot \rho}, \tag{A-3}$$

$$\eta_{p,k} = \sigma + \rho Y_{p,k},\tag{A-4}$$

and the worst-case delay d_k suffered by any voice packet at link k can be bounded by

$$d_k \le \frac{n_k \cdot \eta_k - (C_k - n_k \cdot \rho) \dfrac{n_{j,k} \cdot \eta_k}{C_{j,k} - n_{j,k} \cdot \rho}}{C_k},\tag{A-5}$$

As we described earlier, admission control at run time makes sure that the utilization of link k allocated to voice flows does not exceed u_k

The total number n_k of voice flows from all input links is therefore subject to the following constraint:

$$n_k \le \frac{u_k}{\rho} C_k,\tag{A-6}$$

Then, we are able to maximize the right hand side of (A-5) under the previously stated constrain.

Appendix B

In statistical delay-guaranteed service, all input traffic conforms to a set of random processes. Suppose these processes are independent, if we know the mean value and the variance of each individual traffic random variable, and the number of flows is large enough, then by the Central Limit Theorem, we can approximate the random process of the combined flows. The Central Limit Theorem states that the summation of a set of independent random variables converges in distribution to a random variable that has a Normal Distribution. Actually, using rate-variance envelopes, the traffic arrival rate of each individual flow is a random variable, and the mean rate and the rate-variance of each individual flow can be determined using deterministic traffic models. The following Lemma can be found in [19].

Lemma A.1. With the application of a Gaussian approximation over intervals, the deadline violation probability for a random voice packet is approximately bounded by

$$P\{d_k > D_k\} \le \max_{I < \beta_k} \frac{1}{\sqrt{2\pi}} \exp\left(-\frac{(C(I + D_k) - I\phi_k)^2}{2I^2 RV_k(I)}\right),\tag{B-1}$$

where ϕ_k is the mean rate and $RV_k(I)$ is the rate-variance envelope of voice traffic at link k. β_k is the maximum busy interval.

By this theorem, the deadline violation probability for any random packet can be computed approximately. In the formula given, the final question is how we obtain the values of mean rate and rate-variance envelope. In [19], the rate-variance envelope can be derived with the approximation of a *non-adversarial Mode*: The traffic arrival process conforms to a weighted uniform distribution, where the rate-variance envelope is approximated but non-worst case. Therefore, given the aggregated arrival traffic constraint function $F_k(I)$ of all voice flows, we can specify the *mean rate* and the *rate-variance envelope* as a function of n_k etc. $F_k(I)$ is given as follows:

$$F_k(I) = \begin{cases} C \cdot I, & I \le \tau_k \\ n_k(\sigma + \rho \cdot I), & I > \tau_k \end{cases} \tag{B-2}$$

where

$$\tau_k = \frac{n_k \sigma}{C - n_k \rho}. \tag{B-3}$$

Therefore, we have the following theorem: the mean rate $\phi_{i,j}$ is

$$\phi_k = n_k \rho, \tag{B-4}$$

and the rate-variance envelope is upper bounded by

$$RV_k(I) \approx \frac{1}{12I}(n_k)^2 \rho \, \sigma. \tag{B-5}$$

At this point, the only undetermined variable is the number of flows on each link. In the following, we describe how we eliminate the dependency on the number of flows on each link. The result is a delay formula that can be applied without knowledge of the flow distribution.

As we described earlier, admission control at run time makes sure that the link utilization allocated to each class of flows is not exceeded. The total number n_k of voice flows from all input links is therefore subject to the following constraint:

$$n_k \le \frac{u_k}{\rho} C_k. \tag{B-6}$$

With this constraint, by (B-4) and (B-5), the mean rate and the rate variance can be upper bounded. Therefore, the deadline violation probability can be upper bounded without relying on the run-time information of flow distribution.

References

1. International Telecommunications Union (ITU), "One-way transmission time," Recommendation G.114, 1996.
2. J. Davidson, P. Bailey, T. Fox, and R. Bajamundi, *Deploying Cisco Voice Over IP solutions*, Cisco Press, 2002.
3. S. Wang, D. Xuan, R. Bettati, and W. Zhao, "Providing absolute differentiated services for real-time applications in static-priority scheduling networks," *IEEE/ACM Transactions on Networking*, vol. 12, no. 2, 2004, pp. 326–339.
4. S. Wang, D. Xuan, R. Bettati, and W. Zhao, "Differentiated services with statistical real-time guarantees in static-priority scheduling networks," *Proceedings of IEEE Real-Time Systems Symposium*, December 2001.

5. J. Wu, J. Liu, and W. Zhao, "On schedulability bounds of static priority schedulers," *Proceedings of IEEE Real-Time and Embedded Technology and Applications Symposium (RTAS)*, March 2005.

6. J. Wu, J. Liu, and W. Zhao, "Utilization-bound based schedulability analysis of weighted round robin schedulers," *28th IEEE Real-Time Systems Symposium (RTSS)*, December 2007.

7. S. Wang, Z. Mai, D. Xuan, and W. Zhao, "Design and implementation of QoS-provisioning system for Voice over IP," *IEEE Transactions on Parallel and Distributed Systems*, vol. 17, no. 3, March 2006, pp. 276–288.

8. R. Guerin and V. Peris, "Quality-of-service in packet networks: basic mechanisms and directions," *Computer Networks*, vol. 31, no. 3, February 1999, pp. 169–189.

9. S. Jamin, P. B. Danzig, S. Shenker, and L. Zhang, "A measurement-based admission control algorithm for integrated services packet network SIGCOMM," '95, Cambridge, MA, USA, August 1995.

10. S. Floyd, "Comments on measurement-based admissions control for controlled-load services," *Lawrence Berkeley Laboratory, Tech. Rep.*, July 1996.

11. S. Jamin, S. Shenker, and P. Danzig, "Comparison of measurement-based admission control algorithms for controlled-load service," *Proceedings of INFOCOM '97*, April 1997.

12. C.-N. Chuah, "A scalable framework for IP-network resource provision through aggregation and hierarchical control," PhD thesis, Fall 2001.

13. V. Elek, G. Karlsson, and R. Ronngre, "Admission control based on end-to-end measurements," *Proceedings of the IEEE INFOCOM*, Tel Aviv, Israel, March 2000.

14. Z.-L. Zhang, Z. Duan, L. Gao, and Y. T. Hou, "Decoupling QoS control from core routers: A novel bandwidth Broker achitecture for scalable support of guaranteed services," *Proceedings of ACM SIGCOMM '00*, Sweden, August 2000.

15. Z.-L. Zhang, Z. Duan, and Y. T. Hou, "On scalable design of bandwidth brokers," *IEICE Transactions on Communications*, vol. E84-B, no. 8, August 2001.

16. B.-K. Choi, D. Xuan, R. Bettati, W. Zhao, and C. Li, "Utilization-based admission control for scalable real-time communications," *Real-Time Systems*, vol. 24, March 2003, pp. 171–202.

17. Cisco Documentation, "Gatekeeper external interface reference, version 3.1," 2001.

18. C. Li, R. Bettati, and W. Zhao, "Static priority scheduling for ATM networks," *Proc. IEEE, Real-Time Systems Symposium*, December 1997.

19. E. W. Knightly, "Enforceable quality of service guarantees for bursty traffic streams," *Proc. IEEE INFOCOM*, March 1998.

20. "Abilene network backbone," http://abilene.internet2.edu, 2003.

11

SIP-Based VoIP Traffic Behavior Profiling and Its Applications

Zhi-Li Zhang, Hun Jeong Kang, Supranamaya Ranjan, and Antonio Nucci

CONTENTS

With the widespread adoption of session Session Initiation Protocol (SIP)-based Voice-overIP (VoIP), understanding the characteristics of SIP traffic behavior is critical to problem diagnosis and security protection of IP telephony. In this chapter, we propose a general methodology for profiling SIP-based VoIP traffic behavior at multiple levels: SIP server host, server entity (e.g., registrar and call proxy), and individual user levels. Using SIP traffic traces captured in a production VoIP service, we illustrate the characteristics of SIP-based VoIP traffic behavior in an operational network and demonstrate the effectiveness of our general profiling methodology. In particular, we show how our profiling methodology can help identify performance anomalies through a case study. We also demonstrate the efficacy of our methodology in detecting potential VoIP attacks through testbed experimentation.

11.1 Introduction

Voice over Internet Protocol (VoIP) allows users to make phone calls over the Internet, or any other IP network, using the packet switched network as a transmission medium rather than the traditional circuit transmissions of the Public Switched Telephone Network (PSTN). The maturity of VoIP standards such as SIP [1] and quality of service (QoS) on IP networks opens up new possibilities for carriers as well as enterprises. Consolidation of voice and data on one network maximizes network efficiency, streamlines the network architecture, reduces capital and operational costs, and opens up new service opportunities. At the same time, VoIP enables new multimedia service opportunities, such as Web-enabled multimedia conferencing, unified messaging, while being much cheaper.

Voice over Internet Protocol offers compelling advantages but it also presents a security paradox. The very openness and ubiquity that make IP networks such powerful infrastructures also make them a liability. Risks include Denial of Service (DoS), Service Theft, Unauthorized Call Monitoring, Call Routing Manipulation, Identity Theft, and Impersonation, among others. Not only does VoIP inherit all data security risks, it also introduces new vehicles for threats related to the plethora of new emerging VoIP protocols that have yet to undergo detailed security analysis and scrutiny. There have been several reported incidents and many alerts regarding VoIP attacks or vulnerabilities (e.g., [2,3]). It is therefore imperative for VoIP service operators to deploy scalable monitoring and defense systems to effectively shield their VoIP infrastructure and protect their services and users against potential attacks. In addition, problem diagnosis is also essential to ensure the robustness of VoIP services.

Despite the importance of VoIP problem diagnosis and security, relatively little research has been carried out on analysis of behavior characteristics of SIP traffic—the critical control flow of VoIP services—to help design effective problem diagnosis tools and attack detection mechanisms. This chapter is the first attempt at understanding SIP traffic behavior based on traces from an operational VoIP service. In particular, we develop a novel multilevel profiling methodology for characterizing SIP traffic behavior, with the objective identifying behavior anomalies for problem diagnosis and attack detection. Our methodology characterizes VoIP service activities by extracting and profiling a large variety of traffic features and metrics at three different levels in a progressively refined manner: (i) SIP *server host* characterization, which provides a broad view of their behavior by monitoring and keeping statistics related to only the message types (request vs response) and user activity diversity; (ii) *server entity* characterization, which provides a functional analysis of server activities by separating their logical roles into *registrar*, *call proxy*, and so forth; and (iii) *individual user* characterization, which maintains more detailed profiles of individual user activities. Depending on their needs/requirements, VoIP service operators may choose to profile server/user activities at different levels. In other words, our methodology allows us to balance the speed of profiling, the resource consumption, the desired sophistication of behavior characteristics, and finally the level of security to be offered, based on the specific objectives and needs of the VoIP service operator. Using real-network SIP traffic traces, we illustrate the characteristics of SIP-based VoIP traffic behavior in an operational network and demonstrate the effectiveness of our general profiling methodology. Moreover, we show how our profiling methodology can help identify performance anomalies through a case study. We also develop a profiling-based anomaly detection algorithm and demonstrate its efficacy in detecting potential VoIP attacks in real-time through testbed experimentation.

Related Work While there is a considerable volume of white papers and surveys regarding various vulnerabilities and security threats toward VoIP services (see e.g. [4]), there is relatively few research studies on these topics. Most focus on defense against specific attacks, for example, malformed SIP message format attacks [5,6], DoS and other call disruption attacks [7,8,9], and voice spams [10], albeit these studies are not based on real-network SIP traces. To the best of our knowledge our study is the first analysis of SIP traffic from an operational VoIP service and the first attempt at profiling SIP-based VoIP traffic behavior based on real-network traces.

Chapter Organization Section 11.2 provides some background on SIP, and briefly describes the problem setting and data sets. In Section 11.3, we first introduce a heuristic for discovering SIP servers from passively monitored SIP traffic, and then present our general multilevel profiling methodology for characterizing SIP traffic behavior. Section 11.4 applies our methodology to analyze the SIP traffic behavior using the real-network SIP traces. In Section 11.5, we first use a case study to illustrate how our methodology can help detect performance anomalies; we then present a profiling-based anomaly detection algorithm and demonstrate its efficacy through testbed experimentation. The chapter is concluded in Section 11.6.

11.2 Background and Data Sets

We first provide a quick overview of SIP-based IP telephony. We then briefly touch on the challenges in profiling SIP traffic behaviors based on *passive packet monitoring*, and describe the SIP data sets used in our study.

11.2.1 SIP-based VoIP Service

The session initiation protocol (SIP) [1] is the Internet standard signaling protocol for setting up, controlling, and terminating VoIP sessions.* SIP-based VoIP services require *infrastructure* support from entities such as SIP registrars, call proxies, and so forth

---

* In addition to IP telephony, it can also be used for teleconferencing, presence, event notification, instant messaging, and other multimedia applications.

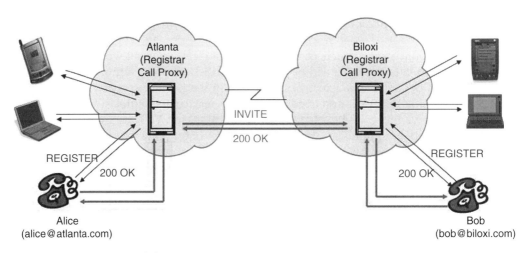

FIGURE 11.1   SIP servers and clients.

(see Figure 11.1)—we collectively refer to these entities as *SIP servers*. A SIP registrar associates SIP users (e.g., names or identities called *SIP URIs*) with their current locations (e.g., IP addresses). A SIP call proxy assists users in establishing calls (called *dialogs* in the SIP jargon) by handling and forwarding signaling messages among users (and other SIP servers). In practice, a physical host (SIP server) may assume multiple logical roles, for example, functioning both as registrars and call proxies.

Session initiation protocol is a text-based request-response protocol, with syntax very similar to HTTP. SIP messages are of either request or response type. The method field is used to distinguish between different SIP operations. The most common methods include REGISTER (for user registration), INVITE, ACK, BYE, CANCEL (these four are used for call set-up or tear-down), SUBSCRIBE, and NOTIFY (for event notification). Response messages contain a response code informing the results of the requested operations (e.g., 200 OK). The FROM and TO fields in an SIP message contain respectively, the SIP URIs of the user where a request message is originated from (e.g., the caller of a call) or destined to (e.g., the callee of a call). In the case of a REGISTER message, both FROM and TO typically contain the SIP URI of the user where the request is originated. Other important fields include VIA and various identifiers and tags to string together various transactions and dialogues. The reader is referred to [1] for details.

11.2.2 Problem Discussion and Data Sets

In this chapter, we focus on characterizing and profiling SIP-based VoIP traffic behavior by using *passive traffic monitoring*, with the objective of identifying anomalies to help diagnose problems and detect potential attacks on critical VoIP services (and their infrastructure). We assume that passive packet monitoring and capturing devices are deployed in the underlying network hosting VoIP services. In addition to the standard layer-3 (IP) and layer-4 (TCP/UDP) header information, portion of layer-7 payload containing appropriate application protocol (SIP) fields are also captured. The captured packet header and payload information is then processed and parsed for our analysis and profiling. Unlike the layer 3/4 header fields which generally have well-defined and *limited* semantics, the layer-7 application protocol such as SIP has a variety of fields, with *rich* semantics that are often context-sensitive and sometimes even implementation-specific. For example, with the SIP protocol itself, the meaning of the same fields may depend on the method used. Hence, a major challenge in performing layer-7 protocol analysis and behavior profiling is to determine how to judiciously incorporate application-specific semantics or "domain knowledge" to select appropriate set of key features to capture the essential behavior characteristics of the application in question. In the following section we present such a general methodology for characterizing and profiling SIP-based VoIP traffic behavior.

Our profiling methodology is motivated and substantiated by an in-depth analysis of SIP traffic traces captured in an operational network of a commercial wireless VoIP service provider. The results reported in this chapter use three SIP traces from this network, referred to as *Trace I* (*13:55–14:30*), *Trace II* (*19:00–19:40*), and *Trace III* (*19:55–20:30*), respectively (the numbers within the parentheses indicate the start and end time of the traces). They are of about 40 min or so long, captured between 13:00 h and 21:00 h within a single day.

11.3 General Methodology

In this section, we present a multilevel profiling methodology for characterizing SIP traffic behavior using layer-3 to layer-7 protocol information obtained from *passive network monitoring*. In order to characterize and profile SIP server behaviors by using passively collected SIP traffic traces, we need to identify the IP addresses associated with SIP servers. We first introduce a simple heuristic for identifying the IP addresses associated with SIP servers (both SIP registrars and call proxies) based on passive monitoring of SIP traffic. We then present a general methodology for characterizing and profiling SIP server behaviors at multiple levels.

11.3.1 Discovering SIP Servers

The key observation behind our heuristics is based on the role of SIP servers in SIP-based VoIP communications: typically, users must register with SIP registrars and users' call signaling must get through SIP call proxies (see Figure 11.1). Hence, the IP address associated with an SIP server will consistently see a large number of SIP messages going through it (i.e., with the said IP address as either the source or destination IP addresses). Furthermore, we will also see a large number of distinct FROM and TO fields in the appropriate SIP messages (e.g., INVITE, REGISTER) associated with this IP address. The baseline algorithm for SIP call proxy discovery is given in Algorithm 1 examining the SIP INVITE messages. By examining the SIP REGISTER messages, we have a similar algorithm for SIP registrar discovery.

In Algorithm 1, for each IP address a in the SIP messages (either as the source or destination IP) we maintain four records, $a.In_{FROM}$, $a.In_{TO}$, $a.Out_{FROM}$, and $a.Out_{TO}$, which maintain, respectively, the set of unique users (or rather their URIs) seen in the FROM and TO fields of the SIP INVITE messages *received* (In) by or *sent* (Out) from a. If the number of distinct users in each of the four records exceeds a threshold α* for an example, then a is

---

* The threshold can be determined, for example, by first plotting In_{FROM} vs. In_{TO} and Out_{FROM} vs. Out_{TO} in a scatter plot, see Figure 1.2.

Algorithm 1 Baseline Algorithm Call Proxy Discovery
_____

```
1:  Parameters: message set M, threshold α;
2:  Initialization: IPSet: = ∅; ProxyIP: = ∅;
3:  for each m ∈ M do
4:    if m.method == INVITE then
5:        x = m.sourceIP; y = m.destinationIP;
6:        from = m.FROM; to = m.TO;
7:        if x ∉ IPSet then
8:            x.Out_FROM = {from}; x.Out_TO = {to};
9:            x.In_FROM = ∅; x.In_TO = ∅;
10:       else
11:           x.Out_FROM = x.Out_FROM ∪ {from};
12:           x.Out_TO = x.Out_TO ∪ {to};
13:       end if
14:       if [|x.In_FROM|, |x.In_TO|, |x.Out_FROM|, |x.Out_TO|] > [α, α, α, α] then
15:           ProxyIP = ProxyIP ∪ {x}
16:       end if
17:       if y ∉ IPSet then
18:           y.In_FROM = {from}; x.In_TO = {to};
19:           y.Out_FROM = ∅; y.Out_TO = ∅;
20:       else
21:           y.In_FROM = y.Out_FROM ∪ {from};
22:           y.In_TO = y.In_TO {to};
23:       end if
24:       if [|y.In_FROM|, |y.In_TO|, |y.Out_FROM|, |y.Out_TO|] > [α, α, α, α] then
25:           ProxyIP = ProxyIP ∪ {y}
26:       end if
27:   end if
28: end for
```
_____

included in the SIP call proxy candidate set *ProxyIP*. By ensuring the diversity of callers (FROM) and callees (TO) in both the SIP INVITE messages originating from and destined to a given IP, we minimize the chance of misclassifying of a user in the *forward* mode in which incoming INVITE messages are forwarded to another location, or similarly, when a user is in a conference mode. In both cases, the TO field of the INVITE messages will contain the URI (or its variants) of the forwarder. Hence, the size of the corresponding In_{TO} and Out_{TO} will be small. We have extended the baseline algorithm to incorporate additional mechanism to address the effect of NAT boxes and other issues, the details of which can be found in [11].

In the following we illustrate the effectiveness of our baseline algorithm using the real SIP traffic traces. Figure 11.2a shows the number of unique FROMs vs. TOs in the SIP REGISTER messages *received* (i.e., In_{FROM} vs. In_{TO}) and *sent* (i.e., Out_{FROM} vs. Out_{TO}) by each IP address seen in the SIP traces. Similarly, Figure 11.2b shows the number of unique FROMs vs. TOs in the SIP INVITE messages *received* and *sent* by each IP address seen in Trace II. It is to be noted that as many hosts (i.e., IP addresses) may have the same number of FROMs and TOs (the labels on the side indicate the number of such hosts). In both cases, only two IP addresses (which are the same two IP entity addresses in Figure 11.2a and b) have significantly more FROMs and TOs than the remaining IP addresses, which have only one or very few distinct FROMs and TOs in a 30–40 min time interval. These two IP addresses are those of two SIP servers (in this case functioning both are registrars and call proxies) in the

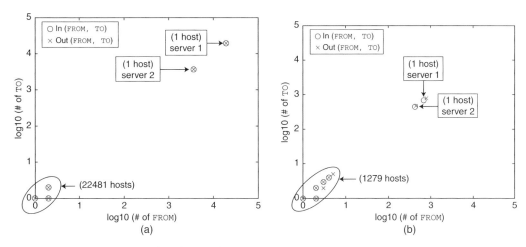

FIGURE 11.2 Hosts corresponding (FROM,TO).

network, one serving more users than the other in this 40 segment SIP trace.* Hence, our baseline algorithm can effectively identify the IP addresses associated with the SIP severs (registrars or call proxies) by appropriately setting the threshold (e.g., $\alpha = 100$).

11.3.2 Profiling SIP Server and User Behaviors

Once we have identified the IP addresses associated with the SIP servers, we characterize and profile the behavior of SIP servers by examining the SIP messages going through them. We characterize and profile the behaviors of SIP servers (and their associated users) at three levels—*server host*, *server entity*, and *(individual) user*—by introducing a range of features and metrics from coarser granularity and finer granularity in terms of the amount of application-specific (i.e., SIP) semantic information. This *multilevel, progressively refined* methodology allows us to balance the speed of profiling, resources required, desired sophistication of behavior characteristics, and level of security, an so forth based on the objectives and needs of a SIP-based VoIP operator.

Figure 11.3 is a schematic depiction of our multilevel profiling methodology. At the *server host* level we maintain only aggregate features and metrics to provide a broad view of a SIP server behavior and its "health" by examining only the message types (`request` vs. `response`) into and out of a SIP server and extracting only coarse-grain user statistics information. At the *server entity* level, we separate the (logical) role of a SIP server into *registrar* and *call proxy*, as these two separate entities require a different set of features and metrics to characterize their respective behavior. On the basis of SIP semantics, we examine the *method* field of a SIP message to attribute it to either the SIP registrar or call proxy (e.g., a SIP `REGISTER` messages and its response are part of a registrar activity while a SIP `INVITE` message and its response are part of a call proxy activity), and compute appropriate features and metrics for the corresponding registrars and call proxies. We also cross-examine the activities of SIP registrars and call proxies to build cross-entity associations. At the (individual) user level, we attribute the SIP messages to individual users, and maintain statistics and features to characterize individual user behaviors. In the following, we provide a more detailed description of our multilevel profiling methodology.

---

* In the Trace III we see that the role of the two SIP servers is reversed, with the latter server serving more users than the former one.

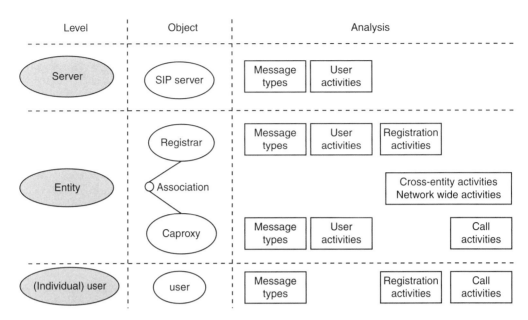

FIGURE 11.3 Multilevel profiling.

a. Server Host-Level Characterization

We characterize the aggregate behavior of a SIP server by maintaining two types of (aggregate) statistics and features: (i) we count the number of `request` and `response` messages received (i.e., *fan-in*) and sent (i.e., *fan-out*) by each SIP server (and derivatively their corresponding ratios) over a given period of time T (say, 5 or 15 min); (ii) we count the number of unique users (URIs) seen in the `FROM` and `TO` fields of SIP `request` messages, and compute an aggregate *user activity diversity* (*UAD in short*) metric from the distribution of such data over T. This UAD metric is computed as follows: Let m be the total number of SIP `request` messages over T, and n the total number of distinct users seen in the message. For each unique user i, m_i is the number of SIP messages with i in either the `FROM` or `TO` field of the messages. Then $p_i = m_i/m$ is the frequency that user i sees in the SIP messages. The user activity diversity metric, *UAD*, is then given by

$$UAD := \left(-\textstyle\sum_i p_i \log p_i\right)/\log m \in [0, 1], \tag{11.1}$$

where the numerator is the entropy of the distribution $\{p_i\}$, while the $\log m$ is its maximum entropy—the ratio of the two is the standardized entropy (or *relative uncertainty*). UAD thus provides a measure of "randomness" of user activities as captured by the distribution $\{p_i\}$: for $n \gg 1$, if $p_i \approx 0$, a few users dominate the SIP activities (in other words, they appear in most of the messages), whereas $p_i \approx 1$ implies that $p_i = O(1/m)$ and thus each user only appears in a few number of SIP messages (hence, overall the user activities appear random).

b. Server Entity-Level Characterization

Registrar: Using the `method` field of SIP messages, we separate registrar-related messages (e.g., the `REGISTER` messages and their responses) and use them to generate statistics and features for registrar behavior profiling. Similar to server-level analysis, we maintain *aggregate statistics* regarding the number (and ratios) of `REGISTER` and other registrar-related requests and responses received and sent by a registrar. In terms of *user activities*, we maintain the number (and list) of users that are successfully registered, and compute a similar user activity diversity (UAD) metric with respect to the registrar. In addition to these aggregate statistics and features regarding the message types and user activities, we also perform a more detailed registration analysis. We examine the response codes in the response messages to maintain statistics about the number of *successful* and *failed* registrations. We also calculate the registration periods of users (i.e., the time lapses between two consecutive REGISTER messages from the same user) and the interarrival times of any two consecutive REGISTER request messages (from different users). From the former we compute the (average) registration period of the registrar and from the latter we derive a (fitted) model for the user REGISTER request arrival process. Together, they not only reveal the configuration of the registrar but also the temporal behavior of the registrar.

Call Proxy: By analyzing the SIP messages related to call activities (e.g., SIP messages with the INVITE, BYE methods, and their responses), we generate statistics and features for call proxy behavior profiling. Similarly, we maintain aggregate statistics regarding the numbers and ratios of various call requests (INVITE, BYE, CANCEL, etc.) and their responses received and sent by a registrar. We maintain several UAD metrics regarding the aggregate user call activities: UAD_{caller}, UAD_{callee}, and $UAD_{caller-callee}$, which measure the UAD of callers, callees, and caller–callee pairs. Each of these metrics is computed using equation (1.1) with appropriately defined parameters: m is the number of SIP call request messages (SIP INVITE, BYE, and CANCEL) requests, and (i) for UAD_{caller}, m_i is the number of SIP call request messages with user i in the FROM field, (ii) for UAD_{callee}, m_i is the number of SIP call request messages with user i in the TO field, and (iii) for $UAD_{caller-callee}$, we replace m_i by m_{ij} where m_{ij} is the number of SIP call request messages with user i in the FROM field and user j in the TO field.

Furthermore, we perform a more detailed call analysis to maintain various call statistics and features of a call proxy. These include the number of ongoing calls, completed calls (calls ended by BYE only), canceled calls (calls ended by CANCEL only), *failed* calls (calls receiving a response with a Request Failure (400–499) response code), and so forth, in a given time period. We also compute statistics (average, standard deviation, or distribution) regarding call durations and call request arrival rates.

Cross-Entity Association: we also correlate statistics and features to generate a cross-entity and network-wide view of the SIP traffic. A detailed description is provided in Kang et al. [11].

c. Individual User-Level Characterization

If needed, we can also maintain statistics and features regarding the individual user activities. For example, from the user call activities we can maintain the (typical or average) number of calls made or received by each user u, and compute the diversity of callees ($UAD_{callee}^{(u)}$) of the calls made by the user as well as the diversity of callers ($UAD_{callee}^{(u)}$) of the calls received by the user u. Other statistics such as (average) call durations may also be maintained. Due to space limitation, we do not elaborate them here.

11.4 Characteristics of SIP Traffic Behavior

In this section, we apply the general profiling methodology presented in the previous section to analyze the SIP traces to illustrate the characteristics of SIP traffic in a real VoIP network and use them to justify the statistics and features we have taken for profiling SIP traffic behavior. In particular, we show that in normal operational environments SIP traffic behavior tends to be very stable both in terms of various SIP message types, user registration, call, and other related activities.

11.4.1 Overall Server-Level Characteristics

Throughout this section, we primarily use *TRACE II* and server-1 as an example to illustrate the results. Figure 11.4a shows the numbers of `request` and `response` messages received (*REQin*, *RESin*) and sent (*REQout*, *RESout*) by server-1 over 5-min time intervals of the SIP traces. We remove the first and last 5 min of the segment to avoid the boundary effect. Figure 11.4b shows, respectively, the ratios of *REQin* vs. *RESout*, *REQout* vs. *RESin*, and *REQin* vs. *REQout* over the same 5-min time interval. We see that overall the total numbers of `request` and `response` messages received and sent by the SIP server do not vary significantly. In particular, for every one `request` message received/sent by the SIP server, on an average there is approximately one `response` message sent/received by it—this is generally expected. There are roughly twice as many `request` messages received by the SIP server than sent by it. As we will see shortly, this is primarily due to the `REGISTER` messages which comprise a large portion of the total `request` messages received by the SIP server. Unlike many SIP `request` messages of other methods (e.g., `INVITE`), a `REGISTER` `request` message does not trigger the SIP server to generate another `request` message except a `response` message.

We break down the SIP `request` messages based on the `method` type and count their numbers over 5-min time intervals. Figures 11.5a and b shows the proportions of `request` messages of each `method` type received and sent, respectively, by the SIP server. As noted earlier, `REGISTER` request messages consist of nearly 60% of the total `request` messages received by the SIP server, while `SUBSCRIBE` `request` messages form 40% of them. In particular, there are no `NOTIFY` `request` messages received by the SIP server. In contrast, the `NOTIFY` messages comprise 90% of the total `request` messages sent by the SIP server, while there is no `REGISTER` `request` messages at all. A more in-depth examination of the `SUBSCRIBE` messages received and `NOTIFY` messages sent by the SIP server reveals that

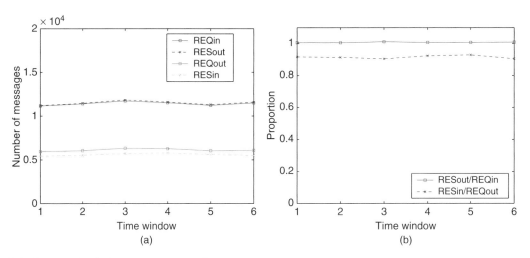

FIGURE 11.4   Analysis on message types for a server.

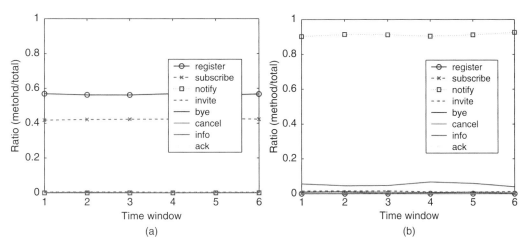

FIGURE 11.5 Analysis on fan-in and fan-out for a server.

there is approximately a one-to-one correspondence between the SUBSCRIBE messages received and NOTIFY sent: this is to be expected, as a SUBSCRIBE received by the SIP server would trigger one (and perhaps a few more) NOTIFY message sent by the SIP server. In both the request messages received and sent by the SIP server, call-related SIP request messages such as INVITE, BYE, and CANCEL consist of only a small portion of the total request messages received/sent by the server.

Figure 11.6 shows the UAD metric of the total SIP messages (both received and sent) by the SIP server over a 5-min time interval, as well as those for SIP request messages received and sent separately. We see that the UAD metrics are close to 1 over all 5-min time intervals and they are fairly stable. As seen in the next subsection this is primarily caused by the periodic exchanges of the REGISTER, SUBSCRIBE, and NOTIFY request messages and their responses between the SIP server and users. Our results show that the aggregate

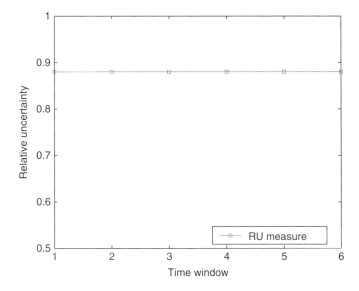

FIGURE 11.6 User activities diversity for a server.

SIP traffic behavior is in general fairly stable and the aggregate statistics/features chosen in our profiling methodology provides a good summary of these stable characteristics. The same observations also hold true for server-2 (which handle a relatively smaller portion of SIP messages in *TRACE II*) as well as for *TRACE III* (where server-2 handles a large portion of SIP messages while server-1 handles a relatively smaller portion of them). *TRACE I*, on the other hand, contains an interesting *anomaly* which is detected by our profiling methodology. We will discuss and dissect this anomaly in more detail in Section 11.5.1.

11.4.2 Registrar Behavior Characteristics

We now focus on the REGISTER request messages and their responses of server-1 (functioning in the role of a registrar), and in particular, examining how REGISTER messages are generated by users. In Figure 11.5a we have shown that REGISTER messages comprises 60% of the total request messages received by the SIP server (registrar). Moreover, the ratio of the number of REGISTER request messages vs. their responses is approximately 1. We observe that the user activity diversity metric for the REGISTER request messages is close to 1, indicating that there are no individual users who dominate the generation of REGISTER messages. The number of *unique* users seen in (the FROM field of) the REGISTER messages over given time intervals are examined. The total number of (distinct) users seen in *TRACE II* is 17 800 that is almost the same as the number of users seen in 15-min intervals and the number of users seen in a time interval of length T (≤ 15 min) is roughly $17\,800 \times (T/15)$. As we will see, this is primarily due to registration periods and a REGISTER arrival process.

To further illustrate how REGISTER messages are generated, we calculate the time lapses between two consecutive REGISTER messages from each user, the distribution of which is shown in Figure 11.7a. The distribution clearly reveals that users generate REGISTER messages roughly periodically with a mean of 15 min. In Figure 11.7b we plot the distribution of the *interarrival* times between two consecutive REGISTER messages (from two different users). The distribution can be well fit into an exponential distribution of the form $p(x) = \lambda e^{-\lambda x}$, where $\lambda = 0.27$. Hence, we see that the number of REGISTER messages seen by the SIP server (registrar) follows approximately a Poisson process. We also study characteristics of SUBSCRIBE and NOTIFY messages which often follow the REGISTER messages of

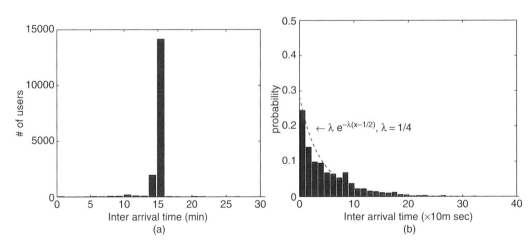

FIGURE 11.7   Analysis on registrar behaviors.

the same users. We perform an in-depth analysis on the number of message types, interarrival times between requests, the SUBSCRIBE periods from each user, and such results are included in [11]. The large number of messages and their regularity lead us to suspect that these SUBSCRIBE and NOTIFY messages are sent to subscribe and notify system resources such as voice mailboxes.

11.4.3 Call Proxy and User Call Behavior Characteristics

We now analyze the characteristics of calls and call-related user activities. Comparing with the number of REGISTER, SUBSCRIBE, and NOTIFY messages, we observe that call-related messages consist of a much smaller portion, indicating that while there are a large number of users (or more aptly, SIP phone devices) in the network, only a very small number of the users actually make phone calls in *a specific* period. This observation is further confirmed in Figure 11.8a which shows the number of unique callers (users seen in the FROM field of INVITE messages) and callees (users seen in the TO field) *in each 5-min interval*. Recall that there are a total of 17 800 unique users in the trace segment. Figure 11.8b is a scatter plot showing the number of calls made vs. calls received per user over 5-min intervals. Again we see that at individual user level, the numbers of calls made and received are generally very small and consistent. In terms of diversity of calls made by users, Figure 11.9a shows the UAD metrics of callers (FROMs), callees (TOs), and caller-callee pairs (FROM-TOs) as defined in Section 11.3.2. We see that the call activities are fairly random, not dominated by any particular user (either as caller or callee).

The number of various call types (ongoing, completed, failed, and canceled calls) over 5-min interval is shown in Figure 11.9b. We see that the number of calls in a typical 5-min interval is fairly small, and the number of *failed* calls is relatively high due to user mobility or receiver statuses (busy or not available). Figure 11.10 shows the call interarrival times are approximately exponentially distributed (the call arrival process is approximately Poisson). We observe that call duration typically lasts between 0 and 3 min while failed and canceled calls tend to last very short. Not surprising, these statistics are similar to traditional telephony, indicating that these call activities are human-generated.

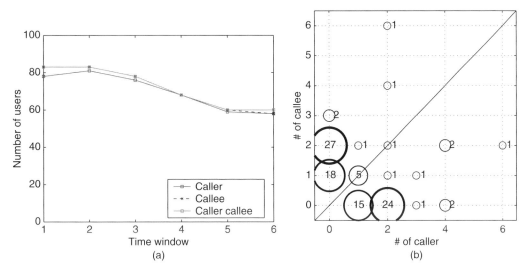

FIGURE 11.8 Analysis of caller and callee behaviors.

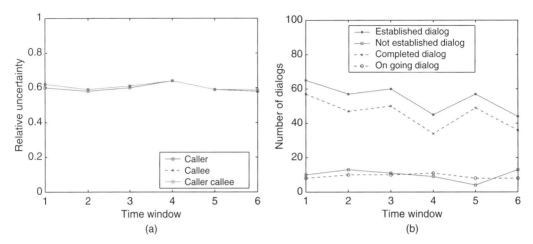

FIGURE 11.9   Analysis on users and calls for a proxy server.

11.5 Applications

In this section, we illustrate the usefulness and applicability of our general profiling methodology in helping diagnose problems and detect potential attacks against VoIP service and infrastructure through a case study as well as testbed experimentation. In particular, we develop a novel profiling-based feature anomaly detection algorithm for these purposes.

11.5.1 Problem Diagnosis: A Case Study

We use a case study—an interesting case uncovered in the analysis of the real-network SIP traces—to illustrate how our methodology can be used to detect and diagnose performance anomalies. As reported earlier, we see that overall the numbers of SIP REGISTER

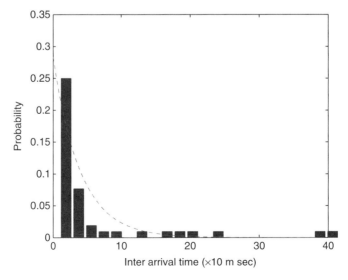

FIGURE 11.10   Analysis on user activities and call types for a proxy server.

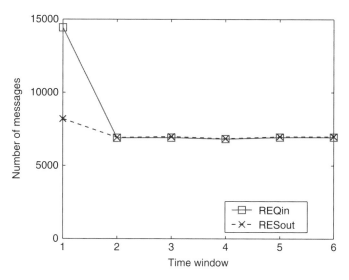

FIGURE 11.11 Analysis on message types for anomaly.

request and *response* messages and their ratios (over 5-min interval) stay fairly stable, and this can be mainly attributed to the fact that users generate REGISTER messages periodically and these messages are generated randomly from the users. These observations hold almost all 5-min intervals for both servers in the traces except for one 5-min interval of server 1 in trace segment 1, where we have found an interesting *"anomaly."* As is evident from Figure 11.11, the number of REGISTER messages received by server 1 in the very first 5-min interval in this trace segment is significantly larger than in other time intervals, and while the number of the responses sent by the server also increases slightly—in particular, the ratio of the numbers of *requests* vs. *responses* increases drastically.

To figure out the cause of this anomaly, we perform a more in-depth analysis of the SIP messages in this anomalous 5-min interval. Figure 11.12a shows the number of REGISTER messages received by server 1 vs. the responses generated by it in each second of the anomalous 5-min interval. We see that between around the 100th second to 160th second

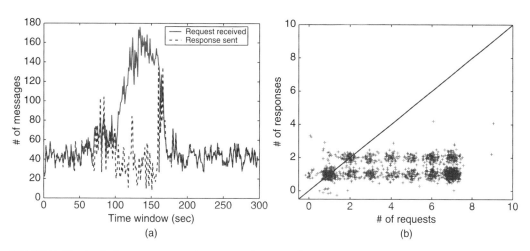

FIGURE 11.12 Analysis on requests and responses for anomaly.

of this 5-min interval, the number of REGISTER *requests* from users shots up quickly, while the *responses* returned by the server first dips for about 50–60 s before it shoots up also, catching up with the number of REGISTER *requests*, after which everything returns to the norm. Figure 11.12b is a scatter showing the number of REGISTER *requests* generated vs. number of *responses* received per user in the 1-min time period from the 100th second to 160th second. To better illustrate the number of data points occupying a particular integer-valued grid (x, y), the data points are "perturbed" slightly around it at random. We see that instead of the normal one REGISTER *request* and *one response* per user, many users send from 2–7 REGISTER requests while receiving one or two *responses*. Closer investigation reveals that the problem is caused by the SIP server not responding to the user registration requests immediately, triggering users to repeatedly retransmit their requests within a few seconds until they either give up or receive a response with either response code 404 Not Found, 408 Request Timeout, or (eventually) 200 OK.

Since all these users were eventually able to successfully register with the SIP server, the surge of the REGISTER *requests* is unlikely to be caused by DoS attacks with spoofed or frivolous REGISTER messages (as were originally suspected by us). That the SIP server failed to respond to the user registration requests in a timely fashion may be caused by delay or slow response from some remote (user/call) database with which the SIP server was interacting.* This performance anomaly can be easily detected using a simple anomaly detection algorithm such as the one described in the following.

11.5.2 Feature Anomaly Detection and VoIP Attacks

Our profiling methodology produces an ensemble of statistics and features over time: for each statistics/feature, a time series is generated. Sudden changes or deviations from expected behavior in a subset of the statistics/feature time series signify anomalies. As has been illustrated later in this section, different VoIP attacks may trigger a different (sub)set of statistics/features to exhibit sudden changes or deviant behaviors. Our profiling-based anomaly detection algorithm consists of the following three key components/phases per feature:

- **Baselining/Profiling**   In this phase, which lasts from $[0 - T_{learn}]$ time windows, we base-line the feature and obtain an estimate of the maximum deviation possible under normal circumstances. We use two sliding windows, one for averaging the feature values and the other for averaging the instantaneous rate-of-change of the feature values, with the windows denoted as T_1 and T_2, where $T_1 + T_2 < T_{learn}$. The output of this phase is the maximum deviation of the rate-of-change from its average. Thus, more precisely: (1) First, for the feature averaging, we use an Exponential Weighted Moving Average (EWMA) with a *beta* = $2/(T_1 + 1)$ to obtain the feature average at time t as: $EWMA(t) = \beta EWMA(t - 1) + (1 - \beta) f(t)$; (2) Thus, we measure the instantaneous rate-of-change as: $s(t) = | [f(t)/EWMA(t - 1)]{-}1 |$; 1; (3) Over the window T_2, we average the instantaneous rates-of-change such that for every time t in the period $[(T_1 + T_2) - T_{learn}]$, we obtain the average slope $s_{avg}(t) = (1/T_2) \sum_{t}^{t-T_2} s(t)$; (4) At a time $t + 1$, we calculate instantaneous deviation of rate-of-change

---

* This problem points to a potential implementation flaw in the SIP client software: when a registration request times out, the client immediately retransmits the request, thereby causing a surge of requests and thus aggra-vating the problem. A better solution would have been to use an exponential back-off mechanism to handle the retransmission of the registration requests.

from its average as: $d(t + 1) = s(t + 1)/s_{avg}(t)$. We learn the maximum of this deviation in the time period $[(T_1 + T_2 + 1) - T_{learn}]$ as: $\alpha = \max_{t=T_1 + T_2 + 1}^{T_{learn}} d(t)$.

- **Alerting along with Lock-in/Lock-out of Averages**   After the learning period is over, we monitor the instantaneous slope $s(t)$ and if it is α times greater than the average slope $s_{avg}(t)$, then we raise an alert and increase the alert level. Otherwise, we update the feature average $EWMA(t)$ and average slope $s_{avg}(t)$. Note that, if an alert was raised, then we *lock in* the values for feature average and average slope, and these values are *locked out* that is, can be updated only when the alert value is reduced to zero. False alarms are avoided by having multiple levels of alerting and the system is said to be in an anomalous state only when the alert level is at its maximum. In particular, we use four levels: Green (alert = 0), Yellow (alert = 1), Orange (alert = 2), and Red (alert = 3).

- **Reporting**   This phase is used for reporting the suspicious elements of a feature that are contributing to the anomaly. This is done only on the features that report the UAD over the feature's distribution, for example, UAD_{caller}, UAD_{callee}. It is applied only when the alert level is updated to its maximum level.

Before we proceed to report the performance of the anomaly detection algorithm stated earlier, we briefly discuss some common potential attacks on VoIP services, and what statistics/features in our SIP behavior profiles may provide alerts about such attacks. Given the generality of our multilevel profiling methodology, we can detect a wide variety of attacks: DoS and DDoS attacks on VoIP infrastructure or users, VoIP spam, and worms that exploit VoIP protocols to spread. The underlying hypothesis toward detecting these attacks is that such an attack introduces either a volume surge, a sudden change or deviation in the ratio/distribution statistics or metrics (e.g., randomness) in one or multiple features. Moreover, the ability to cross-correlate across multiple features also provides us the unique capability to classify the attack. Consider an example of call spam attacks, defined as when a spammer generates many calls, most likely in an automated fashion toward several unsuspecting callees (e.g., automated calls made by telemarketers simultaneously to many callees advertising a product). Thus, by varying the following parameters, a spammer can generate a variety of attacks: (1) number of callers per spammer IP address; (2) whether the IP addresses are legitimate or not; (3) number of IP addresses and; (4) volume of spam calls. The values that each of these features take can be subjectively categorized as "one (or few)" or "many." Thus, there are 2^4 unique spam attacks, two of which are (i) *high volume spam* where one caller per legitimate IP address sends large number of calls to random callees; (ii) and *low volume spam* where one caller per legitimate IP address sends moderate number of calls to random callees.

11.5.3 Experimental Testbed and Results

To validate the efficacy of our detecting algorithm, we use a testbed consisting of two machines, connected via an OC-3 link (1.5 Gbps). Each machine has two Intel(R) Xeon(TM) CPU 3.40 GHz processors and 4 Gbytes memory available to a process running on the Linux kernel 2.4.21. We replay the trace by using *tcpreplay* from one machine while the other one is configured to sniff packets off-the-wire and runs our packet analyzer as well as the anomaly detection module. The packet analyzer can parse layer-7 payloads off-the-wire from network links and emits an annotated vector per packet. Throughout the performance experiments, the capability of our packet analyzer is pegged at a line-rate of

1.5 Gbps while processing VoIP packets up to a maximum of 600,000 concurrent calls with new calls arriving at 1000 calls/second.

For the attack scenario, we take a 3-min sample of clean traffic from our trace and merge the call spam attack toward the end of the trace. In particular, one existing caller URI/IP address from our trace is selected as the spammer. We generate multiple SIP requests from this caller toward randomly generated callees. We generate (i) *high volume spam* that lasts a duration of 25 s, with new calls generated in the first 10 s at the rate of 100 calls/second to yield a total of 1000 spam calls in the trace and (ii) *low volume spam*, consisting of only 10 calls generated by the spammer in the same time duration of 10 s. Each spam call is generated to last the same duration of 15 s, assuming that the spammer is transmitting the same automated message to each caller. In this scenario, we also assume 100% of the callees respond to the caller and none of the callees hang up on the call before the spam call is completed (terminated by BYE). For detecting the attack, we configure the module with a time slot of 2 s, where the learning period T_{learn} lasts for 40 time slots. The averaging time periods T_1 and T_2 are 5 and 15 slots, respectively.

Figure 11.13a and b show two of the several features used in detecting the high volume spam attacks (see [11] for figures of other features). Note that each feature can be observed to be stationary before the attack. Observe that there are two peaks in Figure 11.13a, where the first peak occurs close to the beginning of spam and consists of the flood of INVITE messages and the second peak occurs close to the end of the spam consisting of BYE packets. Our algorithm is able to detect the attack almost instantaneously around time slot 60, when the features of *total requests* and *total unique callee URIs* reach the *Red* alert stage (alert value = 3). Note also that the histogram (distribution) of caller IPs is dominated by the spammer and thereby the corresponding UAD exhibits sharp decreases around the beginning of the attack (Figure 11.13b). Furthermore, at the time when the alerts for these features turn *Red*, comparing the current histogram with the last clean histogram (the one at time slot 60) would reveal the caller URI and caller IP involved in the attack. The efficacy of our anomaly detection in detecting even the *low volume spam* attacks is highlighted via Figure 11.14a and b, which show the performance in the presence of the low volume call spam. In this case, the spammer is able to hide within the background traffic quite well as none of the volume features of *total number of requests* exhibits any significant change during the spam period. However, the intrinsic behavior of the spammer of generating multiple

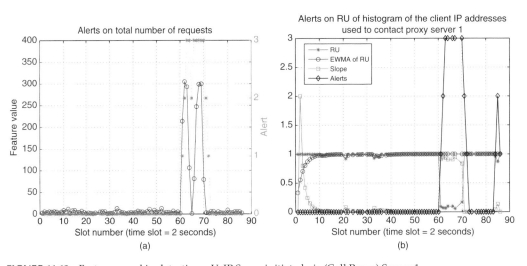

FIGURE 11.13 Features used in detecting a VoIP Spam initiated via (Call Proxy) Server 1.

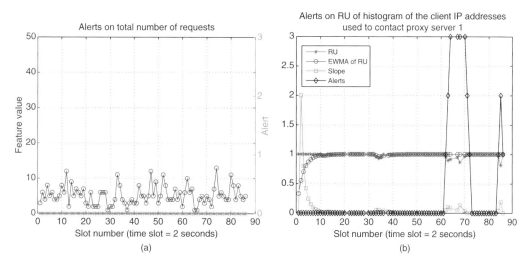

FIGURE 11.14 Features used in detecting a VoIP Low Spam initiated via (Call Proxy) Server.

requests from the same client URI and IP address results in the detection of the attack via the features that track user behavior. Hence, the UAD for caller IPs exhibits three consecutive alerts, leading to the extraction of the spammer's IP address around time slot 63 in Figure 11.14b.

11.6 Conclusions

In this chapter, we have presented a general profiling methodology for characterizing SIP-based VoIP traffic behaviors at multiple levels: the SIP server host, service entity (registrar, call proxy, etc.), and individual user levels. Applying knowledge about application protocol semantics and expected system/user behaviors, an ensemble of statistics and features are selected at each level to capture the essential and stable characteristics of SIP message exchanges, types, volumes, user activities, and so forth. Through our analysis of SIP traffic traces obtained from an operational VoIP service, we show that the overall SIP-based VoIP traffic exhibit stable characteristics and behavior that are well captured by the statistics and features selected in our profiling methodology, thereby justifying the selection of these statistics and features. Finally, we illustrate how our profiling methodology can be used to help identity anomalies for problem diagnosis and attack detection. In particular, we have developed a novel profiling-based anomaly detection algorithm and demonstrate its efficacy in detecting VoIP attacks through testbed experimentation.

Acknowledgment

Hun Kang and Zhi-Li Zhang were supported in part by NSF grants CNS-0435444 and CNS-0626812, a University of Minnesota Digital Technology Center DTI grant, and a Narus Inc. gift grant.

References

1. J. Rosenberg, H. Schulzrinne, G. Camarillo, P. J. Johnston, A. R. Sparks, M. Handley, and E. Schooler, "SIP: Session initiation protocol," RFC 3261, June 2002.

2. N. Wosnack, "A Vonage VoIP 3-way call CID spoofing vulnerability," 2003, http://www.hackcanada.com/canadian/phreaking/voip-vonage-vulnerability.html.

3. A. R. Hickey, "For VoIP, too many threats to count," 2005, http://searchsecurity.techtarget.com/.

4. S. McGann and D. C. Sicker, "An analysis of security threats and tools in SIP-based VoIP systems," *2nd Workshop on Securing Voice Over IP*, Washington DC, USA, June 2005.

5. D. Geneiatakis, T. Dagiuklas, C. Lambrinoudakis, G. Kambourakis, and S. Gritzalis, "Novel protecting mechanism for SIP-based infrastructure against malformed message attacks: Performance evaluation study," *Proceedings of the 5th International Conference on Communication Systems, Networks and Digital Signal Processing (CSNDSP '06)*, Patras, Greece, July 2006.

6. D. Geneiatakis, G. Kambourakis, T. Dagiuklas, C. Lambrinoudakis, and S. Gritzalis, "SIP message tampering: The SQL code injection attack," *Proc. IEEE of the 13th IEEE International Conference on Software, Telecommunications and Computer Networks (SoftCOM '05)*, Split, Croatia, September 2005.

7. B. Reynolds, D. Ghosal, C.-N. Chuah, and S. F. Wu, "Vulnerability analysis and a security architecture for IP telephony," *IEEE GlobeCom Workshop on VoIP Security: Challenges and Solutions*, San Diego, CA, November 2004.

8. B. Reynolds and D. Ghosal, "Secure IP telephony using multi-layered protection," *Proceedings of Network and Distributed System Security Symposium (NDSS '03)*, San Diego, CA, February 2003.

9. Y.-S. Wu, S. Bagchi, S. Garg, and N. Singh, "SCIDIVE: A stateful and cross protocol intrusion detection architecture for voice-over-IP environments," *Proceedings of the 2004 International Conference on Dependable Systems and Networks (DSN '04)*, Split, Croatia, June 2004, pp. 433–442.

10. R. Dantu and P. Kolan, "Detecting spam in VoIP networks," *Proceedings of USENIX, Steps for Reducing Unwanted Traffic on the Internet (SRUTI) Workshop*, Cambridge, MA, July 2005, pp. 31–37.

11. H. J. Kang, Z.-L. Zhang, S. Ranjan, and A. Nucci, "SIP-based VoIP traffic behavior profiling and its applications," *NARUS, Tech. Rep.*, July 2006.

12

VoIP over WLAN Performance

Ángel Cuevas, Rubén Cuevas, Albert Banchs, Manuel Urueña, and David Larrabeiti

CONTENTS

12.1 Introduction

Wireless Local Area Network (WLAN) has been one of the most successful technologies in the recent years. It is a fact that it has been deployed everywhere: companies, educational institutions, airports, cafeterias, homes, and so on. Also, in the press it is frequently reported that the government of a city is to commence a project to offer public Internet access by using WLAN technology.

Two kinds of WLAN architectures or operation modes exist: Infrastructure mode, which is the most common mode and uses a centralized coordination station, usually called an Access Point (AP), for the scheduling of transmissions. All traffic goes via the AP. An example is

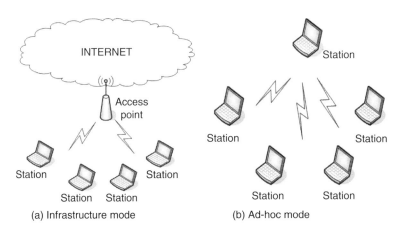

FIGURE 12.1   Wireless LAN architectures.

shown in Figure 12.1a. An ad hoc mode network works without this centralized element and therefore it needs a routing protocol to provide reliable end-to-end communications between users. An example is shown in Figure 12.1b.

The most widely adopted architecture for infrastructure and ad hoc wireless networks is based on the IEEE 802.11 standard [1], which was initially designed for data services. Therefore, it does not have ideal characteristics for real-time communications such as video or voice. The most successful versions of the technology are IEEE 802.11b and IEEE 802.11g. The main difference is the operation rate, that is, the b version has 11 Mbps as upper limit and in the g version the maximum rate is 54 Mbps.

Much research has been carried out with the intention of extending the services offered from data to real-time communications in wireless networks. As a result a new standard, IEEE 802.11e [2], was developed giving support for Quality of Service (QoS), which is needed for real-time communications.

Voice over Internet Protocol (VoIP) [3–5] is currently a predominantly used technology, with a greater number of people using it to telephone across the world. There are many common programmes which make it easy to use VoIP: Skype [6], MSN messenger [7], VoIP cheap [8], etc. They are frequently used every day because they offer a good quality and are cheap. Hence, VoIP is beginning to be a very widely used technology. The principal telephony service provider companies have realized this and most of them are now offering VoIP services, specially to enterprises.

Many organizations are using WLANs as the first hop in their networking strategy, and hence it is important to investigate how VoIP performs over WLAN. An example of a realistic scenario is often the best way to investigate a problem: A university department which uses WLAN as its local area network installs VoIP. Subsequently, all the calls between lecturers in the department will be through VoIP. In this case, it would be interesting to know the performance limitations.

Trying to establish a clear objective in our research, we focused the principal objective as a question: How many VoIP calls can be established over a 802.11b WLAN? To answer this we measured three quantities which are the most important parameters involved in real-time voice communications, assuming that raw data throughput is not a problem. The first parameter is packet delay, which is a measure of the time taken for a packet to travel from an originating node to a destination node. The VoIP maximum one-way delay recommended by ITU-T G.114 is 150 ms [9]. The second parameter is packet jitter, which is a measure of delay variability. For this purpose we measured the time between packet arrivals. We checked the time variability between arrivals, because for jitter the average is not important but the variance is. The last parameter is loss packet rate (LPR). This is a measure of how many packets are dropped in a particular conversation. The results for the LPR are given as a percentage, which will give some clue as to the impact on quality.

In order to answer the question we carried out a large number of simulations of both infrastructure and ad hoc networks. In each case we measured the key performance criteria: packet delay, jitter, and LPR. This chapter presents some of the key results and provides a number of answers, depending on the precise conditions.

Moreover, we provide in this chapter results obtained from real experiments which validate our research. In addition, it will be shown the improvement that 802.11e offers for real-time voice applications in front of 802.11b.

12.2 The Simulated Scenarios

12.2.1 VoIP Traffic

Before working out the simulation the characteristics of the VoIP traffic must be defined. Table 12.1 shows the most popular codecs used for voice.

In order to select one of these codecs to perform the simulations we decided to use codec G.711 because it is the most restrictive. In this way, the obtained results would be an inferior limit. Therefore, we will show the worst case and if other codec is used the number of VoIP calls performing with a good quality will be higher than the number obtained in our work. It must be noticed that in many commercial implementation the codec G.711 is used by sending 160 bytes each 20 ms. However, we use in our simulation the values showed in the table. These values follow the standard PCM codec (an 80 bytes packet generated every 10 ms) and they are more restrictive.

Besides, the way in which the VoIP flows will work over the WLAN must be defined. In our simulations every VoIP call generated two VoIP flows in the WLAN, when the infrastructure mode was used. There was one flow from one of the end nodes involved in the call and the AP and another one from the AP to the other node. That is because the VoIP call was generated as half-duplex traffic which means that only the node which established the call (caller) sends traffic at first. Then, after a random time, this node stops and the node which received the call (callee) starts to send traffic. The time each node spends sending traffic was randomized. Usually, the node which sent traffic changed several times during the call. In the ad hoc mode, each VoIP call generates from one to several flows depending on the number of intermediate nodes which forwards the traffic from the

TABLE 12.1

Most Common Codecs Features

Codec	G.723.1	G.729	GSM	G.736-32	G.711
Bit rate (kbps)	5.3/6.3	8	13.2	32	64
Framing interval (ms)	30	10	20	20	10
Payload (bytes)	20/24	10	33	80	80

originator node to the destination node, but the traffic generation was the same than in the infrastructure mode.

12.2.2 Scenarios Simulated

Five scenarios were simulated: three of them under an infrastructure mode architecture and two using an ad hoc mode network. Some of them are ideal scenarios by which we mean that nodes in these simulations were close together; something that may not happen in a real scenario. The reason to simulate an ideal scenario was to obtain a baseline maximum number of calls value and which we could then compare with some more realistic scenarios. The simulated scenarios were

- Infrastructure mode ideal scenario. AP packet queue length: 50 packets. 20 stations plus 1 AP.
- Infrastructure mode ideal scenario. AP packet queue length: 100 packets. 20 stations plus 1 AP.
- Infrastructure real scenario. 20 stations plus 1 AP (Figure 12.2).
- Ad hoc mode ideal scenario. 31 stations.
- Ad hoc mode real scenario. 31 stations (Figure 12.3).

The key point in the infrastructure mode scenarios is the AP. When more conversations are added the AP has to schedule more packet transmissions, hence its packet queue will lengthen. Those packets which are waiting to be forwarded will be delayed. The AP queue has a limited capacity and when the queue is full, if a new one is received it will be dropped.

In the ad hoc case, a routing protocol has to establish routes among nodes in order to establish end-to-end user conversations. In this case, there will normally be intermediate nodes in a call. Depending on which nodes establish a VoIP call there will be more or less hops needed for a packet to arrive at the destination node. It is possible that some node in a conversation path will become saturated, as it could be a node involved in many conversations. In this case, all the conversations which use that node will be affected, their packets will be delayed and, if the node packet queue is full, some packets will be dropped. This is the normal behavior in an ad hoc network. However, in our simulation there is a

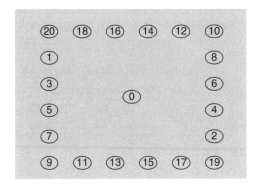

FIGURE 12.2 Infrastructure mode real scenario.

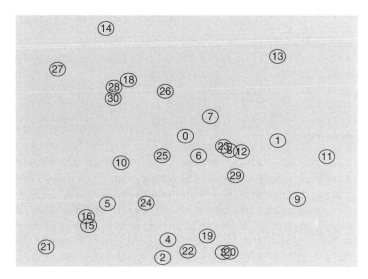

FIGURE 12.3 Ad hoc mode real scenario.

special ad hoc scenario which presents ideal conditions (nodes very close). In this scenario there will not be any intermediate nodes in the conversation path, because the nodes are very close to each other, and only one hop is needed for a packet to reach its destination. Therefore, if there is high delay it is because the environment is saturated and many collisions are happening in the link.

12.2.3 Simulation Features

The simulation time used was 100 seconds in the four first scenarios, and more than 30 seconds in the last one. This time was sufficient to obtain the desired results.

In the infrastructure mode, to obtain results it is sufficient to analyze one call to understand the happenings in that scenario, even if more calls are running at the same time. Since there is a centralized control element, the AP, and there is no quality of service (QoS) mechanism running, all packets receive the same treatment. Therefore, when the AP queue is saturated, packets from all conversations are dropped on a random basis, and packets from all conversations are delayed in the same way. Knowing this, if one conversation starts to present bad results in delay, or dropped packets appear for that conversation, the remaining conversations will also present the same behavior and bad quality. Therefore, it would suffice to check results for only one call in order to achieve our objective.

In the fourth scenario, also, it is enough to obtain results for only one conversation to understand the situation. As it is an ideal ad hoc scenario where all the nodes are very close, the routing protocol establishes one hop path between all the nodes. In other words, one packet needs only one hop to arrive at the destination node. Hence all the conversations suffer the same environment. So if packets from an individual conversation show high delay or are dropped, it is because the link is saturated. It could be concluded that all the conversations must have the same problems. Hence, one conversation is enough to understand what is happening in a particular simulation.

In the ad hoc real scenario we had to obtain data for every VoIP call running, because in this case every call has a different path and different number of hops. It is possible that one intermediate node in a conversation is saturated and this conversation will present bad results, whilst other conversations running at the same time do not cross that node, and will be carried successfully.

12.3 Results

The simulation results are reported in several figures. Figure 12.4 shows average packet delay versus number of calls running over a scenario. Figure 12.5 shows jitter variance versus number of calls. Figure 12.6 shows LPR versus number of calls. These graphs show results for all scenarios except the ad hoc real scenario.

The first scenario shows an abrupt change when the sixth call is added. The LPR changes from 0% to 10%. This is an unacceptable value, because 10% of the information in a phone call cannot be lost, because then the conversation would probably not be comprehendable. Also the delay results indicate bad quality, because the average delay is over 50 ms. This means that the delay of many packets is over this value, whereas in the previous cases it has a very low value. The jitter's variance starts to increase at this point (sixth call) but not too much; but it is a warning showing that something is working worse than in the previous cases. If even more calls are added, all the parameters get worse.

The second scenario results are very similar. We simulated this scenario because in the previous one we obtained an unacceptable LPR when the sixth call was added. The AP drops packets when its queue is full. Trying to reduce this effect we doubled the AP's queue length. We expected to reduce the LPR whilst paying with a higher packet delay. Results obtained showed that when the sixth call is added the average delay is higher than

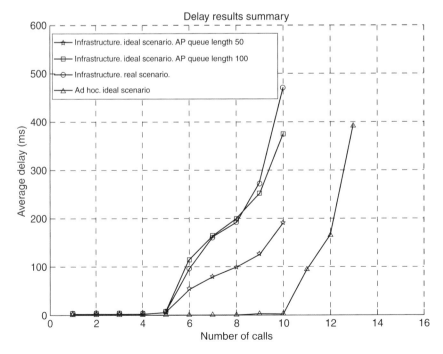

FIGURE 12.4   Average packet delay.

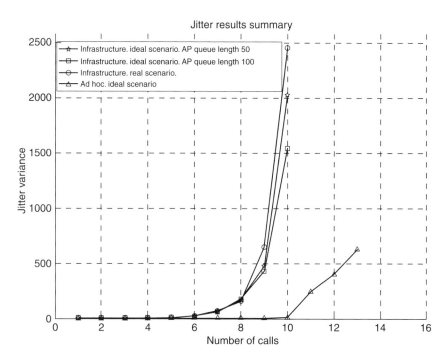

FIGURE 12.5 Variance packet jitter.

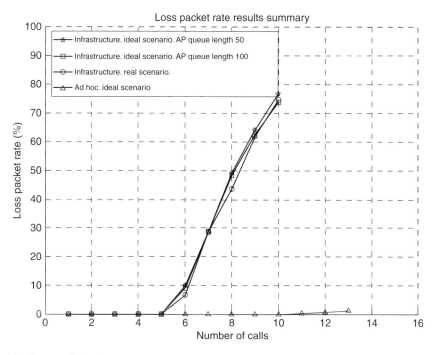

FIGURE 12.6 Loss packet rate.

in the first scenario, more or less double, which means around 100 ms of average delay. This is an unacceptable value because for much of the time the packet delay will be over 150 ms to obtain an average of 100 ms. However, a lower value of dropped packets does not appear. The result is an LPR of 9% instead of 10% with a 50 packet queue in the AP. The commentary for the jitter's variance is also the same than in the previous case. With the sixth call it starts to increase. Again, very bad results are obtained if 6 or more calls are running over this scenario.

The third scenario is a real infrastructure mode scenario in contrast to the previous ideal scenarios. In this case the nodes are located around the AP, as will occur in many real cases. However, results in this case are quite similar to the previous ideal results. Again, when the sixth call is added the call quality becomes unacceptable. Specifically, the average packet delay becomes around 100 ms. The LPR is slightly lower than in the previous cases, around 7% but this is still too much information being lost. The variance in jitter again presents the same bahavior than in the previous cases. When we introduce 7, 8, 9, or 10 calls as in scenario 1 and 2 cases the results are really bad.

Results for the ad hoc mode ideal scenario are reported in Figures 12.4, 12.5, and 12.6. In this case the key step occurs when the eleventh call is added. Delay and jitter are the parameters which show unacceptable values. The average packet delay is around 100 ms, and as we explained in some previous cases this is a bad value because, if the average is 100 ms it is because many packets have a delay over 150 ms which is not acceptable. The jitter's variance presents an abrupt jump between the tenth and eleventh calls. A high variability in the jitter shows bad quality in the communication, in this case very bad quality. The LPR changes from 0% to 0.3%, which is an important change. Since it is an ideal scenario and packets need only one hop to arrive at the destination node, if there are dropped packets it is because in the originating node queue there are 50 packets waiting to be sent, and a new one is generated and dropped because the queue is full. Therefore, this low LPR explains why the delay is so high. A quick conclusion is that eleven or more calls cannot be established in an ad hoc scenario with ideal conditions.

In the ad hoc real scenario we cannot present a summary result, because the results in every simulation were different. It depends on which calls are running in the scenario and which nodes are involved in these calls. To explain this, Figure 12.7 which measures the

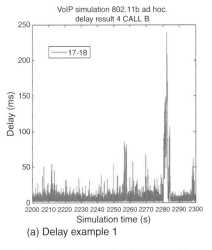

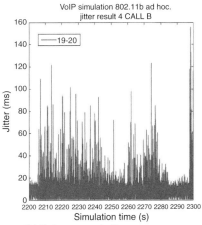

FIGURE 12.7   Ad hoc real scenario with four calls.

instantaneous delay of each packet in the scenario shows the results of two calls running over a four calls scenario, with poor values for the delay. The packet delay value is over 50 ms much of the time in these two calls, and one of them with peaks which reach more than 150 ms. However, we found a scenario where seven calls run properly. In this scenario, the packet delay is always under 40 ms and most of the time under 20 ms for the seven calls, which is an acceptable value.

Other scenarios were simulated, and we found scenarios with four calls with good parameters' values and scenarios with seven calls with unacceptable quality in the conversations. We did not find a simulation with eight calls working properly. We simulated many times with different combinations of calls, but no simulation presented good results. These results are a good example to explain that in a real ad hoc scenario it is impossible to obtain a general result, because it very much depends on which nodes establish a conversation and which nodes are involved in the path of each conversation, whether a particular conversation will have an acceptable or unacceptable quality. The last commentary in this real ad hoc scenario is that we found one case when even one call could not be established, because the routing protocol did not find a path between the caller and the callee. Checking the trace output file we saw that all the packets sent from the originating node were dropped because the originating node did not know the following hop.

12.4 Statistical Analysis

We tried to find if there are some patterns in the delay and jitter statistical distribution. In this way, by just glancing at a distribution from a particular VoIP call we should be able to recognize if the quality of that conversation is good or not. Following this objective we obtained histograms with packet percentage versus time, for packet delay and jitter as we can see in Figures 12.8 and 12.9.

Figure 12.8a shows that when we have a simulation with good parameters, the distribution which describes this situation is an exponential distribution in packet delay, because most part of the packets have a low delay. Obviously, the exponential does not start at time 0, because it is impossible for a packet to experience a delay of 0 seconds. The minimum will be around 1 or 2 ms. If we add more calls and consider the first case with unacceptable values, 6 calls simulation (Figure 12.8b), the distribution has a peak at low values of delay,

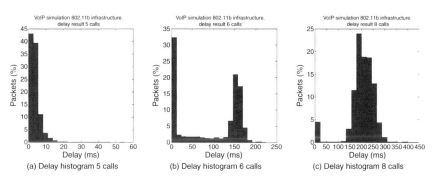

FIGURE 12.8 Delay distribution evolution. All histograms are obtained from infrastructure mode real scenario.

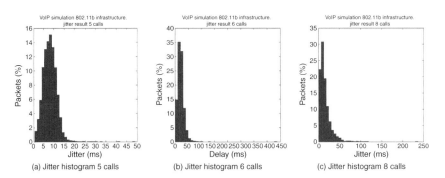

FIGURE 12.9   Jitter distribution evolution. All histograms are obtained from infrastructure mode real scenario.

which includes about 30% of the packets, but there is an important percentage of packets with high delay values, between 100 and 200 ms. The shape of this distribution is something like a U. If more calls are added, the peak in the low values disappears, and more packets are in the high values. At this moment a distribution similar to a normal distribution can be seen, as it is showed in the graph in Figure 12.8c. Evidently, this distribution is centered around a high packet delay value.

A different evolution occurs in packet jitter. With five calls, there is a normal distribution centered around 10 ms, with short and symmetrical queues, as is shown in Figure 12.9a. This is the expected result when the jitter presents a good quality value. In the case with six calls shown in Figure 12.9b, when the quality of the conversation reduces, the queue on the right side is longer and the distribution presents a non-symmetrical shape. With eight calls as shown in Figure 12.9c, the aspect of the distribution is like an exponential distribution starting in low values and with a very long right queue. Since the important factor for the jitter is the variance, obviously, the highest variance is when there are most calls, and the five calls graph is the one which exhibits the lowest variance.

With this statistical analysis an interesting pattern model can be established. Looking at one delay or jitter packet distribution for one conversation we will be able to know what is happening approximately, and if that conversation will have good or bad quality by comparing the distribution with one of the previously explained patterns.

12.5 Related Work

12.5.1 Real Testbed

Much research has been carried out in the field of VoIP over WLAN (e.g., [10–15]). In particular, some research works have deployed real testbed in order to measure the VoIP over WLAN performance [11,13]. Following the question posed in this study, how many VoIP calls can be established over a 802.11b WLAN? Other works, like the one presented below, have provide an answer by using real measurements.

In Garg and Kappes [11], the experimental setup used eight wireless clients associated to a single access point. The access point was connected to an Ethernet LAN network. In this LAN network there were eight PCs which served as end points for the VoIP communications. The clients in the WLAN were located in the same room without obstacles between them. Obviously, the scenario is different from the one posed in our simulations, but the number of VoIP flows within the WLAN will be the same, two VoIP flows per VoIP communication. That is, in this testbed there were two flows per established call at any time, from caller to callee and from callee to caller. Therefore, there were two VoIP flows in the WLAN, one flow from the originating node to the AP and a second one from the AP to the originating node. Hence, the number of flows running on the WLAN is the same than in our simulation. However, the testbed is different because there is a wired part. In any case this scenario was similar to the real infrastructure mode scenario in the simulations.

On completing the experiment, the five first calls experiment were of good quality with reference to communications. When the sixth call was established the authors in [11] say that even if the round-trip time for the packets increased, the quality of the communications was still fine. However, when the seventh call was run all the communications were affected and there were unacceptable loss of quality in all the communications.

The results are quite similar to the ones obtained from our simulation. The difference is that when six calls were running in our simulations, even if the delay and the jitter were acceptable we had a LPR around 7% that we considered unacceptable. In the real testbed work, a loss of performance is perceived when the sixth call was added, but in this case they could check the quality by human intervention and establish that the quality of the sixth conversation was acceptable. In both cases when the seventh call tried to be established there were very bad performance in the communications.

In Garg and Kappes [10], an analytical model is presented. The main results of this analytical model are

- When a G-729 or G-723 codec is used the number of VoIP calls which can be established is higher than using G-711. It is an expected conclusion.
- If you use a different version of the standard, in this case 802.11a with a 54 Mbps transmission rate, you can establish more calls than using 11 Mbps. This being an obvious result, the authors provide the number of calls which can be established in this case. They used the codec G-711, the same that was used in the simulations and the real testbed, and the result is that up to 30 VoIP calls can be established following the analytical model.
- Finally, if the interval frame and the size of every packet is higher, that is sending longer voice fragments in each packet, the number of VoIP calls performing fine is also higher.

12.5.2 VoIP Over 802.11e

Due to the lack of mechanisms to provide QoS in the 802.11 a/b/g standards, the IEEE defined an extension for these standards which provides mechanisms for supporting QoS. This is the standard IEEE 802.11e [2]. The main difference between IEEE 802.11e and the previous standards is that the previous ones gave a fixed value of the involved parameters whereas in 802.11e there are four parameters which are tuneables:

- AIFS: It is the time that the channel must be sensed to be idle before starting a transmission. In the previous standards it was a fixed time called DIFS.
- Contention Window minimum (CW_{min}) and Contention Window maximum (CW_{max}): If at some point the channel is sensed to be busy, the station starts a

backoff period. This backoff period is a number of slots that must be waited before trying a new transmission. The number of slots that a station waits is defined by a uniform distribution in the interval [0,CW]. After the backoff process the station tries to transmit, if a collision occurs, two or more station access the channel simultaneously, the CW is doubled up to a maximum value which is CW_{max}, and the backoff process is restarted. In the previous standards the CW_{min} and CW_{max} values were fixed at 32 and 1024, respectively. In the 802.11e standard this parameter can be established in each station.

- Tx Opportunity (TXOP): Once a station gains access to the channel, this station is allowed to transmit for a duration given by the TXOP parameter. If this parameter is set to 0, it will be allowed to transmit only one packet stored at this moment in its transmission buffer.

Therefore, looking at the previous parameters it is easy to understand that by tuning the different parameters in each station, QoS can be provided. We can give a higher transmission priority by reducing the AIFS or doing $CW_{max} = CW_{min}$, and giving a low value to these parameters. All these techniques cannot be applied to the IEEE 802.11 a/b/g standards where the parameters have fixed values.

Due to the IEEE 802.11e standard much research has been conducted in order to find the optimal configuration to obtain different goals (e.g., [16,17]).

In particular in [18], an experiment to evaluate the improvement of using 802.11e for VoIP traffic instead of using 802.11b is presented. Again, a G-711 codec is used, an 80 bytes packet is generated every 10 ms. In this case the traffic was generated with the tool iperf. In order to develop the testbed the traffic was generated as it is shown in Figure 12.10. The restrictions established in this work were: 50 ms of one-way delay and a maximum loss rate of 5%.

Under these conditions, in this testbed eight VoIP calls could be established by using the standard 802.11b. What this experiment tried was to obtain an optimal configuration of the tuneable parameters in order to improve as much as possible the number of VoIP calls that can be established in a WLAN.

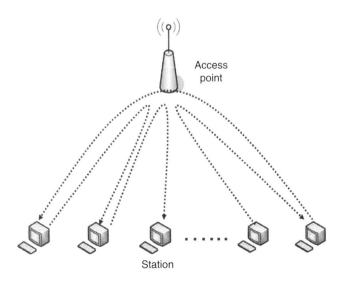

FIGURE 12.10 Testbed traffic generation.

Primarily, it seems obvious that in order to reduce the delay the AIFS parameter should be the smallest possible. Following this, the TXOP should be set to the maximum possible, otherwise the queueing delays would be higher if each packet has to experiment a backoff process. By using the maximum TXOP most of the packets would experiment only one backoff process so that the queueing delay would be lower. Next, CW_{max} should not be of a high value because if a packet suffers several collisions the CW will increase and it means higher delays in queue before transmitting that packet. Therefore, for real-time voice packets the best option is to make CW_{max} equal to CW_{min}.

After the discussion about parameters, three out of four parameters were established. Therefore, in this study the optimum value of CW_{min} in order to obtain a maximum number of VoIP calls within the WLAN was investigated. However, it was taken into account that the configuration for the AP and the station could be different in order to achieve the best result. Hence, they were checked using different values for both CW_{min}^{AP} and CW_{min}^{Sta}.

After developing the test, it was shown that the maximum VoIP calls established with an acceptable performance was 11, improving in three the number obtained by using the 802.11b operation mode.

There were several configurations which provided this call number. However, the configuration which provides the minimum delay was: $CW_{min}^{AP} = 32$ and $CW_{min}^{Sta} = 64$. The average one-way delay obtained under these conditions was 22.56 ms.

Therefore, this experiment proved that the use of IEEE 802.11e provides a better performance than IEEE 802.11b when VoIP traffic is generated over WLAN, and also shows that for each specific case the configuration of the different parameters is different. Hence, if there are data and voice traffic in an IEEE 802.11e WLAN different parameters can be assigned to different stations. In this way, the QoS can be provided, and the real-time traffic such as VoIP traffic can be assigned more priority than the data traffic.

As conclusion of this section, the question posed for the IEEE 802.11b WLANs network can be also answered in relation to the IEEE 802.11e networks. The maximum VoIP calls which can be established in a IEEE 802.11e WLAN is 11.

12.6 Conclusions

In this chapter we have evaluated the performance of VoIP calls over 802.11b and 802.11e WLANs, in infrastructure and ad hoc mode architectures. We have proposed and discussed several scenarios with different features in our simulations as well as presented real testbeds where VoIP has been tested over WLANs.

In the simulation we measured different parameters, involved in every real-time communication, such as packet delay, jitter, and LPR, and on the basis of them we have determined the maximum number of VoIP calls which can be supported by a WLAN.

A statistical analysis is presented, to enable us to look for comparison patterns. In this way, by making a comparison with a given distribution we can know, approximately, the happenings in that conversation.

The most important result obtained in all infrastructure mode scenarios simulated is that five VoIP calls can be established. When the sixth call is added, in the three infrastructure mode cases, the LPR was not 0% anymore and the average delay increases abruptly, in one case up to 50 ms, in the other two up to 100 ms. Jitter's variance presented a slight increase. Worse results were obtained if more calls were added.

A similar conclusion could be made in the ad hoc ideal scenario, when the nodes are very close together. In this case 10 calls can be established at the same time without problems. When the eleventh call was added, average delay and jitter variance grew until they reached unacceptable values. At this point a LPR that was not 0% appeared. It was a low rate of 0.3% but with an important meaning because it expalins why the delay is so high.

There are no conclusions with values in an ad hoc real scenario. The reason is because in this case every simulation presented a different behavior. In a real ad hoc scenario it becomes very important which nodes are establishing a conversation and which are the intermediate nodes. Whether the calls can be supported is a function of node distribution and which nodes are involved in the calls.

We showed that a network scenario with seven calls could be established, whereas a different situation with four calls had bad quality. We did not find any simulation with eight calls running properly over this real scenario.

Finally, the statistical analysis made shows that the evolution in packet delay goes from an exponential distribution, when there are good performance features in a scenario, towards a normal distribution centering around a high delay value, when bad quality is expected. In packet jitter the evolution goes from a normal distribution centering around a known value towards an exponential distribution.

Following, we have presented a real testbed in an infrastructure mode WLAN, where the maximum number of calls achieved in a 802.11b WLAN was six. The methodology and differences between this real testbed and our simulations was explained. And it has been checked that the results are quite similar.

Finally, the IEEE 802.11e standard extension has been introduced to provide QoS. Another testbed related to this has been presented, and it has been shown that the usage of IEEE 802.11e outperforms the results obtained in the IEEE 802.11b. In particular, under this testbed condition the improvement goes from 8 to 11 calls.

Acknowledgment

This work was partially supported by the Spanish government through the Project IMPRO-VISA TSI2005-07384-C03-027.

References

1. IEEE Std 802.11, "IEEE standard for wireless LAN medium access control (MAC) and physical (PHY) layer specification," 1999.
2. IEEE 802.11e, "Part 11: Wireless LAN medium access control (MAC) and physical (PHY) layer specification; Medium access control (MAC) enhancements for quality of service (QoS)," Supplement to IEEE 802.11 Standard, 2005.
3. J. Davidson, *Voice Over IP fundamentals*, Cisco Press, 2007.
4. U. D. Black, *Voice Over IP*, Prentice-Hall, 1999.
5. M. Goncalves, *Voice Over IP networks*, McGraw-Hill, 1999.
6. http://www.skype.com

7. http://www.msn.com

8. http://www.voipcheap.com

9. ITU Rec. G.114, "One-way transmission time," *International Telecommunications Union*, February 1996.

10. S. Garg and M. Kappes, "Can I add a VoIP call?," *ICC 2003*, vol. 2, 2003, pp. 779–783.

11. S. Garg and M. Kappes, "An experimental study of throughput for UDP and VoIP traffic in IEEE 802.11b networks," *WCNC 2003*, vol. 3, 2003, pp. 1748–1753.

12. W. Wang, S. C. Liew, and V. O. K. Li, "Solutions to performance problems in VoIP over a 802.11 wireless LANs," *IEEE Transactions on Vehicular Technology*, vol. 54, no. 1, January 2005, pp. 366–384.

13. G. L. Agredo and J. A. Gaviria, "Evaluacion experimental de la capacidad de IEEE 802.11b para soporte de VoIP," *i2Comm 2006*, vol. 1, 2006, pp. 125–151.

14. A. Cuevas, "VoIP over WLAN ns-2 simulations for infrastructure and *ad hoc* architectures," Master thesis, University of Reading (UK) and Universidad Carlos III de Madrid (Spain), July 2006.

15. V. J. Garcia, "Estudio experimental de VoIP en un escenario 802.11e EDCA," Master thesis, Universidad Carlos III de Madrid (Spain), November 2005.

16. A. Banchs, A. Azcorra, C. Garca, and R. Cuevas, "Applications and challenges of the 802.11e EDCA mechanism: An experimental study," *IEEE Network*, vol. 19, no. 4, July 2005, pp. 52–58.

17. R. Cuevas, "Soporte QoS en redes WLAN 802.11e: análisis práctico de prestaciones e implementación de una arquitectura QoS," Master thesis, Universidad Carlos III de Madrid (Spain), December 2005.

18. P. Serrano, A. Banchs, J. F. Kukielka, G. d'Agostino, and S. Murphy, "Configuration of IEEE 802.11e EDCA for voice and data traffic: An experimental study," *Internal Report*, Universidad Carlos III de Madrid (Spain).

13

Burst Queue for Voice over Multihop 802.11 Networks

Xinbing Wang and Hsiao-Hwa Chen

CONTENTS

13.1 Introduction

In the last few decades, VoIP over wireless has received a considerable amount of attention. Since wireless networks based on the IEEE 802.11 standard are popular and have been widely deployed, it is natural to study the transmission of VoIP traffic over 802.11 networks. Numerous studies have been carried out to evaluate the performance of VoIP traffic over 802.11 networks. However, most of these studies are concentrated on one-hop situations; only a few works have addressed the topic of supporting audio application over multihop, ad hoc networks. Multihop networks are more flexible and have many applications, and it would be valuable to investigate VoIP performance in these situations.

VoIP traffic has two significant features that are different from the no-voice flows. The first difference is the payload size. As is well known, a network packet consists of two parts: header and payload. The packet header records how the data are to be transmitted while the payload has the real information that has to delivered. In each network, the header size is almost fixed for each packet, so it is common to use a big payload size to increase network reuse. But for voice flows, the payload size is always small (even smaller than the header size) as the information to be delivered depends upon the speed of human speech. This leads to the low channel reuse of the networks. The second difference comes from the fact that the hearing process can in general tolerate a certain amount of loss ratio (less than 4%) [1]. VoIP traffic should meet the delay requirement (usually less than 200 ms) [1], although it does not require 100% reliable connection.

This chapter focuses on the VoIP traffic over multihop 802.11 networks. Through a simulation study, we identify the bottleneck in the network, which lies in the media access control layer. The queue overflow of the busiest node is a major factor that degrades the performance of VoIP traffic. Thus, we propose a burst queue (BQ) scheme to improve the performance. The idea is that busy nodes try to pack several packets in their queue to increase channel efficiency and use broadcast to save protocol overhead. Each node monitors its queue length and decides whether to use BQ to pack several packets into a big one and broadcast it. We simulate various traffic patterns to examine the effectiveness of the BQ scheme, and our simulation results indicate the superiority of our scheme over legacy 802.11.

The remainder of this chapter is organized as follows. The background and related previous work are presented in Section 13.2. The problem description of the multihop 802.11 networks are discussed in Section 13.3. Then we propose our BQ scheme in Section 13.4. The simulation results of the our scheme over various traffic patterns are presented in Section 13.5. Finally, we conclude our work in Section 13.6.

13.2 Background and Related Work

The ultimate objective of VoIP is to deliver a high-quality voice service comparable to that provided by traditional circuit-switching networks. There are numerous challenges in transmitting VoIP traffic over wireless networks. For instance, due to the deficiency in the wireless media access methods, the delivery of VoIP often leads to unpredictable delay and packet-loss [2] performances.

In recent years, extensive research and development efforts have been conducted on IEEE 802.11 networks [3–5]. Garg and Kappes [5] present experimental studies on the throughput of IEEE 802.11b wireless networks and point out that inherent channel inefficiency limits the maximum number of voice flows. Their experiments revealed that the aggregated bandwidth of the networks is diminished by ongoing VoIP connections. Bianchi [6] proposed a Markov chain model for the binary exponential backoff procedure [6]. Based on the saturated throughput derived from the model, Foh et al. obtained the mean packet delay [7]. In [8], the authors use a p-persistent backoff strategy to approximate the original backoff in the protocol, which could have potential implications in the P2P system as well [9,10].

Since 802.11b can only support the best-effort traffic, 802.11e is designed to make up for the deficiency. The creation of 802.11e EDCA is a consequence of the extensive research that aimed to support prioritized service over 802.11 DCF [11,12]. There are even some works [13,14] that try to enhance the performance of 802.11e. However, most of these works are focused on the priority for different types of traffic, which means that if we just transmit voice flows over the networks, there would be no difference between DCF and EDCA. Therefore, it is necessary to study the networks that transmit only voice packets. The idea is that if a scheme can support more VoIP flows in a voice-only environment, its performance should be better in no-voice-only (with EDCA) conditions.

For voice flows, delays longer than 100 ms require echo cancelation and long delays would lead to cross-talk. The regular retransmission mechanism also results in excessive end-to-end delays. However, voice can afford a small amount of packet loss since the ear–brain is less sensitive to short drops in received speech [1]. Keeping in mind this characteristic of voice, we developed the BQ scheme by which delay and throughput are dramatically improved while the loss ratio is still kept within an acceptable level.

13.3 Problem Description

The most significant difference between VoIP traffic and other flows is the payload size. The payload size of VoIP ranges from 33 bytes (GSM) to 160 bytes (G.711), while the TCP payload size is generally 1460 bytes. Since the transmission overhead is fixed for each packet, a lower payload size results in lower channel utilization. Consider a VoIP packet with a 33-byte payload, that transmits at 11 Mbps at 24 µs ($33 \times 8 \div 11$). The transmission time for the 40-byte IP/UDP/RTP header is 29 µs ($40 \times 8 \div 11$). However, the 802.11 MAC and PHY layers have an additional overhead of more than 800 µs, attributed to the physical preamble, MAC header, MAC backoff time, and MAC ACK [15]. Thus, the overall channel efficiency is less than 3%. Note that this calculation already eliminates the CTS/RTS handshake overhead of the 802.11 MAC of unidirectional VoIP flows as shown in Table 13.1.

Another problem in multihop networks is that its topology may affect the performance of VoIP calls. When we simulated 10 VoIP traffics in a flat area with randomly positioned nodes, the total throughput differed for each random topology. In some conditions, the throughput degraded dramatically, while in others the performance was much better. Investigating the poor performance conditions, we found that the busiest node in the whole network affects performance the most. Figure 13.1 shows a common situation: when calls are established between nodes L_i and R_j, nodes A, B, and C will become the busiest nodes. If we do not consider the influence of node C, any end of nodes A, B, L_1, L_2, L_3 has

TABLE 13.1

Parameter Values of 802.11b DCF

DIFS	50 µs
SIFS	10 µs
Slot time	20 µs
CWmin	32
CWmax	1024
PHY header	192 µs
MAC header	34 bytes
MAC ACK	248 µs

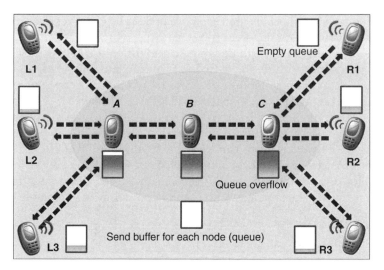

FIGURE 13.1   Bottlenecks in a common network for VoIP.

equal probability of accessing the shared channel. In saturated conditions, in which each node always has packet to send, the packet arrival rate at node *A* is four times the packet sending rate. This finally leads to queue overflow. Even in unsaturated conditions, when node *A* can send additional packets, when the other nodes' send queue is empty, the channel access overhead of node *A* results in a higher delay.

Thus, if we increase the number of calls, the send buffer (queue) of node *A*, *B*, and *C* will first overflow, which results in an unacceptable packet loss ratio. This is due to the extreme low channel utilization and rather high packet arrival rate. Nodes *A*, *B*, and *C* are identified as the bottlenecks in the whole network in Figure 13.1. If we want to provide more VoIP calls in multihop conditions, we must first solve this problem. In the rest of this chapter, we take nodes *A*, *B*, and *C* the nodes in our basic bottleneck model.

13.4 Burst Queue Scheme

As in the bottleneck model, busy nodes have a huge backlog, and each packet in the queue is sent in a channel that is efficient only to the extent of 3%, which finally results in that queue overflow. To improve channel utilization, we propose a Burst Queue scheme that increases channel efficiency and reduces delay as shown in Figure 13.2. Since the MAC and PHY overheads are too high, we combine several transmissions into one burst broadcast, to save this overhead. That is, when the backlog is too long, the MAC warned of either a packet overflow or an undesirably long delay. Then the MAC packs the packets in the queue, creating a packet of size not exceeding 1200 bytes. Figure 13.3 shows the details of the scheme. Queue length represents the business of the node. If the queue length is kept at a low level, the MAC performs the default operations. If the queue length exceeds the WATER_MARK (we set it to 10), the queue notifies the MAC and the BQ scheme starts up. The BQ broadcast packet contains just one MAC header, but each packet is packed with its destination MAC address ahead. The receiving MAC uses this information to decide whether to deliver the packet to the up layer or just discard it.

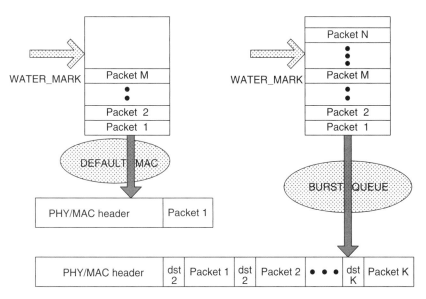

FIGURE 13.2 Burst Queue packet structure.

If we pack *n* packets each time, we save about 40*n* bytes extra MAC and PHY header, *n* − 1 backoff time, and *n* MAC ACK time at the cost of a lower broadcast transmission rate and a higher collision probability. Our simulation results show that the cost is worth paying. With the BQ scheme, the bottleneck model could support more VoIP calls and obtain a lower delay without queue overflow. For non-bottleneck nodes that have a short queue, burst could never happen. Such nodes preserve the reliability of the 802.11 retransmission and ACK mechanism. Our idea is that if the node is not busy, reliability is our concern; otherwise, we pack several packets in the queue and make a burst broadcast to increase the channel reuse as shown in Figure 13.2. This is a tradeoff between contention and collision that prevents the queue from overflowing and acquiring a lower delay to satisfy the VoIP traffic.

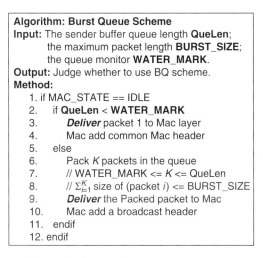

FIGURE 13.3 The Burst Queue Scheme implementation.

13.5 Simulation and Results

13.5.1 Simulation Setup

We simulate the bottleneck nodes in Figure 13.1 and develop a test bed with nodes in a chain topology: each node is 200 m away and has a 250 m transmit range and a 550 m sensing range. To simulate the bottleneck condition, we send multiple aggregated flows from each end of the node; thus the middle node would become the bottleneck node. We use CBR traffic as our VoIP flow, the payload size is 33 bytes, the interval is 20 ms, and the data rate is 13.2 kbps as specified in the GSM. We run a TCP flow of the Reno version over the test model whose packet size is set to 1024 bytes. All simulations last 100 s in NS-2 [16].

13.5.2 BQ Over 802.11b/g

We first simulate the unidirectional condition. We transmit N aggregated VoIP flows from node A to node C (as shown in Figure 13.4).

As shown in Figure 13.5, when the number of flows is less than 5, broadcast does not happen, which means that the queues of three nodes never exceed 10. In these unsaturated networks, our scheme acts the same as legacy 802.11b/g protocol. Then we increase the number of the flows. When N reaches 8, the burst broadcast starts to occur under the BQ scheme as displayed in Figure 13.5. We can see that the average delay of BQ is much lower than that of legacy 802.11b/g. In addition, we observe that the average delay of legacy 802.11b/g remains almost constant (106 ms) as shown in Figure 13.6 from 9 to 20 flows. The reason is that when the network is saturated, the packet transmission rate reaches its

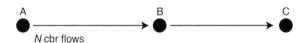

FIGURE 13.4   Unidirectional VoIP flow over bottleneck model.

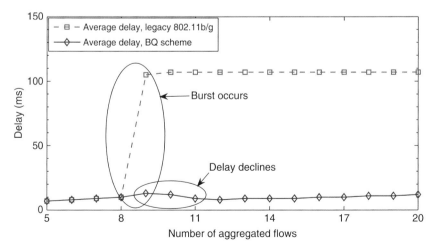

FIGURE 13.5   Packet delay of unidirectional VoIP flows over bottleneck model.

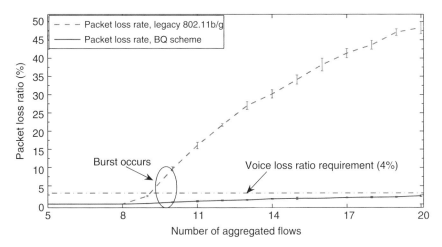

FIGURE 13.6 Packet loss ratio with 95% confidence interval of unidirectional VoIP flows over bottleneck model.

maximum capacity, and the extra packets are dropped due to the queue overflow. Thus, the average duration of each packet in the queue is the same. For BQ, we further observe that the average delay for 11 voice flows is even lower than that of 9-and 10-flows. This is because the busier the network, the more the BQ scheme takes place. For the 11-flow condition, the queue length of each node almost always exceeds the WATER_MARK, while for the 9- and 10-flow conditions, BQ does not occur as frequently. Moreover, the increasing trend of a short BQ delay for $N > 12$ comes from the packets leftover in the queue even after the broadcast.

Regarding the packet loss ratio, Figure 13.6 shows that when the number of flows is small (i.e., $N \leq 9$), there is no queue at the individual nodes. When the number of voice flows reaches 10, due to the queue overflow, legacy 802.11b/g yields a loss ratio of 9.9%, which exceeds the voice transmission requirement (4%). While in the BQ scheme, with 2043 times burst broadcast within the simulation period of 100 seconds, the queue in each node never overflows. The resulting 0.06% loss ratio due to the collision is much smaller than the voice transmission requirement (4%). We then increase N to 20, which is rather large, and see that with the queue overflow, legacy 802.11b/g suffers from a high packet loss rate, reaching up to 46.5%, while BQ maintains the loss ratio within the voice transmission requirement. Note that although BQ decreases the chances of queue overflow, if we keep increasing the number of flows, the packet loss ratio of BQ will also increase and the queue of our scheme will eventually overflow (e.g., 50 flows).

We now examine the simulation results for *bidirectional* flows. There are M voice flows transmitting two at a time from each end to the other end as shown in Figure 13.7, resulting in a total number of flows of $2M$.

Like the unidirectional condition, Table 13.2 shows the superiority of the BQ scheme as well. When $M = 4$, the network is unsaturated and no burst happens. When M reaches 5,

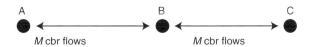

FIGURE 13.7 Bidirectional VoIP flow over bottleneck model.

TABLE 13.2

Loss Rate, Burst Broadcast Times Comparison

		M	4	5	6	7
Legacy 802.11b	Loss rate (%)		0	1.5	24.1	34.8
BQ over 802.11b	Loss rate (%)		0	0.1	0.06	3.0
	Burst times		0	153	2615	4717

TABLE 13.3

Average Delay and Maximum Delay Comparison

		M	5	6	7
Legacy 802.11b	Average delay (ms)		96	268	275
	Maximum delay (ms)		156	353	364
BQ over 802.11b	Average delay (ms)		18	23	21
	Maximum delay (ms)		71	92	92

bursts occur 156 times and BQ reduces the average delay by 81.3%. Legacy 802.11b/g could support only 5 pure voice pairs, and its delay is rather high, while BQ over 802.11b/g could support 7 pure voice pairs and still achieve a low average delay at 21 ms as shown in Table 13.3. The nodes A, B and C become the bottleneck when M reaches 6. The queue overflow results in the high loss ratio of 802.11b/g, and we also change the queue length to 500, but it becomes unhelpful and the delay exceeds 400 ms, which is apparently not acceptable. We note that the improvement in delay leaves sufficient margin for further transmission, as the nodes in both ends may not be the terminals of the whole transmission. This saved delay could be used for the transmission over Internet, etc.

Finally, we add a background TCP flow to the bidirectional condition. The results are shown in Table 13.4. Without the support of priority service, the performance of the VoIP degrades dramatically. The aggressiveness of TCP leads to greater contentions in such a wireless circumstance. By removing MAC layer ACK, the BQ scheme reduces the contention to some extent, and hence achieves the balance between contention and reliability. With slightly more unreliability due to the lack of ACK for broadcasting, BQ reduces contention and flow delay as shown in Table 13.5.

We observe that legacy 802.11b/g MAC has a lower packet loss ratio, but the delay is too high to support voice. The throughput achieved by TCP and the average delay establish that BQ is superior. It is the TCP that affects the performance of voice. Aggressive TCP flow

TABLE 13.4

Loss Rate, Burst Broadcast Times Comparison

		M	3	4
Legacy 802.11b	Loss rate (%)		1.8	3.6
	TCP throughput (kbps)		265.25	136.99
BQ over 802.11b	Loss rate (%)		3.1	3.9
	Burst times		3761	5466
	TCP throughput (kbps)		400.91	365.45

TABLE 13.5

Average Delay and Maximum Delay Comparison

		M	3	4
Legacy 802.11b	Average delay (ms)		117	111
	Maximum delay (ms)		303	362
BQ over 802.11b	Average delay (ms)		45	39
	Maximum delay (ms)		182	133

results in more contentions and consequently, in more broadcast failure. (Some TCP packet is packed with voice packets and broadcasted out.)

13.5.3 BQ Over 802.11e

With a TCP flow, 802.11b becomes unsuitable for voice services, for which reason we implemented the BQ scheme in 802.11e. We deployed it precisely in the AC0 queue, which is designed to transmit voices. We assign VoIP to AC0, and TCP to AC3 and run the simulation of Figure 13.8 over 802.11e again.

The results in Tables 13.6 and 13.7 show that 802.11e can support VoIP better than 802.11b. When $M = 4$, the loss ratio of legacy 802.11e is zero. For BQ, 37 burst times result in a

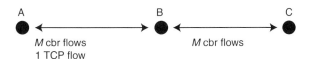

FIGURE 13.8 Bidirectional VoIP flow and a TCP flow over bottleneck model.

TABLE 13.6

Loss Rate, Burst Broadcast Times Comparison

		M	4	5	6
Legacy 802.11e	Loss rate (%)		0	0.6	20.0
	TCP throughput (kbps)		161.38	17.15	4.33
BQ over 802.11e	Loss rate (%)		0.07	1.9	4.1
AC0	Burst times		37	664	2830
	TCP throughput (kbps)		163.13	59.43	69.08

TABLE 13.7

Average Delay and Maximum Delay Comparison

		M	4	5	6
Legacy 802.11e	Average delay (ms)		13	55	248
	Maximum delay (ms)		66	193	363
BQ over 802.11e	Average delay (ms)		13	19	22
AC0	Maximum delay (ms)		58	70	72

slightly higher loss ratio (0.07%), while improving the TCP throughput by 2 kbps. As M increases, BQ always achieves a small delay, and the TCP throughput is much better than that of legacy 802.11e. As shown in Table 13.6, for $M = 5$ and 6, BQ reduces the average delay by 65% and 91.1%, respectively, and improves the TCP throughput by 246% and 1494%. We can see that BQ sacrifices a little packet loss rate to obtain a lower delay and the TCP throughput of BQ is always higher than that of legacy 802.11e.

13.6 Conclusion

In a multihop wireless network with heavy load, contention becomes the one of the most serious problems. Voice oriented network a in particular, are filled with huge amounts of small packets, all of which contend for the shared network resources. To alleviate this situation and improve network efficiency, we first make the voice packet more efficient to the extent that each packet has a higher information overhead ratio. We pack several MAC packets together and merely hold their MAC destination address. Second, we reduce the protocol overhead, including hand shaking and backoff time through burst broadcast. For small voice packets, 802.11 four way hand shake is too wasteful and burst broadcast becomes an efficient option. We use burst broadcast to relieve a part of contention consumption by the way that it reduces the total amount of DIFS, increases the efficiency of the channel and prevents more contention. Without MAC layer ACK, broadcasts incur more unreliability on one hand, and on the other hand, can reduce contention and delay, and improve background TCP throughput. Therefore, our BQ scheme achieves the trade-off between contention and reliability.

Acknowledgment

The works included in this chapter are supported by NSF China (No. 60702046, 60832005); China Ministration of Education (No. 20070248095); Sino-Sweden Joint Project (No. 2008 DFA11630); Shanghai Jiaotong University Pre-Research Funding; Qualcomm Research Grant; Huawei Research Grant; Shanghai PUJIANG Talents (08PJ14067); Shanghai Innovation Key Project (08511500400); Shanghai Jiaotong University Young Faculty Funding (No. 06ZBX800050); and Taiwan National Science Council research grants (NSC96-2221-E-110-035 and NSC96-2221-E-110-050).

References

1. D. E. McDysan and D. Spohn, *ATM Theory and Applications*, McGraw-Hill, Signature Edition, 1999.
2. W. C. Hardy, *VoIP Service Quality: Measuring and Evaluating Packet-Switched Voice*, McGraw-Hill, New York, December 2002.

3. G. Anastasi, E. Borgia, M. Conti, and E. Gregori, "IEEE 802.11 *ad hoc* networks: Performance measurements," *Proceedings of MWN 2003*, Providence, USA, May 2003.

4. H. Zhai, Y. Kwon, and Y. Fang, "Performance analysis of IEEE 802.11 MAC protocols in wireless LANs," *Wireless Communications and Mobile Computing*, vol. 4, December 2004, pp. 917–931.

5. S. Garg and M. Kappes, "An experimental study of throughput for UDP and VoIP traffic in IEEE 802.11b networks," *Proceedings of WCNC 2003*, New Orleans, USA, March 2003.

6. G. Bianchi, "Performance analysis of the IEEE 802.11 distributed coordination function," *IEEE Journal on Selected Areas in Communications*, vol. 18, no. 3, March 2000, pp. 535–547.

7. C. H. Foh and M. Zukerman, "Performance analysis of the IEEE 802.11 MAC protocol," *Proceedings of European Wireless 2002*, Florence, Italy, February 2002.

8. F. Cali, M. Conti, and E. Gregori, "Tuning of the IEEE 802.11 protocol to achieve a theoretical throughput limit," *IEEE/ACM Transcriptions Networking*, vol. 8, no. 6, December 2000, pp. 785–799.

9. Y. Liu, L. Xiao, X. Liu, L. M. Ni, and X. Zhang, "Location awareness in unstructured peer-to-peer systems," *IEEE Transactions on Parallel and Distributed Systems (TPDS)*, vol. 16, no. 2, February 2005, pp. 163–174.

10. Y. Liu, L. Xiao, and L. M. Ni, "Building a scalable bipartite p2p overlay network," *IEEE Transactions on Parallel and Distributed Systems (TPDS)*, vol. 18, no. 9, September 2007, pp. 1296–1306.

11. J. L. Sobrinho and A. S. Krishnakumar, "Real-time traffic over IEEE 802.11 medium access control layer," *Bell Labs Tech. J.*, Autumn 1996.

12. I. Ada and C. Castelluccia, "Differentiation mechanisms for IEEE 802.11," *Proceedings IEEE INFOCOM '01*, Anchorage, Alaska, April 2001.

13. Y. Xiao, "Enhanced DCF of IEEE 802.11e to support QoS," *Proceedings IEEE WCNC '03*, New Orleans, Lousiana, March 2003.

14. X. Chen, H. Zhai, and Y. Fang, "Enhancing the IEEE 802.11e in QoS support: Analysis and mechanisms," *Proceedings Second International Conference on Quality of Service in Heterogeneous Wired/Wireless Networks (QShine05)*, August 2005.

15. IEEE Std 802.11, "Part 11: Wireless medium access control (MAC) and physical layer (PHY) specifications," 1999.

16. "The network simulator NS-2," Web site: http://www.isi.edu/nsnam/ns/index.html.

14

Radio Access Network VoIP Optimization and Performance on 3GPP HSPA/LTE

Markku Kuusela, Tao Chen, Petteri Lundén, Haiming Wang,
Tero Henttonen, Jussi Ojala, and Esa Malkamäki

CONTENTS

14.1 Introduction

Circuit switched (CS) voice used to be the only way to provide voice service in the cellular networks, but during the past few years there has been growing interest to use cellular networks for real-time (RT) packet-switched (PS) services such as Voice over Internet Protocol (VoIP) to provide voice service without circuit switched service. The main motivation for the operators to use VoIP instead of CS voice are the savings, that could be achieved when the CS related part of the network would not be needed anymore. It is also expected that VoIP can bring better capacity than CS voice due to more efficient utilization of resources. Supporting VoIP in any radio access technology faces certain challenges due to VoIP traffic characteristics (strict delay requirements, small packet sizes), which make the efficient exploitation of radio interface capacity difficult due to control channel constraints. The solutions to these challenges vary for different technologies.

The introduction of 3G networks with integrated IP infrastructure [1] included in WCDMA Release 99 made it possible to run VoIP over cellular networks with reasonable quality, although with lower spectral efficiency than the circuit switched voice [2]. 3GPP Release 5 and 6 have brought High Speed Packet Access (HSPA) [3] to WCDMA downlink (DL) and uplink (UL). HSPA consists of High-Speed Uplink Packet Access (HSUPA) [4,5] in UL direction and High Speed Downlink Packet Access (HSDPA) [6] in DL direction, and was originally designed to carry high-bit rate delay tolerant non-real time (NRT) services like web browsing. Even though HSPA is not originally designed to support RT services, with careful design of the system, the RT services can be efficiently transported over HSPA, and number of new features have been introduced to 3GPP Release 6 and 7 to improve the efficiency of low-bit rate delay sensitive applications like VoIP. It has also been shown that VoIP can provide better capacity on HSPA than CS voice [7].

Long-term evolution (LTE) of 3GPP [8] work (targeted to 3GPP Release 8) has been defining a new packet-only wideband radio-access technology with a flat architecture, aiming to develop a framework for a high-data-rate, low-latency, and packet-optimized radio access technology called E-UTRAN. As E-UTRAN is purely a PS radio access technology, it does not support CS voice, which stresses the importance of efficient VoIP traffic support in E-UTRAN.

This chapter provides an overview of the challenges faced in implementing VoIP service over PS cellular networks, in particular 3GPP HSPA and LTE. Generally accepted solutions by 3GPP leading to an efficient overall VoIP concept are outlined and the performance impact of various aspects of the concept is addressed. The chapter concludes with a system simulation-based performance analysis of the VoIP service, including a comparison between HSPA and LTE.

14.2 System Description

14.2.1 HSPA

High-Speed Packet Access, a 3GPP standardized evolution of Wideband Code Division Multiple Access (WCDMA), has become a huge success as the world's leading third-generation mobile standard. HSPA consists of HSUPA and HSDPA which enhance WCDMA in the uplink and the downlink separately. Several key techniques such as HARQ, fast base station-controlled scheduling, and the shorter frame size were introduced into HSPA targeting the higher data rates with more efficient spectrum usage and lower transmission delay (see Figure 14.1).

In HSDPA, high-speed downlink shared channel (HS-DSCH) is associated with a 2 ms frame size, in which the users share at most 15 fixed spreading factor 16 (SF 16) high-speed physical downlink shared channels (HS-PDSCHs) by code multiplexing. In this way, radio resource can be utilized in a more efficient manner than earlier dedicated resource allocation in WCDMA. Furthermore, HARQ in HSPA can provide physical layer (L1) retransmissions and soft combining, which reduce the amount of higher layer

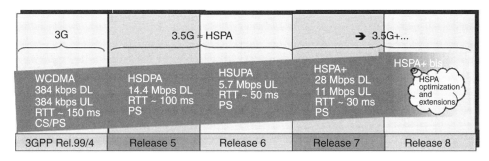

FIGURE 14.1 WCDMA/HSPA key performance indicators.

ARQ transmissions and frame selections (i.e., hard combining). This also improves the efficiency of the Iub interface between base station and radio network controller (RNC) significantly.

Rather than the conventional dedicated transport channel (DCH), the enhanced DCH (E-DCH) mapping on the enhanced dedicated physical data channel (E-DPDCH) is used in HSUPA to further improve the uplink data rate up to 5.76 Mbps. The 2 ms TTI frame length introduced into HSUPA can further reduce the transmission delay in the air interface: the minimum Round Trip Time (RTT) is decreased from about 150 ms in WCDMA to 50 ms in HSPA. Besides, base station controlled fast scheduling can select the user in a good channel condition for the transmission reaching the higher throughput due to the multiuser diversity gain. This is impossible in WCDMA system, where the reaction of the RNC-controlled scheduling is too slow to follow the fast changing channel conditions. This is because of the transmission delay between UE and the RNC via the base station.

14.2.2 LTE

To ensure competitiveness of 3GPP radio access technologies beyond HSDPA and HSUPA, a LTE of the 3GPP radio access was initiated at the end of 2004. As a result, the Evolved Universal Terrestrial Radio Access Network (E-UTRAN) is now being specified as a part of the 3GPP Release 8. Important aspects of E-UTRAN include reduced latency (below 30 ms), higher user data rates (DL and UL peak data rates up to 100 Mbps and 50 Mbps, respectively), improved system capacity and coverage, as well as reduced CAPEX and OPEX for the operators [9].

In order to fulfill those requirements, 3GPP agreed on a simplified radio architecture. All user plane functionalities for the radio access were grouped under one entity, the evolved Node B (eNB), instead of being spread over several network elements as traditionally in GERAN (BTS/BSC) and UTRAN (Node B/RNC) [8]. Figure 14.2 depicts the resulting radio architecture where:

- The E-UTRAN consists of eNBs, providing the E-UTRA user plane (PDCP, RLC, MAC) and control plane protocol terminations (RRC) towards the UE and hosting all radio functions such as Radio Resource Management (RRM), dynamic allocation of resources to UEs in both uplink and downlink (scheduling), IP header compression and encryption of user data stream, measurement and measurement reporting configuration for mobility and scheduling.

- The eNBs may be interconnected by means of the X2 interface.

- The eNBs are also connected by means of the S1 interface to the Evolved Packet Core (EPC), which resides in MME/SAE gateway.

Figure 14.3 shows the E-UTRA frame structure, which consists of 10 sub-frames, each containing 14 symbol blocks. In E-UTRAN system, one sub-frame (1 TTI) of length 1.0 ms is regarded as the minimum time allocation unit. In frequency domain, a minimum allocation unit is a Physical Resource Block (PRB), which consists of 12 subcarriers (each subcarrier is 15 kHz). In DL, OFDMA scheme is designed to allow signal generation by a 2048-point FFT. This enables scaling of system bandwidth to the needs of an operator. At least the bandwidths of {6, 12, 25, 50, 100} PRBs in the range of 1.4 MHz to 20 MHz are available yielding a bandwidth efficiency of about 0.9. In DL, user data is carried by downlink shared channel (DL-SCH), which is mapped to physical downlink shared channel (PDSCH).

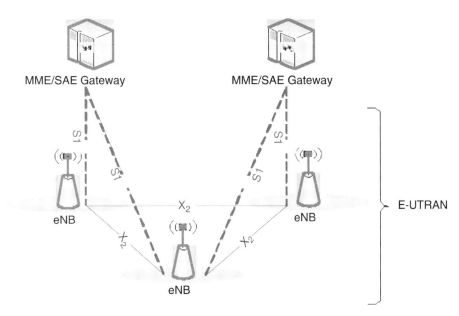

FIGURE 14.2 E-UTRAN architecture. (*Source:* 3GPP TS 25.308, "HSDPA; Overall description; Stage 2.")

The UL technology is based on Single-Carrier FDMA (SC-FDMA), which provides excellent performance facilitated by the intra-cell orthogonality in the frequency domain comparable to that of an OFDMA multicarrier transmission, while still preserving a single carrier waveform and hence the benefits of a lower Peak-to-Average Power Ratio (PAPR) than OFDMA. In UL direction, user data is carried by uplink shared channel (UL-SCH), which is mapped to physical uplink shared channel (PUSCH).

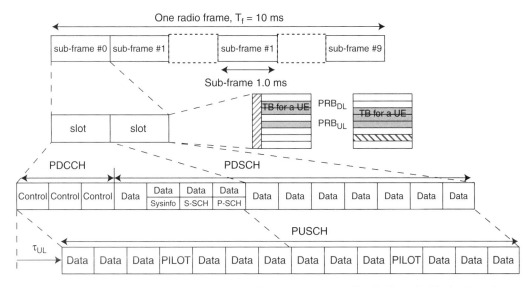

FIGURE 14.3 Frame structure in E-UTRAN system. (*Source:* M. Rinne, K. Pajukoski et al., "Evaluation of recent advances of the Evolved 3G (E-UTRA) for the VoIP and best effort traffic scenarios," IEEE SPAWC, June 2007.)

With all the radio protocol layers and scheduling located in the eNB, an efficient allocation of resources to UEs with minimum latency and protocol overhead becomes possible.
This is especially important for RT services such as VoIP, for which, high capacity requirements were set: at least 200 users per cell should be supported for spectrum allocations up
to 5 MHz, and at least 400 users for higher spectrum allocations [9].

14.3 Properties and Requirements for Voip

This section illustrates the characteristics of VoIP traffic by describing the properties and
functionality of the used voice codec in HSPA/LTE (Section 14.3.1) and by presenting the
requirements and the used quality criteria for VoIP traffic in Sections 14.3.2 and 14.3.3,
respectively.

14.3.1 Adaptive Multirate (AMR)

Adaptive Multirate (AMR) is an audio data compression scheme optimized for speech
coding. AMR was adopted as the standard speech codec by 3GPP in October 1998 and is
now widely used. The AMR speech codec consists of multirate speech codec, a source
controlled rate scheme including a voice activity detector, a comfort noise generation system, and an error concealment mechanism to combat the effects of transmission errors
and lost packets.

The multirate speech codec is a single integrated speech codec with eight source rates
from 4.75 kbps to 12.2 kbps, and a low rate background noise encoding mode. For AMR
the sampling rate is 8 kHz. The usage of AMR requires optimized link adaptation that
selects the best codec mode to meet the local radio channel and capacity requirements.
If the radio conditions are bad, source coding is reduced and channel coding is increased.

In addition to AMR audio codec, an extension of it is AMR-wideband (AMR-WB) audio
codec. Sampling rate for AMR-WB is 16 kHz, which is double the sampling rate of AMR.
Therefore, AMR is often abbreviated as AMR-NB (narrowband). AMR-WB is supported if
16 kbps sampling for the audio is used in the UE. Similarly, like AMR-NB, AMR-WB operates with various bit rates. For AMR-WB the bit rates range from 6.60 kbps to 23.85 kbps.
It is emphasized that both AMR-NB and AMR-WB are used in HSPA and LTE radio
systems. A more detailed description of AMR-NB and AMR-WB can be found in [10].

General functionality of the AMR codec is illustrated in the Figure 14.4. During voice
activity periods there is one voice frame generated every 20 ms (number of bits/frame

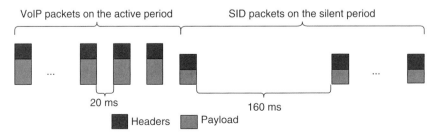

FIGURE 14.4 Illustration of VoIP traffic.

depending on the AMR codec mode), whereas during the silent periods Silence Descriptor (SID) frames are generated once in every 160 ms.

14.3.2 Delay Requirements

Voice over Internet Protocol (VoIP) is a conversational class service and the packet delay should be strictly maintained under reasonable limits. The maximum acceptable mouth-to-ear delay for voice is in the order of 250 ms, as illustrated in the Figure 14.5. Assuming that the delay for core network is approximately 100 ms, the tolerable delay for MAC buffering/scheduling and detection should be strictly below 150 ms. Hence, assuming that both end users are (E-)UTRAN users, the tolerable one-way delay for MAC buffering and scheduling should be under 80 ms, which is the used air interface delay for HSPA. To improve voice quality further on LTE, tolerable air interface delay was reduced to 50 ms for LTE [11]. For LTE FDD the average HARQ RTT equals 8 ms implying that at most 6 HARQ retransmissions are allowed for a VoIP packet when air-interface-delay of 50 ms is used. For HSPA the HARQ RTT would be 12 ms in downlink and 16 ms in uplink with 2 ms TTI, which means at most 6 HARQ retransmissions in downlink and 4 HARQ retransmissions in uplink with 80 ms HSPA delay budget. Similarly, at most 1 HARQ retransmission could be allowed in the uplink, corresponding to 40 ms HARQ RTT with 10 ms TTI.

14.3.3 Quality Criteria

Considering the nature of radio communication it is not practical to aim for 100% reception of all the VoIP packets in time. Instead, certain degree of missing packets can be tolerated without notably affecting the QoS perceived by the users. For VoIP traffic, the system capacity is measured as the maximum number of users who could be supported without exceeding 5% outage. During the standardization phases of HSPA and LTE in 3GPP,

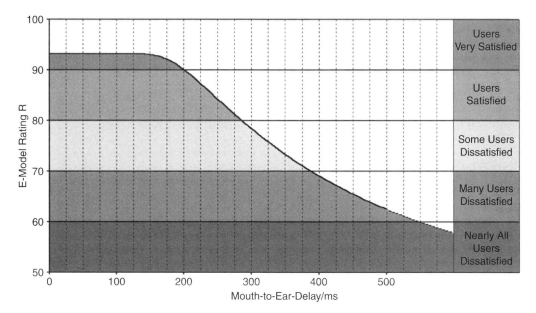

FIGURE 14.5 E-Model rating as a function of mouth-to-ear delay (ms) [8]. (*Source:* ITU-T Recommendation G.114, "One way transmission time." With permission.)

slightly different outage criteria were used when evaluating VoIP system performance. Used outage criteria are defined according to system as follows: For HSPA, a user is in outage if more than 5% of the packets are not received within the delay budget when monitored over a 10-second window. For LTE [11], an outage is counted if more than 2% of the packets are not received within the delay budget when monitored over the whole call. It is emphasized that even though the used outage criteria differ slightly between HSPA and LTE, the performance difference originating from different outage criteria is rather small being inside the error margins of the simulations.

In UL direction there also exists an alternative method to measure VoIP performance, which is used in the context of this chapter for HSPA. That is, UL VoIP capacity can be measured as the allowed number of users in a cell with an average noise rise of X dB measured at the base station. Optimal value for X may depend on the used system, and for HSPA the used value was 6 dB.

14.4 Radio Interface Optimization for VoIP

The most important mechanism to optimize air interface for VoIP traffic in (E-)UTRAN are presented in this section.

14.4.1 Robust Header Compression (ROHC)

In streaming applications such as VoIP, the overhead from IP, UDP, and RTP headers is 46 bytes for IPv4, and 66 bytes for IPv6. For VoIP traffic, this corresponds to around 144–206% header overhead, assuming that the most typical VoIP codec (AMR 12.2 kbps) is used with 31 bytes payload. Such a large overhead is excessive in wireless systems due to the limited bandwidth and would lead to serious degradations in the spectrum efficiency.

In order to overcome this, a robust header compression (ROHC) technique [12] was adopted as a nonmandatory feature to 3GPP specifications from Release 4 onwards—hence ROHC feature is supported in HSPA and LTE radio systems as well. With ROHC, the IP/UDP/RTP header is compressed down to few bytes with the minimum size being 3 bytes, which is also the most commonly used header size. Hence, with ROHC the header overhead may be compressed down to 25%, implying that the utilization of ROHC leads to significant improvements in the VoIP air interface capacity.

14.4.2 Multiuser Channel Sharing

Due to the characteristics of the VoIP traffic (strict delay constraints, regular arrival of small size packets) multiuser channel sharing plays an important role when optimizing radio channel interface for VoIP traffic. With multiuser channel sharing, multiple users can be multiplexed within the same TTI, and this will lead to significant improvements in VoIP capacity. To stress the importance of multiuser channel sharing as a building block to optimize radio interface for VoIP traffic, it is concluded that VoIP capacity is directly proportional to the average number of scheduled users per TTI. In HSPA, multiuser channel sharing is done by separating users multiplexed into the same TTI in Orthogonal Variable Spreading Factor (OVSF) code domain, whereas in LTE user separation on PDSCH is done in frequency domain.

An important issue related to multiuser channel sharing is how the transmission resources are divided between simultaneously scheduled users. Conventional methods use an even-share method, where each multiplexed user is given the same amount of resources. The main shortcoming of this method with VoIP traffic is that it wastes resources: The size of the resource allocation has to be done according to the needs of the weak users to guarantee sufficient QoS at cell border as well. If users in good channel conditions are assigned too much transmission resources, part of the resources are wasted as the delay requirements do not allow to buffer enough packets to take full benefit of the allocated transmission resources. At the same time users in unfavorable channel conditions might not have enough transmission resources to transmit even one VoIP packet at target block error rate (BLER).

A more sophisticated method to assign transmission resources for the scheduled users is to determine the size of the resource assignment dynamically based on the channel quality and on the supportable payload of a user. With dynamic approach each user is allocated the amount of transmission resources that is necessary to reach target BLER. This can be estimated by link adaptation unit based on reported channel quality indicator (CQI) and selected modulation and coding scheme (MCS). With dynamic approach the size of the resource allocation matches more accurately to the requirements of the user. Thus the same total amount of transmission resources can be used more efficiently which increases the resource utilization efficiency. Moreover, weak users can be boosted with a higher share of transmission resources, which also improves the DL coverage of the VoIP service.

The impact of used resource allocation algorithm to the VoIP system performance on HSDPA was studied in [13]. The results indicated that the dynamic resource allocation on top of the VoIP-optimized scheduling provides substantial (up to 20%) capacity gain over the even share approach.

14.4.3 Utilization of VoIP-Specific Packet Scheduling (PS) Algorithm

Perhaps the most essential building block for the efficient support of VoIP traffic in HSPA/LTE is the packet scheduling (PS) algorithm. This algorithm at (e)NodeB is responsible for user selection and the resource allocation for the selected users. Considering VoIP traffic characteristics, PS buffer status and individual user's channel conditions when making scheduling decisions and resource allocation for the selected users, VoIP capacity can be boosted up to 115% [13] when compared to blind Round Robin (RR) scheduling method without any VoIP specific user prioritization. In this section, a brief description of the proposed VoIP optimized PS algorithm is given (details on the algorithm and corresponding performance analysis on HSDPA and LTE is given in [7,13–15]). Moreover, it is assumed that the proposed algorithm schedules each transmission of data by L1 control signaling. Henceforth, this kind of PS method is referred to as fully dynamic PS.

14.4.3.1 Design Criteria of the Packet Scheduling Algorithm for VoIP

The following issues should be considered when designing PS algorithm for VoIP traffic:

- *Strict delay requirements of VoIP service*—As described in the Section 14.3, the tolerable air interface delay for VoIP traffic (including packet scheduling and buffering) should be strictly within 80 ms. When this delay is exceeded, the quality of the voice perceived by the user starts to deteriorate rapidly. Due to the stringent delay limit, time domain scheduling gains for VoIP traffic are very limited, as the buffered VoIP packet(s) should be scheduled within the given delay budget in order to sustain good speech quality. For users whose channel is in a deep fade, the delay limitation implicitly means that there is not enough time for PS to wait until the conditions become more favorable. Instead, the data should be scheduled early enough to allow enough time for the needed retransmissions. This implies that weak users, that is, users relying on HARQ retransmissions, should be prioritized in the user selection.

- *Low user data rate of VoIP service*—The size of the VoIP packet is typically very small, for example, for AMR 12.2 kbps codec the size of the most common ROHC compressed packet is 40 bytes (every 20 ms). Together with strict delay requirements of VoIP traffic this leads to frequent transmission of small payloads, as it is not possible to wait long enough to send more than few VoIP packets at a time. Therefore, when high capacities are targeted, it is necessary for several users' transmissions to share the DSCH resources in each TTI and each user needs to be scheduled fairly often.

- *Managing the control overhead*—Due to the low user data rates, the number of scheduled users per TTI can become quite large with multiuser channel sharing. When each of these users is scheduled by L1 control signaling, the control channel overhead can become the bottleneck of VoIP system performance. In that case the dynamic packet scheduler is unable to fully exploit PDSCH air interface capacity and some of the PDSCH resources remain unused while the control channel capacity is already exhausted. One attractive technique with fully dynamic scheduler to improve VoIP capacity within control channel limitations is to bundle multiple VoIP packets together at L1 in one transmission for a user. Packet bundling is CQI based, which means that PS decides whether to utilize it or not for a user based on the user's channel conditions. Packet bundling is described in more detail in Section 14.6.2.1. An alternative way of addressing the control overhead problem is semi-persistent scheduling, which instead of scheduling each transmission individually, relies on long-term allocations. Semi-persistent scheduling is discussed in more detail in Section 14.6.2.2.

14.4.3.2 Proposed Fully Dynamic Packet Scheduling Algorithm for VoIP

The following packet scheduling algorithm aims to address the particulars of the VoIP service listed in Section 14.4.3.1. The proposed user selection procedure relies on the concept of scheduling candidate set (SCS) [13,15], which classifies the schedulable users into three groups. These three groups in the SCS are dynamically updated in each TTI. The groups are further ordered in the following priority.

1. Users with pending retransmissions in their HARQ manager.
2. Users whose head-of-line (HOL) packet delay is close enough to the delay budget, where HOL delay is the delay of the oldest unsent packet in the (e)NodeB buffer.
3. Remaining users (other than the retransmission users and HOL users).

The priority metric used for ordering users within each group is a relative CQI metric. The relative CQI metric for each user is given by CQI_inst/CQI_avg, where CQI_inst is

TABLE 14.1

Packet Scheduling and Resource Allocation Algorithm Alternatives

Scheduler	Description
RR	RR scheduling with even share resource allocation
PF	Proportional fair scheduling with even share resource allocation
PF + SCS	SCS based VoIP optimized scheduling algorithm using PF metric to order the users in the SCS. Even share resource allocation
PF + SCS + RA	As PF + SCS, but with dynamic resource allocation

the instantaneous CQI calculated over the full bandwidth and *CQI_avg* represents the average wideband CQI value in the past. The relative CQI metric reflects user's instantaneous channel condition compared to the average and therefore can indicate a favorable moment for scheduling the user. Of course, the time domain scheduling gains are in the end somewhat limited because of the tight delay budget.

As a further improvement, compared with sending one packet per TTI, bundling multiple packets together in one transmission can reduce related control channel signaling overhead and thus improve air interface efficiency (see Section 14.6.2.1 for further details). Therefore, in the user selection within each group, priority is given to users who can support bundling.

Further, in Table 14.1 the abbreviations of considered scheduler alternatives are listed and explained.

Table 14.2 illustrates the relative HSDPA capacity improvements that can be obtained with a packet scheduler specifically designed for VoIP service over blind RR scheduling method without any VoIP specific user prioritization. The table shows capacities for 80 ms delay budget assumption. The selected set of scheduler alternatives compares the enhancements incrementally.

These numbers verify that packet scheduling and allocation of transmission resources play a key role in achieving good VoIP capacity. There is a large difference between the best and the worst scheduler option. PF gives gain over RR as it is on average able to schedule users at better channel conditions. Introducing SCS is an important improvement, as it considers the delay requirements of VoIP traffic by prioritizing urgent transmissions and retransmissions. Moreover, SCS makes the packet bundling more likely thus improving the overall efficiency. The capacity gain of dynamic resource allocation (PF + SCS + RA) over even share resource allocation is approximately 20% to 30% for the PF + SCS scheduler.

TABLE 14.2

Relative Capacity Gain with Different Scheduler Improvements Compared to Round Robin Scheduler with 80 ms Delay Budget

Scheduler	Capacity Gain (%)
RR	0
Proportional fair (PF)	20
PF + SCS	42
Dynamic resource allocation (PF + SCS + RA)	80

Source: P. Lundén and M. Kuusela, "Enhancing performance of VoIP over HSDPA," VTC 2007 Spring.

14.4.3.3 RNC Controlled Nonscheduled transmission in the Uplink for VoIP Over HSUPA

There are two scheduling schemes defined for HSUPA: Node B controlled scheduled transmission and RNC controlled nonscheduled transmission. The former scheme does not guarantee a minimum bit rate since the scheduling algorithm decides whether to allocate any power to the user or not. The latter scheme (nonscheduled transmission) defines a minimum data rate at which the UE can transmit without prior request. When nonscheduled transmission is configured by the serving RNC (SRNC), the UE is allowed to send E-DCH data at any time, up to a configured bitrate, without receiving any scheduling command from Node B. Thus, signaling overhead and scheduling delay are minimized. Therefore, RNC controlled nonscheduled transmission is the most suitable choice for VoIP traffic.

14.4.4 Advanced Receivers

By using advanced receiver techniques at UE the performance of VoIP can be significantly improved. The following advanced receiver techniques were introduced to HSDPA terminals: 2-antenna Rake receiver with Maximum Ratio Combining (MRC) and 1-antenna LMMSE time domain chip level equalizer were adopted by 3GPP Release 6 onwards, and 2-antenna LMMSE chip level equalizer was adopted by 3GPP Release 7. When comparing the performance improvements achieved with 3GPP Release 6 advanced receiver techniques, it was concluded that the biggest improvement is achieved with plain receive diversity. By having additional receive antenna at terminal 3 dB array gain is added to the realized signal-to-noise ratio (SNR) for all terminals, due to which the VoIP DL capacity is significantly boosted. Further, the utilization of LMMSE chip equalizer improves VoIP system performance as well by reducing the intracell interference of the users in the cell, but the achieved performance improvement is much smaller compared to utilization of receive diversity, as equalizer is not able to improve the performance of the cell edge users, that is, the users that most probably are in the outage, as the performance of those users is mainly limited by the intercell interference. On the basis of system level simulations the utilization of receive diversity provides approximately 80% gains in HSDPA capacity compared to the case with 1-antenna RAKE receivers at terminals. For 1-antenna LMMSE chip level equalizer the corresponding gain is in the order of 20–30%.

In LTE, the baseline receiver technique is 2-antenna receiver with either MRC combining or Interference Rejection Combining (IRC), where the diversity antennas are used to reject the co-channel interference by estimating the interference covariance matrix and selecting combining weights accordingly to maximize SNR. According to [16], IRC combining provides approximately 0.5 dB gain in the received SNR over MRC combining, and this would correspond to 10–15% gains in VoIP LTE DL capacity.

14.5 Practical Constraints in VoIP Concept Design and Solutions

When optimizing air interface for VoIP traffic, several practical constraints are to be considered in VoIP concept design on HSPA and LTE radio systems. This section provides description for the most important practical limitations in radio interface optimization. Existing solutions (if any) to avoid these limitations are covered in Section 14.6.

14.5.1 Control Channel Limitations

With VoIP traffic the number of supported users may increase drastically. To support VoIP transmission of DL and UL transport channels, there is a need for certain associated DL and UL control signaling. Therefore, the overhead from associated control signaling in UL and DL may become a bottleneck for the VoIP system performance. Sections 14.5.1.1 and 14.5.1.2 describe the main contributors for control channel overhead for HSPA and LTE radio systems. Methods to reduce control channel overhead are described in Sections 14.6.1 and 14.6.2.

14.5.1.1 HSPA

For VoIP traffic, the packet size is considerably small corresponding to a low data rate transmission on the data channel, for example, 32 kbps with 10 ms TTI and 160 kbps with 2 ms TTI in HSUPA. In other words, it implies that the overhead from control channels would consume relatively more power for the transmission of VoIP traffic than the other packet data traffics transmitted with the high data rate. On the other hand, the control channels cannot be subject to HARQ retransmissions, that is, the information carried by the control channels cannot be soft combined. Thus, more retransmissions may lead to the higher overhead from the control channels associated with data transmission. It makes sense to study the effect of the control channel overhead to performance and investigate how the overhead could be reduced.

In HSPA, there are mainly three types of dedicated physical control channels in the UL and one in the DL: the UL Dedicated Physical Control Channel (UL DPCCH), the E-DCH Dedicated Physical Control Channel (UL E-DPCCH), the UL Dedicated Physical Control Channel associated with HS-DSCH transmission (UL HS-DPCCH), and the DL Shared Control Channel for HS-DSCH (HS-SCCH). Their constraints on the VoIP performance would be addressed in the following sections.

DPCCH overhead in the uplink transmission

The uplink DPCCH in 3GPP Release 6, that is, HSUPA, is used for carrying physical layer control information which mainly consists of known pilot bits to support channel estimation for coherent detection and transmit power-control (TPC) commands to be used for the DL transmission. When the high capacity (in terms of users in a cell) is targeted for VoIP on HSDPA and HSUPA, the interference caused by continuously transmitted UL DPCCHs becomes the limiting factor for capacity.

To illustrate this, let us calculate the UL capacity by using the following UL load formula [1]:

$$NR_{dB} = -10\log 10\left(1 - \frac{\rho}{W/R} N \cdot v \cdot (1+i)\right), \qquad (14.1)$$

where ρ is the $E_b N_0$ target after antenna combining, W is the chip rate, R is the E-DPDCH bit rate, N is the number of users, v is the equivalent activity factor, i is the other cell to own cell interference ratio, and NR_{dB} is the noise rise in decibels. When the UL load formula (14.1) is used, the overhead from DPCCH, E-DPCCH, and the retransmissions are included in the equivalent activity factor v.

Further, it is also possible to divide this formula into components as shown in (14.2) and then have separate requirements for E_c/N_0 (i.e., the signal to noise target per chip after the antenna combining) and activity factors for each channel. Here E-DCH is assumed to include both E-DPCCH and E-DPDCH, which would always be transmitted together thus having the same activity factor.

$$NR_{dB} = -10\log 10\left(1 - \left[\left(\frac{E_c}{N_0}\right)_{E-DCH} v_{E-DCH} + \left(\frac{E_c}{N_0}\right)_{DPCCH} v_{DPCCH} + \left(\frac{E_c}{N_0}\right)_{HS-DPCCH} v_{HS-DPCCH}\right]N\cdot(1+i)\right)$$

(14.2)

For DPCCH gating (see Section 14.6.1.1), it is pessimistically assumed that E-DPDCH and HS-DPCCH transmissions never overlap and thus the DPCCH activity was obtained as the sum of activities on E-DPDCH and HS-DPCCH.

Thus, the noise rise (NR) contributed by inactive UEs (i.e., UEs with inactive packet transmission) can be calculated as shown in Figure 14.6 assuming that i is 0.65 for the typical 3-sector macro cell scenario and the required DPCCH E_c/N_0 is about −18 dB for the inactive UE [17]. Obviously, up to 30 inactive users with the continuous DPCCH transmission in this case can eat up all reserved radio resource, that is, 6 dB noise rise target. However, the less DPCCH activity or the less inactive UEs, the less interference is experienced in UL.

With VoIP over E-DCH, the UL transmission of voice is not continuous as the CS voice over Dedicated Channel (DCH). One voice frame is transmitted every 20 ms or two voice frames every 40 ms if bundling of two voice packets is allowed. With VoIP, the continuous transmission of the UL control channel DPCCH would consume the scarce radio resource and become inefficient.

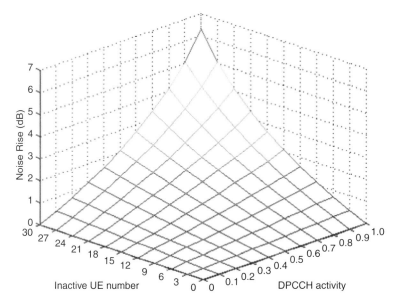

FIGURE 14.6   Noise rise as a function of the inactive UE number and the DPCCH activity factor. (*Source:* Tao Chen et al., "Uplink DPCCH gating of inactive UEs in continuous packet connectivity mode for HSUPA," IEEE WCNC, March 2007. With permission.)

Overhead from HS-DPCCH carrying CQI report

Except for DPCCH channel, there are certain other control channels such as HS-DPCCH which may consume the UL radio resource. HS-DPCCH, carrying CQI information for DL channel sensitive schedulers and adaptive modulation scheme selections, may be transmitted periodically according to a CQI reporting period controlled by RRC. Correspondingly, the effect of CQI reporting period can be studied according to the earlier-stated equations as well. Reducing CQI activity (i.e., increasing CQI reporting period) from once per 10 ms to once per 20 ms increases the VoIP capacity by about 10% and going down to one CQI transmission per 80 ms increases the VoIP capacity by 20% (from CQI once per 10 ms). On the other hand, the larger transmission interval of CQI may worsen the downlink performance due to less reliability of CQI. Therefore, the CQI reporting period should be leveraged for the UL and DL performance.

HS-SCCH and E-DPCCH overhead associated with the data transmission

Other control channel constraints for VoIP over HSPA would be the overhead from the VoIP packet transmission associated control channels such as HS-SCCHs in the DL and E-DPCCHs in the UL. The HS-SCCH is a DL physical channel used to carry DL signaling related to HS-DSCH transmission, whereas the E-DPCCH is a physical channel used to transmit control information associated with the E-DCH. When a large number of users are served at a time, the control overhead (i.e., HS-SCCH overhead and E-DPCCH overhead) increases and the respective share of resources for the VoIP packet transmission of the user is reduced. Otherwise, more HARQ retransmissions would also increase these overhead. Thus, there would be a tradeoff between the allowed retransmission number and the control overhead.

14.5.1.2 *LTE*

To support the transmission of DL and UL transport channels, certain associated UL and DL control signaling is needed. This is often referred to as L1/L2 control signalling, as the control information originates from both physical layer (L1) and MAC (L2). The DL L1/L2 control signaling, transmitted by using Physical Downlink Control Channel (PDCCH), includes both PDSCH scheduling related information for each scheduled terminal and PUSCH related scheduling messages (a.k.a scheduling grants). PDSCH scheduling information includes information about the used resource assignment and MCS, and it is necessary to the scheduled terminal to receive, demodulate, and decode PDSCH properly. PDCCH overhead limitations to VoIP DL capacity are discussed in "PDCCH overhead" subsection.

Like LTE DL, LTE UL needs for certain associated L1/L2 control signaling to support the transmission of DL and UL transport channels. The L1/L2 control signaling is carried by Physical Uplink Control Channel (PUCCH), and the carried L1/L2 control signaling includes HARQ acknowledgments for scheduled and received PDSCH resource blocks in DL, CQI information indicating the channel quality estimated by the terminal and UL scheduling requests indicating that the terminal needs UL resources for PUSCH transmissions. The biggest part of UL L1/L2 control signaling originates from CQI signaling, which is required by channel aware scheduler at eNodeB to exploit the time and frequency domain scheduling gains of PDSCH. With VoIP traffic the CQI signaling overhead may start to limit VoIP UL capacity, as the number of supported users in E-UTRA is large. CQI overhead issue is handled in "CQI overhead" subsection.

PDCCH overhead

As in HSPA, the baseline scheduler for VoIP traffic is the fully dynamic packet scheduling method, where each packet is transmitted by L1 control signaling. As described in Section 14.4.2, multiuser channel sharing plays an important role when optimizing air interface capacity for VoIP traffic. As PDCCH should be transmitted to each scheduled terminal, PDCCH overhead may become a limiting factor for VoIP capacity. Due to this reason there is not enough control channel resources to schedule all physical resource blocks (PRBs) and hence part of PDSCH capacity is wasted. For example, at 5 MHz bandwidth with AMR 12.2 kbps codec the PDSCH utilization rate for fully dynamic PS is only 35%.

One way to avoid control channel limitations with fully dynamic PS, several of a user's VoIP packets can be bundled at L1 for one transmission, as described in Section 14.6.2.1. The main benefit from bundling is that more users could be fitted to the network with the same control channel overhead as good users are scheduled less often. This will lead to significant capacity improvements in the control channel limited situation. For example, at 5 MHz bandwidth with AMR 12.2 kbps codec VoIP DL capacity is improved by 75–80% and PDSCH utilization rate is increased from 35% to 70% when packet bundling (up to two packets bundled together) is used.

An alternative method to avoid control channel limitations for VoIP performance is to utilize semi-persistent packet scheduling method [8,18–20]. This approach has been heavily studied in 3GPP. With this method, savings in control channel overhead are realized as the initial transmissions of VoIP packets are scheduled without L1 control signaling by using persistently allocated time/frequency resources. Semi-persistent PS method works very well in control channel limited circumstances, which can be seen as increased PDSCH utilization rate. As an example, by using semi-persistent PS the utilization rate of PDSCH is approximately 90% at 5 MHz bandwidth with AMR 12.2 kbps codec. Semi-persistent resource allocation is described in more detail in Section 14.6.2.2.

CQI overhead

As EUTRA is based on OFDMA scheme, in LTE system, the channel-dependent scheduling can be conducted both in time and frequency domain. To achieve maximal frequency domain scheduling gains in DL, each UE has to measure CQI in frequency domain for all PRBs across the whole bandwidth and report the information to eNodeB for link adaptation and scheduling decisions. In practice, the feasible CQI feedback resolution in frequency domain is limited by the UL control overhead related with CQI feedback, which may become prohibitively large especially with VoIP traffic, where the number of supported users may be large. Therefore, CQI feedback should be reduced both in time and in frequency domain to keep the UL control overhead at reasonable level. Quantity of the reduction of CQI feedback in time domain depends on the velocity of an user, for example, at 3 km/h the feasible CQI update rate for VoIP traffic is 5–10 ms. For this traffic, the impact of CQI update rate is not that crucial, as due to the delay limitations of the traffic the time domain scheduling gains are rather limited.

When reducing CQI feedback in frequency domain, the challenge is how much it could be reduced whilst maintaining the benefits of frequency domain scheduling. A simple method to reduce CQI feedback in frequency domain is to relax CQI measurement granularity in frequency domain and calculate one CQI feedback over N consecutive PRBs. Additionally, frequency selective CQI feedback schemes have been adopted by 3GPP, such as the best-M average scheme, where the UE reports only the average CQI for the M CQI blocks having the highest CQI value and indicates the position of those M blocks within the bandwidth (wideband CQI for the whole bandwidth is reported as well in this scheme). Detailed description of the various CQI reporting options available in LTE can be found in [21]. The impact of CQI feedback reduction on the performance of fully dynamic PS with VoIP traffic is analyzed in [22].

Due to the nature of the VoIP traffic the realized time and frequency domain scheduling gains are lower than, for example, the best effort (BE) traffic. This implies that VoIP traffic is rather robust against reduced CQI feedback. This is also verified by the results presented in [22]—according to the results, 84% reduction in CQI overhead implied only 7% loss in capacity compared to the case, where full CQI (one CQI per PRB) was used. On the other hand, this compressed CQI overhead still corresponds to 4 kbps UL channel bit rate, which might be too much considering the limited capacity of PUCCH.

As the number of supported users in LTE with VoIP traffic is high, it may necessitate the usage of wideband CQI in order to keep the overhead from CQI feedback at reasonable level. This would mean lowest possible UL signaling overhead from CQI feedback at the cost of reduced capacity, as all frequency domain scheduling gains will be lost. In order to keep the capacity reduction at a minimum, the impact of lost frequency domain scheduling gains to the performance should be compensated with more efficient utilization of frequency diversity. Means to achieve this are described in Section 14.6.2.3.

14.5.2 Co-channel Interference and Noise Limitations for LTE UL

When the frequency re-use factor is set to one (1) for LTE, co-channel interference is generated in the network degrading the UL performance of LTE significantly. In order to mitigate the impact of co-channel interference in LTE UL, interference aware scheduling [23,25], described in Section 14.6.2.5, could be used.

Further, in extreme noise limited scenarios such as macro-cell scenario case 3 [26], LTE UL may suffer from UL coverage problems. This is caused by poor UE power utilization from the small 1 ms TTI duration. To overcome this, a TTI bundling technique [27,28] was adopted as a part of LTE 3GPP standard. TTI bundling technique is described in Section 14.6.2.6.

14.6 VoIP Solutions to Avoid System Related Limitations

14.6.1 Solutions in HSPA

14.6.1.1 Uplink Gating Associated with the Packet Bundling

To address the practical constraints for VoIP over HSPA, several solutions have been proposed to improve VoIP performance.

In 3GPP Releases up to the Release 6, DPCCH channel is transmitted continuously regardless of whether there is actual user data to be transmitted or not, thereby highly loading the cell (see Section 14.5.1.1). The idea of gating, introduced in 3GPP Release 7 and presented in [29] is to stop the transmission of DPCCH when there is no data to be sent on E-DCH and no L1 feedback signaling on HS-DPCCH (see Figure 14.7). This would reduce interference to other users and increase the UL capacity. Besides, it can increase UE standby time efficiently because of less power consumption.

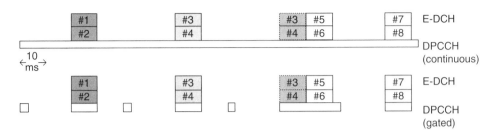

FIGURE 14.7 Illustration of VoIP transmission on E-DCH assuming 10 ms TTI and bundling of two VoIP packets into one TTI with continuous and gated DPCCH transmission.

DPCCH gating might pose difficulty for the network to distinguish an inactive period from a lost connection. Hence it is desirable that DPCCH transmission is not totally silent but transmitted periodically following a predefined pattern. Another issue to consider is transmission power. After a gating period without tracking the received SIR on DPCCH, the channel response variations could degrade the transmission and increase the interference between the UEs. Thus, in the 3GPP specifications for DPCCH gating the optional power control preambles are defined to be sent prior to the data channel reactivation. As a further improvement, compared with sending one packet per TTI, bundling multiple packets together in one transmission can reduce related control channel signaling overhead and thus improve air interface efficiency. More details about packet bundling can be found from Section 14.6.2.1. Packet bundling is typically utilized in the downlink direction only, due to the UE power shortage. However, packet bundling is applicable in UL direction as well if 10 ms TTI is used [30].

14.6.1.2 HS-SCCH-Less Operation

3GPP Release 7 also allows HS-SCCH-less operation in the DL to minimize the interference from HS-SCCH. The high VoIP capacity typically requires 4–6 code multiplexed users on HS-DSCH in the DL of Release 6. Each code multiplexed user requires HS-SCCH channel. HS-SCCH-less operation can be applied for VoIP when the packet interarrival time is constant. In HS-SCCH-less operation it is possible to send HS-PDSCH without associated HS-SCCH by using predefined transmission parameters (e.g., modulation and coding scheme, channelization code set, transport block size) configured by higher layers and signaled to UE via RRC signaling. HS-SCCH-less operation implies that UE has to blindly decode the HS-PDSCH that has been received without associated HS-SCCH. The interference from HS-SCCH transmissions can be avoided with HS-SCCH-less operation leading to higher VoIP capacity.

The dynamic HS-DSCH power allocation on HS-PDSCH and HS-SCCH channels in the case of HS-SCCH-less operation for VoIP service enables supporting more than four code-multiplexed users on HS-DSCH due to the more efficient use of the scarce power resource. This improves performance especially with the short 80 ms delay requirement, as buffering of VoIP packets in MAC-hs is becoming very restricted leading to the higher power requirement.

14.6.2 Solutions in LTE

14.6.2.1 Packet Bundling

The size of the VoIP packet is typically very small, for example, for AMR 12.2 kbps codec the size of the most common ROHC compressed packet is 40 bytes. This, together with strict delay requirements of VoIP traffic necessitate the use of PDSCH multiuser sharing when high capacities are targeted. As each user is scheduled by L1 control signaling, with multiuser channel sharing the control channel overhead becomes the limiting factor for VoIP system performance and the dynamic packet scheduler is not able to fully exploit PDSCH air interface capacity. One attractive technique with baseline scheduler to improve VoIP capacity within control channel limitations is to bundle multiple VoIP packets (of one user) together and transmit the bundled packets within the same TTI to a scheduled user. Packet bundling is CQI-based meaning that it is used only for such users whose channel conditions are favorable enough to support bundling. With packet bundling, good users are scheduled less frequently implying that transmission resources can be more efficiently used and L1/L2 control overhead can be reduced. Hence packet bundling can be seen as an attractive method in DL to improve VoIP capacity for fully dynamic PS whilst keeping L1/L2 control overhead at a reasonable level. From the perspective of voice quality it is important to keep the probability of losing consecutive packets as small as possible. This can be achieved by making the link adaptation for the TTIs carrying bundled packets in a more conservative manner leading to a reduced packet error rate for the first transmission.

In UL, packet bundling could be used as well but it is not an attractive technique because of the limited UE transmission power and non-CQI-based scheduling.

The impact of packet bundling on PDSCH utilization rate in macro cell case 1 [26] with AMR 12.2 kbps codec at 5 MHz bandwidth is presented in the Figure 14.8, which contains cumulative distribution function for scheduled PRBs per OFDM subframe. Here, it is assumed that the statistics are plotted with the load corresponding to 5% outage point–this assumption is valid for the other statistics presented in this chapter as well. In Case 1, approximately 70% of the users use bundling (see Figure 14.9), so approximately 70–80% more users can be fitted to the network without exceeding outage, as is shown in Section 14.7. Due to the higher load compared to nonbundling case, the average PDSCH utilization rate is approximately doubled with bundling.

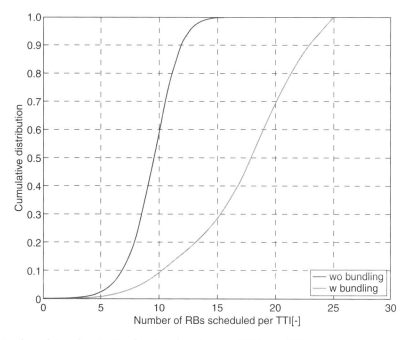

FIGURE 14.8 Cumulative distribution function for scheduled PRBs per TTI.

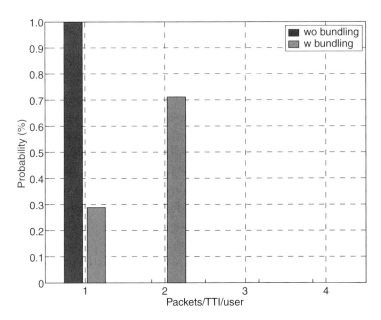

FIGURE 14.9   Probability distribution function for bundled packets/TTI/user.

14.6.2.2 Semi-persistent Resource Allocation

In order to avoid control channel limitations for VoIP traffic in E-UTRAN, a concept of semi-persistent packet scheduling (PS) [8,18] was adopted in 3GPP for LTE. Semi-persistent PS can be seen as a combination of dynamic and persistent scheduling methods, where initial transmissions of VoIP traffic are scheduled without assigned L1/L2 control information by using persistently allocated time and frequency resources, whereas the possible HARQ retransmissions and SID transmissions are scheduled dynamically. For VoIP traffic the used BLER target in DL is of the order 10–20%, and this together with SID transmissions implies that only 20–30% of the transmissions require L1/L2 control signaling in DL. As initial transmissions are scheduled by using persistently allocated time and frequency resources, savings in control channel overhead are achieved with the cost of reduced time and frequency domain scheduling gains.

The semi-persistent resource allocation method adopted in 3GPP LTE is talk spurt based persistent allocation, and in DL direction the method works as follows. At the beginning of a talk spurt, a persistent resource allocation is done for the user and this dedicated time and frequency resource is used to transmit initial transmissions of VoIP packets. At the end of the talk spurt, persistent resource allocation is released. Thus, the released resource can be allocated to some other VoIP user, which enables efficient usage of PDSCH bandwidth.

The persistent allocation at the beginning of the talk spurt is signaled by using L1/L2 signaling. RRC signaling is used to signal some semi-static parameters such as the periodicity of the allocation. A slow, talk spurt based link adaptation is possible since the modulation scheme as well as the amount of resources are signaled with L1/L2 signaling.

The release of the persistent allocation at the end of the talk spurt is done via explicit signaling from the eNodeB with L1/L2 signaling. It is still open whether implicit release (without signaling) of DL semi-persistent resources is supported, too. DL operation for the talk spurt based persistent allocation is illustrated in Figure 14.10.

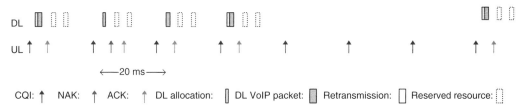

FIGURE 14.10 Talk spurt-based persistent allocation in downlink.

In UL direction, the UE should send a scheduling request (SR) to the eNodeB at the beginning of a talk spurt to get the radio resource allocated for it. Similarly, as in DL, the allocated radio resource is used to carry initial transmissions of VoIP packets without any scheduling related L1/L2 signaling in DL. SID frames can be allocated dynamically (thus requiring SR for each SID). The required signaling in the UL direction is similar to the DL. The persistent UL allocation at the beginning of the talk spurt is signaled by using L1/L2 signaling.

When the talk spurt ends, the resource is released either explicitly with release signaling or implicitly by noticing that no more data is coming. Thus, the released resource can be allocated to some other VoIP user.

Figure 14.11 shows the required UL and DL signaling as well as the transmission resources. Here we assume that SID frames are scheduled dynamically.

The impact of semi-persistent resource allocation method on DL control overhead is illustrated in Figure 14.12, which contain cumulative distribution function (CDF) of the dynamically scheduled users per TTI (out of maximum 10 users). As is evident from the figure, with semi-persistent resource allocation method DL control overhead is approximately 40% of the control channel overhead of fully dynamic PS. In Figure 14.13 CDF for the number of scheduled PRBs per OFDM subframe is presented. Performance of fully dynamic PS is control channel limited, that is, there is not enough control channel resources to utilize the total transmission bandwidth: with fully dynamic PS only 70% of the total transmission bandwidth (5 MHz ~ 25 PRBs) is used on average. Here we assume that packet is used for dynamic scheduler. Semi-persistent PS does not suffer from control channel limitation, but its performance is data limited: on average only 10% of the total transmission bandwidth is unused.

14.6.2.3 Means to Overcome CQI Imperfections

CQI feedback is utilized by the channel aware scheduler at eNodeB when selecting the scheduled users and allocating resources for them. Additionally, CQI reports are used by

FIGURE 14.11 Talk spurt-based persistent allocation in uplink.

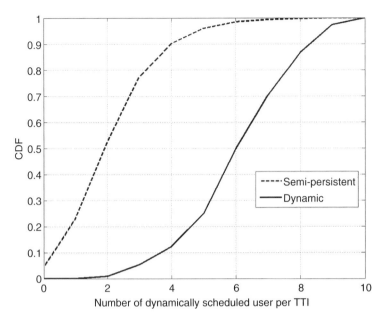

FIGURE 14.12 Cumulative distribution function for the dynamically scheduled users per TTI.

Link adaptation algorithm when selecting the used modulation and coding scheme for the scheduled users. CQI feedback is subject to reporting delays and UE measurement/ estimation imperfections, and these are compensated by the Outer Loop Link Adaptation (OLLA) algorithm, which uses HARQ ACK/NAK feedback from the past to maintain the average BLER at the target level. OLLA has been shown to provide good mitigation mechanism for the LA errors caused by the imperfect CQI feedback information [31–33].

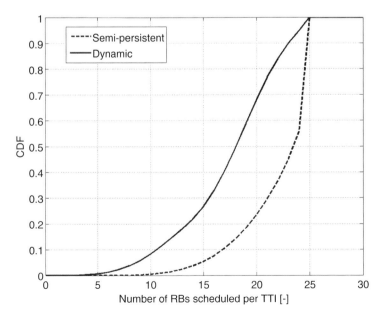

FIGURE 14.13 Cumulative distribution function for scheduled PRBs per TTI.

As described earlier, wideband CQI may be required for VoIP traffic in order to keep the UL control overhead at a reasonable level. This implies that frequency domain packet scheduling gains are lost: the eNodeB packet scheduling can no longer track the frequency selective fading and will have to rely on a transmission scheme that offers maximal frequency diversity. Or similarly, frequency diversity transmission should be favored if due to the increased velocity frequency dependent CQI information is unreliable (velocity >25 km/h). Frequency diversity transmission can be achieved by one of the following methods [34]:

(1) Using localized transmission where a user is scheduled on multiple PRBs that are scattered over the full system bandwidth to offer maximal frequency diversity.

(2) Using distributed transmission where a number of PRBs (scattered over the full system bandwidth) are shared by a set of users.

Distributed transmission refers to the case where one PRB of 12 contiguous subcarriers is shared between multiple users on a subcarrier resolution. Distributed transmission feature was adopted to 3GPP Release 8 with the restriction that a user scheduled with distributed transmission shall always be scheduled on groups of $Nd \times 12$ subcarriers, where Nd is limited to two (2).

Method (1) offers a high degree of frequency diversity for cases where there is sufficient data for a user to be scheduled on multiple PRBs, that is, at least 3–4 PRBs. Unfortunately, with VoIP traffic the data amount for one user typically requires at most 2 PRBs, and therefore it is difficult to achieve efficient frequency diversity with localized transmission. Frequency diversity offered by method (1) can be further improved if HARQ retransmissions are scheduled in different locations in frequency than previous transmission(s).

VoIP LTE DL performance of semi-persistent PS with wideband CQI was analyzed in [34] by comparing localized transmission with method (1) (with HARQ retransmissions scheduled so that frequency diversity is maximized) and distributed transmission with $Nd = 2$. It was shown that distributed transmission provided only 2% higher capacity than localized transmission with enhanced method (1).

In Table 14.3 relative losses in capacity due to utilization of wideband CQI instead of narrowband CQI are given for fully dynamic and semi-persistent PS methods. Here it is assumed that localized transmission with method (1) assuming diversity enhancement over HARQ retransmissions is used. As can be seen from the results, when the frequency domain scheduling losses are compensated with efficient utilization of frequency diversity, degradation in performance due to usage of wideband CQI stays tolerable. Without frequency diversity enhancements corresponding losses would be higher. Moreover, the relative losses in performance realized by the semi-persistent PS are only half of the corresponding losses for fully dynamic PS, as only HARQ retransmissions and SID transmissions benefit from FD scheduling gains.

TABLE 14.3

Impact of Wideband CQI to Capacity

Used Packet Scheduling Method	Dynamic	Semi-persistent
Relative loss in capacity (%)	14	7

14.6.2.4 Adaptive HARQ in UL

For tradeoff purpose, the currently agreed proposal in 3GPP is to use synchronous nonadaptive HARQ protocol as much as possible for LTE UL. It means adaptive HARQ will be enabled for use if fully nonadaptive HARQ is impossible to be used. Such decision is due to the drawbacks of fully nonadaptive HARQ. The nature of nonadaptive solution is that retransmissions occur at a predefined (normally fixed) time after the previous (re)transmission using the same resources. The benefit of synchronous nonadaptive HARQ is that the UL control signaling can be minimized since only a NAK need to be signaled back to start a HARQ retransmission. However, due to orthogonality and single carrier requirements in LTE UL, there are some obvious problems for nonadaptive HARQ retransmissions combined with semi-persistent allocation:

(a) Resource fragmentation: When several UEs are scheduled in one TTI and some users require retransmissions whereas others do not, the UL resources can be fragmented and the scheduling of new users becomes more difficult or the required resources cannot be scheduled to a user due to single carrier requirement [36].

(b) Low resource efficiency: For semi-persistent allocation, part of the resources is allocated persistently and part is scheduled dynamically. A nonadaptive HARQ sets unnecessary restrictions and leads to poor resource utilization. Since with synchronous nonadaptive HARQ, the possible retransmission would overlap with the persistent allocation, a persistent allocation in a given time-frequency resource implies that nothing can be scheduled on the same frequency resource at a HARQ RTT earlier (or a few HARQ RTTs earlier).

(c) Separate ACK/NAK channel in DL: With synchronous nonadaptive HARQ, a separate ACK/NAK has to be sent strongly coded (repetition) and with high power in order to guarantee low error probability.

Synchronous adaptive HARQ can then naturally be used to solve the problems: when resource fragmentation happens, an adaptive HARQ allows moving of the retransmissions to the edges of the band thus avoiding the fragmentation; when the collision happens between initial persistent transmissions and retransmissions, by allowing adaptive HARQ, the possible retransmission can be sent on different frequency resource thus avoiding the problem; adaptive HARQ implies that all the retransmissions are scheduled. This has the advantage that no separate ACK/NAK channel is needed in the DL. Retransmissions are requested with UL allocation (implicit NAK) and an UL allocation for a new transmission implies that retransmissions are not continued. Though ACK/NAK channel still exists in terms of 3GPP agreement, it is overridden by UL allocation signaling.

14.6.2.5 Interference Aware Packet Scheduling for LTE UL

As proposed in [23,25], an interference aware packet scheduling method to alleviate the co-channel interference for larger cells (e.g., macro cell Case 3 [26]) could be used. Briefly, the method performs as follows: First, users in the same sector are sorted on the basis of DL pathloss measurement. Second, UEs closer to BS are allocated into middle frequencies and UEs at cell edge are allocated into edge frequencies. SNR target is first set to cell edge users, and the SNR target of users closer to BS may be gradually increased. In other words, the basic idea in this method is that users with similar pathloss use same frequency in all sectors/cells. Hence, with this method the interference level is

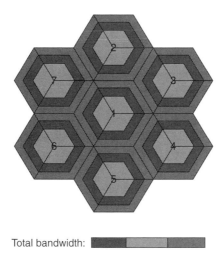

Total bandwidth:

FIGURE 14.14 IC scheme for large cell.

comparable to the useful signal level, but as can be seen from Figure 14.14, the cell edge users may interfere with each other. Therefore, for small cells (e.g., macro cell Case 1 [26]) different re-use pattern to minimize intercell interference can be used. This is depicted in Figure 14.15. Similarly, as for the interference control scheme for larger cells, UEs are sorted according to DL pathloss measurements. Then, the required transmission power levels of UEs are figured out according to SNR target. The SNR target is the same for all UEs in the network, that is, for re-use factor three power sequence $[0\ -4\ -4]$ is used. The key point here is that users with similar pathloss use different frequency in different sectors within one cell. Hence, the cell edge users in different sectors use different sub-bands implying that they do not interfere with each other. On the other hand, the interference among the three sectors within one cell seems to be a bit larger.

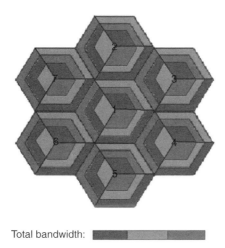

Total bandwidth:

FIGURE 14.15 IC scheme for small cell.

Of course, some kind of combined scheme by utilizing the advantages of the two IC schemes can be designed to further enhance the performance as well.

14.6.2.6 TTI Bundling for LTE UL

A natural solution to improve the coverage in noise limited scenario would be to use RLC layer segmentation of VoIP packets. The drawback of this method is mainly the significantly increased overhead. To overcome this problem, TTI bundling method is adopted in 3GPP [27,28], in which a few consecutive TTIs are bundled together and a single transport block is first coded and then transmitted by using this bundled set of consecutive TTIs. With TTI bundling only one L1/L2 grant is needed to schedule the transmission and only one HARQ feedback signal is sent from the eNodeB for the bundled subframes. Moreover, the same hybrid ARQ process number is used in each of the bundled subframes. The operation is like an autonomous retransmission by the UE in consecutive HARQ processes without waiting for ACK/NAK feedback. The redundancy version on each autonomous retransmission in consecutive subframes changes in a predetermined manner. There is a possibility that the code rate might be larger than one for the first subframe, but decoding becomes possible at the eNodeB as soon as the code rate becomes less than or equal to one. TTI bundling is a technology that improves the data rate, reducing overheads and leading to a packet-optimized transmission.

14.7 Network Performance of VoIP

14.7.1 HSPA

14.7.1.1 Simulation Methodology and Assumptions

The quasi-static system level simulators for HSUPA and HSDPA, where all necessary RRM algorithms as well as their interactions are modeled, are used to investigate the performance of VoIP on HSPA. These tools include a detailed simulation of the users within multiple cells. The fast fading is explicitly modeled for each user according to the ITU Vehicular-A profile. Regarding the methodology, this kind of quasi-static simulator is based on descriptions in [3]. A wrap-around multi-cell layout modeling several layers of interference is utilized in this study.

   The main parameters used in the system simulation are summarized in Table 14.4, where UE moving speeds other than 3 kmph are mostly applied to investigate the effect of the mobility on the performance of the DL.

14.7.1.2 Simulation Results and Analysis

VoIP performance in HSUPA

The UL capacities for VoIP over HSUPA are presented to evaluate the performance of some advanced solutions as described in Section 14.6.1.1. For the 2 ms TTI without UL packet bundling, Figure 14.16 shows UE power distribution for 60 UEs/cell for gating DPCCH and continuous DPCCH. It may be noted that there can be huge power savings by applying DPCCH gating. The exact power saving level would depend on the activity of DPCCH, which is affected by the transmission of E-DCH and HS-DPCCH. It also implies that the frequent

TABLE 14.4

Simulation Parameters Settings

Parameter	Value
System Configuration	
Inter-site distance	2.8 km
Cell configuration	ITU Veh-A, Macrocell
Voice call mean length	60 seconds
UE speed	3 (30, 50, 90) kmph
Outage observation window length	10 seconds
Cell outage threshold	5%
Residual FER	1%
HSUPA Specific Configuration	
Receiver	2 antenna RAKE with MRC combining
Frame size	2 ms/10 ms
Channels	E-DCH/DPCCH
Number of HARQ channels	8 (2 ms TTI)/4 (10 ms TTI)
Max number of L1 transmissions	4 (2 ms TTI)/2 (10 ms TTI)
UL delay budget	80 ms
Scheduling algorithm	Non-scheduled
HSDPA Specific Configuration	
Receiver	1 antenna RAKE with MRC combining
Frame size	2 ms
Channels	HS-DSCH/CPICH/HS-DPCCH
Number of HARQ channels	6
Max number of L1 transmissions	4
Max code-multiplexing users	4
Number of codes	10
Total BS transmission power	20 W
HS-DSCH power	10 W
DL delay budget	80 ms

CQI reporting carried on HS-DPCCH would reduce the power saving gain because of having fewer opportunities to use DPCCH gating. Similarly, if the DL data transmission over HS-DSCH is quite active, the transmission of ACK/NAK feedback information carried over HS-DPCCH in the UL would be quite frequent, which will also constrain the gating gain.

Table 14.5 summarizes the VoIP capacity results obtained with a quasi-static system simulator for both 2 ms and 10 ms TTI in the case of 3 kmph UE moving speed. The capacity is calculated with two different criteria presented in Section 14.3.3: the number of VoIP users per sector not exceeding the 5% FER target measured over 10 s or the number of VoIP users allowed per sector with an average noise rise of 6 dB. Here the delay budget of 80 ms is assumed for HSUPA. This would not be the bottleneck at the performance since HARQ retransmissions will be completed within this delay budget when the maximum allowed HARQ retransmission number is 1 for 10 ms TTI and 3 for 2 ms TTI. The results show that VoIP capacity over HSUPA is comparable or to a great extent better than CS voice over DCH (about 65 voice users per sector) [3] with continuous DPCCH and gating gives 40–50% further gain. Otherwise, fewer transmissions can increase the DPCCH gating gain in the

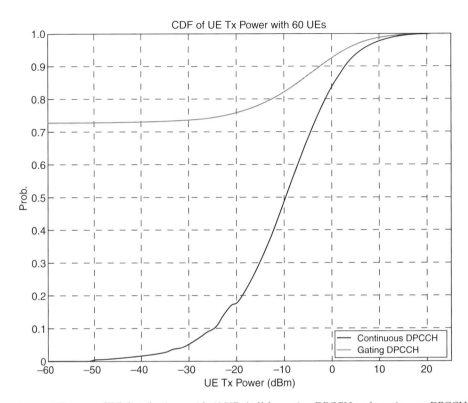

FIGURE 14.16   UE power CDF distributions with 60 UEs/cell for gating DPCCH and continuous DPCCH.

capacity from 50% to 77% with 10 ms TTI because of the increasing opportunities for DPCCH gating from fewer transmissions on E-DCH. (For more results and simulation assumptions, see e.g. [3,29,30,37].)

VoIP performance in HSDPA

VoIP system performance in HSDPA achieved with the best mode scheduling alternative ("PF + SCS + RA" described in the Section 14.4.3.2) is shown in Figure 14.17 for both Release 6 and Release 7. Release 6 results are achieved with 1-antenna LMMSE chip equalizer whereas Release 7 results are achieved with 2-antenna RAKE with MRC combining.

TABLE 14.5

Summary of the UL VoIP Capacity Results without and with UL DPCCH Gating

TTI	Average Number of Transmissions	Capacity Criteria: 5% FER Over 10 s			Capacity Criteria: Noise Rise 6 dB		
		Continuous DPCCH (Users)	Gated DPCCH (Users)	Gating Gain (%)	Continuous DPCCH (Users)	Gated DPCCH (Users)	Gating Gain (%)
2 ms	~3	82	123	50	75	106	41
10 ms*	1.25	65	115	77	61	93	52
	1.67	80	120	50	73	103	41

*   10 ms TTI results using bundling of 2 VoIP frames in a single TTI.
Source: 3GPP TR 25.903 V1.0.0 (2006–05), "Continuous connectivity for packet data users."

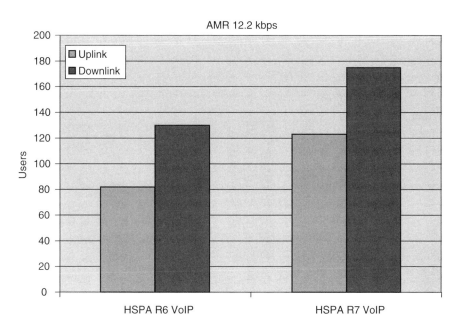

FIGURE 14.17 VoIP HSPA capacity comparison between downlink and uplink.

Corresponding UL capacities for Release 6 and 7 are also presented in the figure for comparison. As the results show, receive diversity provides superior performance compared to 1-antenna LMMSE chip equalizer. Furthermore, results indicate that HSPA capacity is uplink limited.

Effect of the mobility on the performance

In the real world, UE is always moving at a varying speed. Therefore, it makes sense to understand the effect of UE velocity and handover delay in practice. VoIP cell capacities are presented in Figure 14.18. The figure shows the achieved capacities for 100 ms, 200 ms, and 400 ms handover delays.

From the results it can be seen that the VoIP capacity is highly sensitive to UE velocity. Higher velocities lead to lowered VoIP capacity: the capacity drops by roughly 20% when going from 3 km/h to 30 km/h and with higher velocity the capacity drops even further. This is because the radio channel is changing more rapidly and scheduling is thus more challenging. Moreover, the number of handovers is increased significantly when UE velocity increases.

The results show that the handover delay is critical to the performance of VoIP service over HSDPA, especially when the UE velocity gets higher. This is caused by the fact that with high UE velocities the performance is already greatly affected by the radio channel conditions and thus it is required to have fast response to the handover procedures.

14.7.2 LTE

14.7.2.1 Simulation Methodology and Assumptions

The performance evaluation for LTE is conducted with a quasi-static system level simulator [11]. The simulated network is composed of 19 cell sites with 3 sectors each. The scheduling algorithm and other radio resource management (RRM) functions are explicitly modeled

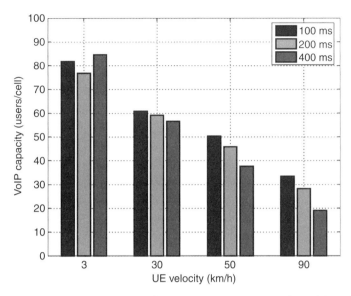

FIGURE 14.18   Average VoIP cell capacity with different UE velocities and handover delays. (*Source:* P. Lundén, J. Äijänen, K. Aho, and T. Ristaniemi, "Performance of VoIP over HSDPA in mobility scenarios," VTC 2008 Spring. With permission.)

in 3 center cells and statistics are also gathered from these 3 center cells. The remaining cells generate intercell interference with full load. The propagation characteristics of the physical link as a distance dependent pathloss, shadowing and frequency selective fast fading are modeled in the system simulator, whereas the link performance is fed to the system simulator through an Exponential Effective SNR Mapping (EESM) interface [24], or alternatively through an Actual Value Interface (AVI). Furthermore, for LTE the Spatial Channel Model [38] SCM-C shall be employed. The main simulation parameters are based on [16] and are also listed in Table 14.6. UEs are uniformly distributed in 3 center cells and assigned with VoIP traffic. The VoIP traffic is modeled as in [11] and the most important characteristics are summarized in Table 14.7. Moreover, the main DL related parameters and UL related parameters are summarized in Table 14.8 and Table 14.9, respectively.

TABLE 14.6

Summary of the Main Simulation Parameters

Description	Settings
Number of cells	19 sites with 3 cells
Carrier center frequency	2.0 GHz
Simulation scenarios	Macro Case 1 and Macro Case 3
System bandwidth	5 MHz
Pathloss	According to [26], minimum coupling loss (MCL) 70 dB
Std of log-normal shadowing	8 dB
Velocity	3 km/h
Channel	SCM-C with correlations for spatial channels
Link to system model	DL EESM (realistic ChEst), UL AVI (realistic ChEst)
PDCCH modeling	Realistic

TABLE 14.7

Main Traffic Related Assumptions

Description	Settings
Codec	AMR 12.2 kb/s, AMR 7.95 kb/s
Header compression	ROHC
Payload including all overhead	40 bytes for AMR 12.2 kbps 28 bytes for AMR 7.95 kbps
SID payload including all overhead	15 bytes
Voice activity	50%
Talk spurt	Negative exponential distribution, mean 2.0 s
One-way delay (ms)	50

TABLE 14.8

Main DL Related Simulation Parameters

Description	Settings
Multiple access	OFDMA
eNode-B Tx power	43 dBm (5 MHz)
Max C/I limit in the receiver	22 dB
Transmission scheme	2×2 Space-Time Transmit Diversity (STTD)
Receivers	2-antenna with MRC combining
CQI settings	Zero mean i.i.d Gaussian error with 1 dB std, 5 ms measurement window, 2 ms reporting delay
DL L1/L2 control channel overhead	4 OFDM symbols per TTI including pilot overhead
MCS set	As in [16]
HARQ	8 HARQ SAW channels. Asynchronous, adaptive HARQ with chase combining. Maximum 6 retransmissions
Packet scheduler	VoIP optimized PS, semi-persistent PS
Packet bundling	Can be used for fully dynamic PS (up to 2 packets could be bundled per TTI)
PDCCH assumption	10 CCEs reserved for DL traffic scheduling per TTI (max 10 users could be scheduled dynamically per TTI)

TABLE 14.9

Main UL Related Simulation Assumptions

Description	Settings
Multiple access	SC-FDMA
UE transmission power	Max 24 dBm
Power dynamics of UEs at eNodeB	17 dB
Transmission scheme	1×2
Receivers	2-antenna with MRC combining
UL control overhead	2 PRBs per TTI
MCS set	As in [16]
HARQ	8 HARQ SAW channel. Synchronous, adaptive HARQ with incremental redundancy (IR). Maximum 6 retransmissions
Packet scheduler	Dynamic PS, semi-persistent PS
PDCCH assumption	10 CCEs reserved for UL per TTI (max 10 users could be scheduled dynamically per TTI)

The performance is evaluated in terms of VoIP capacity. VoIP capacity is defined according to the quality criteria given in Section 14.3.3, that is, capacity is given as the maximum number of users that can be supported without exceeding 5% outage. User is in outage, if more than 2% of the packets are lost (i.e., lost or erroneous) during the whole call.

14.7.2.2 Simulation Results and Analysis

VoIP performance in LTE

VoIP capacity numbers for LTE are summarized in Table 14.10 for both DL and UL directions. Let us analyze DL results first.

As described in Section 14.5.1.2, the performance of the fully dynamic PS without packet bundling is limited by the available PDCCH resources due to which savings in VoIP packet size are not mapped directly to capacity gains. This can be verified from the results showing almost identical performance for AMR 7.95 and AMR 12.2. With packet bundling, control channel limitations can be partly avoided and hence the VoIP capacity with fully dynamic PS can be boosted up to 90%. With bundling, the performance of fully dynamic PS starts to suffer from data limitation as well, due to which small gains in capacity (8%) can be achieved when AMR 7.95 is used instead of AMR 12.2. On the other hand, the performance of semi-persistent PS is data limited instead and hence savings in VoIP packet size are nicely mapped to gains in capacity—approximately 35% higher capacity is obtained if AMR 7.95 is used instead of AMR 12.2. When comparing the performances of fully dynamic PS and semi-persistent PS with each other, it is concluded that if packet bundling is not allowed then the performance of fully dynamic PS is badly control channel limited and hence semi-persistent PS is able to have 50–100% higher capacities than fully dynamic PS. As the performance of semi-persistent PS suffers from data limitation, gains in capacity over fully dynamic PS are smaller for the AMR 12.2, which has higher data rate and hence bigger VoIP packet size. When packet bundling is used with AMR 12.2, the control channel limitation for the performance of fully dynamic PS is not that significant compared to the corresponding case with AMR 7.95 due to lower number of supported users at 5% outage. Therefore, fully dynamic PS is able to have 15% gains in capacity over semi-persistent PS if AMR 12.2 is used with packet bundling. When bundling is used with AMR 7.95, control channel limitation again starts to dominate the performance of fully dynamic PS due to high number of supported users at 5% outage, and hence semi-persistent PS achieves 8% higher capacities than fully dynamic PS. However, in practice it will be most likely

TABLE 14.10

VoIP Capacity in LTE

VoIP Codec	AMR 7.95	AMR 12.2
Downlink Capacity		
Dynamic scheduler, without bundling	210	210
Dynamic scheduler, with bundling	400	370
Semi-persistent scheduler	430	320
Uplink Capacity		
Dynamic scheduler	230	210
Semi-persistent scheduler	320	240

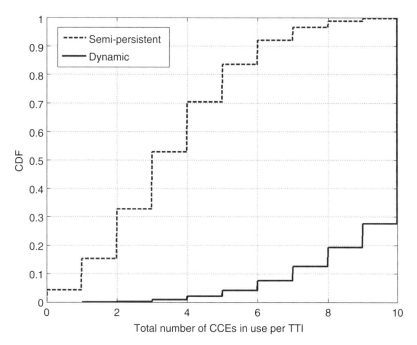

FIGURE 14.19 Distribution of used CCEs per TTI.

necessary to use wideband CQI with VoIP traffic to keep the CQI related UL signaling load reasonable. This will reduce the performance of the dynamic PS more than that of the semi-persistent PS (as the dynamic PS relies more on the frequency selective CQI information). Therefore, the semi-persistent PS can be seen as an attractive option for supporting high number of simultaneous VoIP users.

When analyzing VoIP results for LTE UL, it is emphasized that similar to DL, the semi-persistent scheduling suffers much less from control channel limitations, whereas the performance of dynamic scheduling greatly relies on the number of control channels per TTI as illustrated in Figure 14.19, which contains cumulative distribution function of the used control channel elements (CCEs) per TTI. As control channel limitation becomes a bottleneck for VoIP UL system performance with fully dynamic PS especially if very high number of users are accessed into system, that is, as is the case with AMR 7.95. Due to the control channel limitations fully dynamic PS can support only 230 users per sector in Case1 with AMR 7.95, whereas with semi-persistent scheduling control channel limitations can be avoided and 320 users with a gain of 40% can be supported. However, with AMR 12.2, the impact of control channel limitation is much smaller for the fully dynamic PS due to bigger VoIP payload size and hence smaller number of supported users at 5% outage. Therefore, the capacity gain of semi-persistent PS over fully dynamic PS is only about 14%.

Finally, when comparing the performances of DL and UL together, it is concluded that the VoIP performance in LTE is UL limited, similar to HSPA. This is mainly due to the coverage limitation in UL, which is caused by relatively low maximum transmit power of the UEs. The extra DL capacity is not wasted, but can be used, for example, for supporting additional best effort traffic (web browsing etc.), which is typically asymmetric and geared more towards DL.

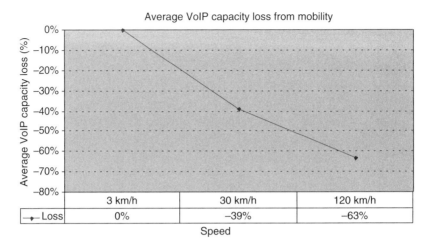

FIGURE 14.20 Average VoIP cell capacity with different UE velocities and handover delays. (*Source:* T. Henttonen, K. Aschan, J. Puttonen, N. Kolehmainen, P. Kela, M. Moisio, and J. Ojala, "Performance of VoIP with mobility in UTRA long term evolution," IEEE VTC Spring, Singapore, May 2008.)

VoIP performance with mobility in LTE

Figure 14.20 shows the average behavior of different UE mobility scenarios in LTE. The simulations were run with dynamic scheduling, but it is assumed that similar behavior would result with semi-persistent scheduling.

Mobility clearly degrades VoIP capacity. With high speed the loss (compared to case with the pedestrian speed of 3 km/h) is about 63%, and so the capacity is roughly only one-third of the nonmobility case. The loss in capacity is caused be several factors: First, the CQI measurements become less accurate as the speed increases, which causes the scheduler to lose most of the efficiency of frequency-dependent scheduling. Link adaptation also performs badly as it is tied to the CQI reports, leading to increase in retransmissions. And the faster the users move, the more handovers are done, with all the resulting effects (L1 packets flushed when handover is done, delays due to handover execution delay) also affecting the VoIP capacity [43].

14.7.3 VoIP Performance Comparison Between HSPA and LTE

The voice capacity comparison among WCDMA, HSPA, and LTE is summarized in Figure 14.21 for 5 MHz bandwidth. From the comparison, it is obvious that HSPA Release 7 provides clear improvement to the voice capacity compared to the CS voice. When comparing voice capacities between HSPA and LTE, it is observed that LTE achieves more than 100% higher capacity than HSPA Release 7. Furthermore, it can be noted that both VoIP capacity is UL limited both in HSPA and LTE.

14.8 Summary and Conclusion

The potential for simplifications in the core network and the cost benefits that an "All IP network" can provide are clear. However, the VoIP solution has to work well to enable a

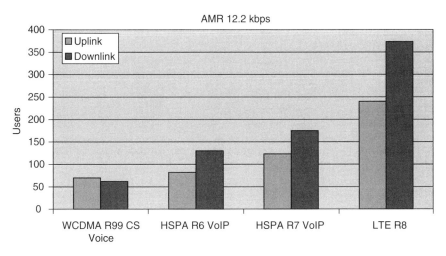

FIGURE 14.21 Voice capacity with WCDMA/HSPA/LTE.

standalone cellular system with all the commonly used services. The network has to be able to guarantee the same service quality and higher voice capacity with VoIP than has been required from CS-based voice systems. The principles of capacity and coverage gaining mechanisms, explained in Section 14.4, are: Limiting the IP-header overhead, multiuser channel sharing, VoIP specific packet schedulers and advanced receivers. ROHC, a solution for the first mechanism, enables the system to perform more efficiently by reducing the overhead and has been widely studied and recognized as an essential optimization for VoIP. The other gaining mechanism enhance the capacity and the coverage by helping to guarantee the quality also at the cell edge.

Control channel overhead and the co-channel interference, as explained in Section 14.5, are severe obstacles to the VoIP performance. To enable the whole potential offered by the different gaining mechanisms in the networks with the practical limitations and further gain in capacity and coverage, several methods were explained in Section 14.6. These solutions (packet bundling, UL gating, HS-SCCH-less operation, semi-persistent allocation, distributed allocation, adaptive HARQ, TTI bundling) with VoIP optimization provide optimistic sight for the VoIP success. The simulation results of Section 14.7 shows that especially, the control channel overhead reduction solutions adopted by 3GPP are attractive as they provide very good performance.

References

1. H. Holma, A. Toskala, "WCDMA for UMTS–HSPA evolution and LTE," Fourth edition, Wiley, 2007.
2. R. Cuny and A. Lakaniemi, "VoIP in 3G networks: An end-to-end quality of service analysis," IEEE, VTC Spring, vol. 2, April 2003, pp. 930–934.
3. H. Holma and A. Toskala, "HSDPA/HSUPA for UMTS," John Wiley, 2006.
4. 3GPP TS 25.309, "FDD enhanced UL; Overall description; Stage 2."
5. 3GPP TR 25.896, "Feasibility study for enhanced UL (FDD)."

6. 3GPP TS 25.308, "HSDPA; Overall description; Stage 2."

7. H. Holma, M. Kuusela, E. Malkamäki, K. Ranta-aho, and T. Chen, "VoIP over HSPA with 3GPP, Release 7," PIMRC 2006, Helsinki.

8. 3GPP TS 36.300, "Evolved universal terrestrial radio access (E-UTRA) and evolved universal terrestrial radio access network (E-UTRAN); Overall description; Stage 2 (Release 8)."

9. 3GPP TR 25.913 V7.1.0, "Requirements for evolved UTRA (E-UTRA) and evolved UTRAN (E-UTRAN)."

10. 3GPP TS 26.071, v7.0.1, "Mandatory speech CODEC speech processing functions; AMR speech CODEC; General description (Release 7)."

11. 3GPP R1-070674 "LTE physical layer framework for performance verification," Orange, China Mobile, KPN, NTT DoCoMo, Sprint, T-Mobile, Vodafone, Telecom Italia.

12. IETF RFC 3095: "RObust Header Compression (ROHC): Framework and four profiles: RTP, UDP, ESP, and uncompressed."

13. P. Lundén and M. Kuusela, "Enhancing performance of VoIP over HSDPA," VTC 2007 Spring.

14. Y. Fan, M. Kuusela, P. Lunden, and M. Valkama, "Downlink VoIP support for evolved UTRA," *IEEE Wireless Communication and Networking Conference (WCNC) 2008*, Las Vegas, United States.

15. B. Wang, K. I. Pedersen, T. E. Kolding, and P. Mogensen, "Performance of VoIP on HSDPA," VTC 2005 Spring, Stockholm.

16. "Next generation mobile networks (NGMN) radio access performance evaluation methodology," *A White paper by the NGMN Alliance*, January 2008.

17. T. Chen, E. Malkamäki, and T. Ristaniemi, "Uplink DPCCH gating of inactive UEs in continuous packet connectivity mode for HSUPA," IEEE WCNC, March 2007.

18. 3GPP TS 36.321, "Evolved Universal Terrestrial Radio Access (E-UTRA); Medium Access Control (MAC) protocol specification (Release 8)."

19. 3GPP R1-070475, "Downlink scheduling for VoIP," Feb 12–16, 2007.

20. 3GPP R1-070476, "Uplink scheduling for VoIP," Feb 12–16, 2007.

21. 3GPP TS 36.213 V8.3.0 (2008-05), "TSG-RAN; Evolved universal terrestrial radio access (E-UTRA); Physical layer procedures (Release 8)."

22. Y. Fan, P. Lundén, M. Kuusela, and M. Valkama, "Performance of VoIP on EUTRA DL with limited channel feedback," IEEE ISWCS 2008, Reykjavik, Iceland.

23. 3GPP R1-050813, "UL interference control considerations," Nokia.

24. K. Bruninghaus et al., "Link performance models for system level simulations of broadband radio access systems," *Proceedings of IEEE International Symposium on Personal, Indoor and Mobile Radio Conference (PIMRC-2005)*, Berlin, September 2005.

25. 3GPP R1-060298, "UL inter-cell interference mitigation and text proposal," Nokia.

26. 3GPP TR 25.814 v7.0.0: "Physical layer aspect for evolved universal terrestrial radio access (UTRA)."

27. 3GPP R2-074061, "HARQ operation in case of UL power limitation," Ericsson.

28. 3GPP R2-074359, "On the need for VoIP coverage enhancement for E-UTRAL UL," Alcatel-Lucent.

29. 3GPP TR 25.903 V1.0.0 (2006–05), "Continuous connectivity for packet data users."

30. O. Fresan, T. Chen, K. Ranta-aho, and T. Ristaniemi, "Dynamic packet bundling for VoIP transmission over Rel'7 HSUPA with 10ms TTI length," ISWCS 2007.

31. A. Pokhariyal et al., "HARQ aware frequency domain packet scheduler with different degrees of fairness for the UTRAN long term evolution," *IEEE Proceedings of the Vehicular Technology Conference*, May 2007.

32. K. I. Pedersen et al., "Frequency domain scheduling for OFDMA with limited and noisy channel feedback," *IEEE Proceedings of the Vehicular Technology Conference*, October 2007.

33. I. Z. Kovács et al., "Effects of non-ideal channel feedback on dual-stream MIMO OFDMA system performance," *IEEE Proceedings of the Vehicular Technology Conference*, October 2007.

34. 3GPP R1-074884, "On the impact of LTE DL distributed transmission," Nokia, Nokia-Siemens-Networks.

35. 3GPP TS 36.201, "TSG-RAN; Evolved universal terrestrial radio access (E-UTRA); LTE physical layer—General description, Release 8."

36. 3GPP R2-071841, "Resource fragmentation for LTE UL," May 7–11, 2007.
37. O. Fresan, T. Chen, E. Malkamaki, and T. Ristaniemi, "DPCCH gating gain for voice over IP on HSUPA," IEEE WCNC, March 2007.
38. 3GPP TR 25.996, "Spatial channel model for MIMO simulations," v.6.1.0, September 2003.
39. 3GPP TS 36.211, "TSG-RAN; Evolved universal terrestrial radio access (E-UTRA); LTE physical layer—Physical channels and modulation, Release 8."
40. P. Lunden, J. Äijänen, K. Aho, and T. Ristaniemi, "Performance of VoIP over HSDPA in mobility scenarios," *Proceedings of the 67th IEEE Vehicular Technology Conference*, Singapore, May 2008.
41. A performance summary for the evolved 3G (E-UTRA); VoIP and best effort traffic scenarios.
42. ITU-T Recommendation G.114, "One way transmission time."
43. T. Henttonen, K. Aschan, J. Puttonen, N. Kolehmainen, P. Kela, M. Moisio, and J. Ojala, "Performance of VoIP with mobility in UTRA long term evolution," IEEE VTC Spring, Singapore, May 2008.
44. M. Rinne, K. Pajukoski et al., "Evaluation of recent advances of the Evolved 3G (E-UTRA) for the VoIP and best effort traffic scenarios," IEEE SPAWC, June 2007.

15

Emerging Methods for Voice Transport over MPLS

Junaid Ahmed Zubairi

CONTENTS

15.1 Introduction

This chapter describes the emerging methods and protocols for transport of voice over Multi Protocol Label Switching (MPLS) networks. Internet Engineering Task Force (IETF) and International Telecommunications Union-Telecommunication (ITU-T) protocols and mechanisms implement packet-based voice over the Internet. Voice over IP (VoIP) deployment has experienced exponential growth over the last four years and this pace of growth is expected to continue in the future. As MPLS provides effective Traffic Engineering (TE) and quality of service (QoS) mechanisms, it is the natural choice for carriers who wish to offer their customers toll-quality voice services. MPLS configures Label Switched Paths (LSPs) through the access and core networks and LSPs so that voice can be tagged with stringent service requirements to prioritize traffic. We look at the protocols and standards that deal with the voice deployment over MPLS networks, and the challenges and solutions that are being developed.

Voice deployment over packet networks has experienced tremendous growth over the last four years. The number of worldwide VoIP customers reached 38 million by the end of 2006, and is projected to grow to approximately 250 million by the end of 2011 [1]. The quad play (voice, video, wireless, and data) services are taking shape with MPLS as the core technology for integration. The standards and protocols for voice over MPLS have been developed by the IETF, IP/MPLS Forum (formerly MFA Forum/MPLS Forum), and the ITU-T.

In this chapter, we focus on the transport of voice over MPLS. There are some specific delay and loss limits for voice traffic. The recommended maximum one-one-way delay for interactive real-time voice transmission is 150 ms as per ITU-T recommendation G.114. The loss of some amount of voice packets is compensated using redundant data to recover lost content and silence intervals to emerge from problems. As the TCP/IP network cannot guarantee the delay bounds of service-aware traffic, new protocols have been developed that ensure the provisioning of required QoS using bandwidth allocation, traffic management, and fault handling. MPLS is the technology of choice for integration and provisioning of voice, video, data, and wireless services, and is poised to replace legacy technologies in the mobile backhaul networks [2] with MPLS Mobile BackHaul Initiative (MMBI). MPLS may be used to resolve interprovider technical and commercial issues with MPLS-Inter Carrier Connection (MPLS-ICI) specification as outlined in [3]. In the next section, we define the service requirements of voice traffic. Later, the key concepts of MPLS are described, and traffic treatment in an MPLS domain is discussed. MPLS serves as an enabling technology for QoS, and it works with Differentiated Services (DiffServ) for prioritized handling of the QoS traffic. DiffServ is introduced, and the interworking of the MPLS–DiffServ network is explained. ITU-T has defined Transport MPLS (T-MPLS) as a simplified version of MPLS for carrier transport networks. T-MPLS and IEEE Provider Backbone Transport (PBT) are discussed briefly. Next, we discuss the protocols and standards for voice over MPLS (VoMPLS) that evolved through carrier implementation agreements facilitated by IP/MPLS Forum. The last section sums up the VoMPLS deployment methods and related issues.

15.1.1 Voice Traffic Requirements

Real-time two-way interactive voice service places several demands on the network, including limits on packet loss, delay, jitter, and near zero call completion rate change [4]. Table 15.1 summarizes the key requirements for voice calls of acceptable quality over the network.

There are other secondary requirements that include protection against power failure, authentication, privacy security against intruders, and high availability for call placement. QoS can be defined as a combination of the upper limits of delay, loss, and jitter, and the level of protection against link and node failures. LSPs woven within the MPLS fabric can

TABLE 15.1

Key Voice Call Requirements

S. No.	Factor	Requirement
1.	Latency	Less than or equal to 150 ms (ITU-T G.114) 1-way
2.	Jitter	Generally minimized to values around ETE. Delay 1–50 ms
3.	Loss	Less than or equal to 3–5%
4.	Equipment failure	Fast recovery within 50 ms

be tagged with the required QoS by combining MPLS and DiffServ protocols. We describe the key protocols that meet the voice traffic QoS requirements including MPLS, DiffServ, T-MPLS, and PBT.

15.2 MPLS and DiffServ Protocols

With statistical multiplexing, the datagrams do not follow a fixed path and may arrive at the destination out of order. Moreover, under congested conditions, there is increased jitter. These problems make conventional IP networks largely unsuitable for connection-oriented applications such as interactive real-time voice. MPLS [5] has emerged as the key integration technology for carrying voice, video, and data traffic over the same network. In an MPLS-enabled network, LSPs are installed from an ingress node to an egress node prior to start of transmission,enabling connection-oriented applications to take advantage of the "virtual connections" set by MPLS. There may be several ingress–egress pairs in an MPLS domain, resulting in a complex management challenge. The most popular deployment of MPLS is the Virtual Private Network (VPN), a technology for establishing encrypted private tunnels across the Internet between clients and corporate networks [6].

Since the LSPs are stackable, traffic from different flows sharing common characteristics can be aggregated on an LSP. These characteristics may include common egress and identical QoS and protection requirements. "Stackable" refers to the fact that each packet is treated according to the topmost label in the label stack. Two LSPs can therefore be merged by pushing the same label on top of the label stack for packets belonging to both LSPs. Various LSPs may also be aggregated and segregated based on the loading and the status of the network links. For example, in case of link failure, the Fast Reroute (FRR) protection trigger would reroute the traffic around the faulty link, and merge it with the original path ahead of the downstream node of the faulty link. In this case, the downstream node would act as the Protection Merge LSR (PML) for the rerouted traffic, and would swap labels to ensure that the rerouted traffic would be handled transparently after the fault location. A few key definitions are summarized as in Table 15.2.

MPLS operates by defining a label inside the MPLS "shim header," which is placed on the packet between the layer-2 and layer-3 headers. The 32-bit shim header is organized as shown in Figure 15.1. It is referred to as "shim" header as MPLS allows layer-2 and layer-3 to fit together firmly.

The MPLS header fields include an unstructured 20-bit label that identifies the LSP to which the packet belongs. The label is assigned based on the forwarding equivalency class

TABLE 15.2

MPLS Key Definitions

S. No.	Definition	Explanation
1.	LER	Label Edge Router; Ingress or Egress router in an MPLS domain
2.	RSVP-TE	A signaling protocol to distribute labels for setting up an LSP
3.	CR-LDP	A signaling protocol to distribute labels for setting up an LSP
4.	LSR	Label Switched Router; core router inside MPLS domain that forwards packets based on labels
5.	LSP	Label Switched Path; established with LDP or RSVP signaling across a domain
6.	FEC	Forwarding Equivalence Class; a subset of packets that can be grouped together under the same label
7.	FRR	Fast recovery (within 50 ms) from localized node or link failures

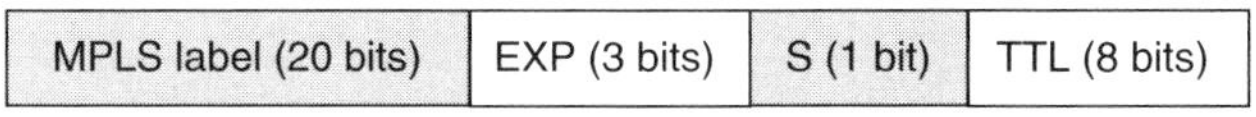

MPLS label (20 bits)	EXP (3 bits)	S (1 bit)	TTL (8 bits)

FIGURE 15.1 MPLS header fields.

(FEC) of the traffic trunk, and is followed by a 3 bits experimental (EXP) field and a 1 bit indicating whether it is the bottom of the label stack. The Time to Live (TTL) field has the same use as the TTL field in an IP header.

15.2.1 Constrained LSP Routing

The MPLS network runs constrained routing against traffic demands to find the most suitable LSP. Constrained routing involves computing the shortest path after pruning the network links that do not meet requirements that may include reliability, bandwidth availability, cost, and delay. Constrained routing requires extended interior gateway protocols (IGP) such as Open Shortest Path First–Traffic Engineering extensions (OSPF-TE) or Intermediate System to Intermediate System protocol–Traffic Engineering extensions (ISIS-TE). This capability to find routes other than the IGP shortest paths is known as MPLS Traffic Engineering (MPLS-TE) [7]. Optimum utilization of the whole network is gained as a result of the MPLS-TE achieving uniform distribution of traffic across it. Once an LSP is installed, its minimum bandwidth is guaranteed on original and protection paths, and the status of all the links in the network is updated to reflect the bandwidth booking by the new LSP. LSPs are installed using Label Distribution Protocol (LDP) or traffic engineering extensions of the resource reservation protocol (RSVP-TE). MPLS allows seamless integration with conventional IP networks because the MPLS header is removed by the egress router before forwarding the packet to the conventional IP network. The scalability offered by LSPs is unmatched even by Asynchronous Transfer Mode (ATM) networks as the ATM offers only two levels of hierarchy as opposed to an arbitrary number of levels in MPLS.

15.2.2 DiffServ Types of Service

In general, traffic can be categorized as elastic best effort and inelastic QoS traffic. Best effort traffic does not need any bandwidth, delay, and loss guarantees. On the other

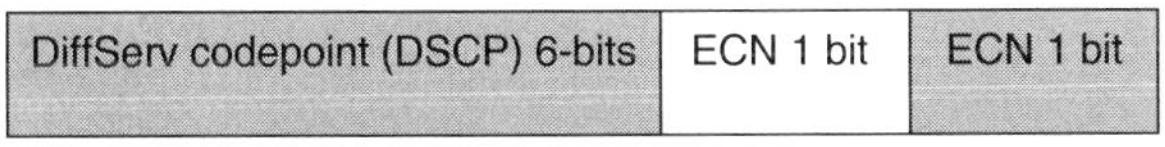

DiffServ codepoint (DSCP) 6-bits	ECN 1 bit	ECN 1 bit

FIGURE 15.2 DiffServ mapping in IPv4 ToS field.

hand, QoS-sensitive traffic such as VoIP has certain minimum bandwidth and delay requirements. Therefore, the core routers should offer appropriate services based on the type of traffic involved. Class based queuing (CBQ) can be invoked for routers that implement the DiffServ protocol to create separate queues for each class of traffic. DiffServ divides the traffic into Expedited Forwarding (EF) and Assured Forwarding (AF) Per-Hop Behavior (PHBs). EF is the premium service offered under DiffServ, and is suitable for low latency and low jitter flows that maintain almost constant rates. EF flows receive a minimum guaranteed bandwidth even under the most congested conditions. AF PHB is subdivided into four classes, each receiving a minimum bandwidth, and being subdivided into three drop preferences. The packets in microflows falling into one of the subclasses cannot be reordered where a microflow refers to individual user-application generated traffic. The router can choose to discard packets based on their drop preferences. Best effort traffic receives default forwarding (DF) behavior from the network with no bandwidth or delay guarantees. Each packet carries a PHB identifier that is used by the router to place the packet in an appropriate queue, and queues are individually serviced according to priority. DiffServ provides a mechanism to control the amount of traffic entering a domain per PHB, through the use of shaper, marker, and dropper entities operating within the ingress node. The core routers identify the packet PHBs and offer appropriate services accordingly. The Type of Service (ToS) fields in an IP packet header is modified to carry the DiffServ class information as shown in the Figure 15.2.

Differentiated Services Code Point (DSCP) is a field in the IP packet header for classifying the packet in one of the DiffServ PHBs. As seen in Figure 15.2, the DSCP extends to 6 bits, allowing a total of 64 PHBs to be defined. However, only 14 PHBs have actually been defined because of varying drop precedences, with one EF, one DF, and twelve AF subclasses. To indicate drop precedences, the DSCP field is subdivided into PHB Scheduling Class (PSC) and Drop precedence. The two explicit congestion notification (ECN) bits are for explicit congestion notification. With two ECN bits, only four values are possible for congestion notification. Values 10 and 01 are set by the sender to indicate ECN-capability. The value 00 indicates that the packet is not using the ECN mechanism, and the value 11 is used by a router to indicate congestion to the end nodes [8]. The DiffServ protocol follows the Internet model of keeping the network core simple. The core routers merely determine the DSCP and serve each packet according to the PHB to which it belongs. The complex functions of admission control, shaping, marking, and dropping are delegated to the edge routers.

15.2.3 MPLS-DiffServ Network

In an MPLS-DiffServ network, the routers jointly implement various MPLS and DiffServ functions. The ingress router is responsible for determining an LSP for a new flow request. The QoS requirements of the new flow can be translated into a DiffServ class assignment at the ingress, for which purpose the EXP field in the MPLS shim header is used. As the EXP field is 3 bits in length, it can represent only eight different scheduling

MPLS label (20 bits) (FEC destination)	EXP (DiffServ PSC) 3 bits	S (1 bit)	TTL (8 bits)

FIGURE 15.3 DiffServ PSC placement in MPLS E-LSP header.

and drop precedences. Under MPLS-DiffServ [9], two types of LSPs are defined. Explicit LSP (E-LSP) interprets the label field of the MPLS shim header as the egress identifier, and the EXP field as the DiffServ PSC combined with the drop precedence. On the other hand, Label only inferred PSC LSP (L-LSP) interprets the label field as the DiffServ scheduling priority and the destination. The EXP field in L-LSP is used to indicate only the drop precedence of the packet. The main difference between E-LSP and L-LSP is the aggregation feature in E-LSP resulting in scalability; however, some PSCs in an E-LSP may suffer because the bandwidth is reserved for the entire LSP [10].

When the transmission starts, the packets belonging to an assigned flow are marked according to the DiffServ PSC and labeled as according to the LSP installed. When a core router receives a packet belonging to this LSP, the label of the LSP is swapped based on the preinstalled label table, and the packet is placed in an appropriate outgoing queue depending on the DiffServ PSC identifier found in the packet. Figure 15.3 shows the placement of the DiffServ class identifier and MPLS label for an E-LSP.

Priority flows must be protected against link and node failures. In an MPLS domain, protection schemes are implemented to provide backup paths. The next section discusses some MPLS protection schemes. Security and portability are other requirements that can be ensured by using MPLS-based VPN tunnels.

In summary, MPLS-DiffServ networks offer the following services that are of particular importance to VoIP and VoMPLS:

- Advance reservation of bandwidth through installation of LSPs
- Aggregation (Multiplexing) of similar flows by the use of stackable LSPs
- Priority handling through the use of DiffServ identifiers and CBQ on routers
- Fast and efficient fault handling by the use of MPLS-based protection schemes
- Security and portability through the use of MPLS-based VPN tunnels

15.3 T-MPLS Protocol

Two new technologies were recently introduced to provide QoS in packet switched networks. The first is the T-MPLS (ITU-T G.8110.1) protocol, which implements a simplified and stripped down MPLS architecture for carrier-class networks. The second technology is known as PBT (Provider Backbone Transport) or PBB-TE (Provider Backbone Bridging—Traffic Engineering). PBT extends the proposed IEEE 802.1ah Provider Backbone Bridges standard, and addresses the problems involving traditional Ethernet. It enlarges the Virtual LAN (VLAN) address space and includes a new service tag to indicate end-to-end connections. The two approaches attempt to combine the traffic engineering capabilities of Synchronous Optical Network/Synchronous Digital Hierarchy (SONET/SDH) and layer 2 network designs [11]. Both T-MPLS and PBB provide point-to-point primary and backup tunnels with 50ms protection switching, thus fulfilling the need to migrate from SONET/SDH transport to a network that is fully packet switched, with comparable reliability [12].

T-MPLS was defined in 2006 to support connection-oriented streams such as VoIP. The standard documents approved were G.8110.1 Architecture of T-MPLS layer network, G.8112 Interfaces for the T-MPLS hierarchy (TMH), and G.8121 Characteristics of T-MPLS equipment. T-MPLS is described as a connection-oriented packet transport technology that uses the MPLS label swapping and forwarding, carrier networks performance monitoring, simplified protection switching, and Automatic Switched Optical Network/ Generalized MPLS (ASON/GMPLS) control and management functions [13]. It involves setting up unidirectional and bidirectional Point-to-Point (P2P) connections that take identical network paths in both directions and receive TE support and management. T-MPLS assumes that the end-to-end Operation, Administration and Management (OAM) can be greatly simplified by removing the IP-specific functionality. For example, T-MPLS removes the Penultimate Hop Popping (PHP) feature that would require IP processing of the packet at the end router, and the Equal Cost Multi Path routing (ECMP) feature is similarly eliminated in order to avoid possible source identification confusion. The T-MPLS tunnel can carry multiple L2 and L3 services including IP/MPLS LSPs and Pseudo Wire Emulation Edge to Edge (PWE3) pseudowires. Companies that deploy T-MPLS are expected to find it easier and more economical in terms of man-hours of staff training, as compared with IP/ MPLS networks. This is because of the fact that the architecture of the T-MPLS network is similar to the SONET/SDH network.

Figure 15.4 shows a T-MPLS network configured with primary and backup LSPs, which provides connection-based services. A fish topology diagram shows T-MPLS nodes connected with IP/MPLS nodes that send their IP/MPLS tunnels across the T-MPLS network. Besides IP/MPLS tunnels, T-MPLS networks serve PWE3 and other L2/L3 services.

Early in 2008, ITU-T and IETF deliberated to resolve the inconsistencies between MPLS and T-MPLS, and a consensus is emerging to define separate code points for MPLS and

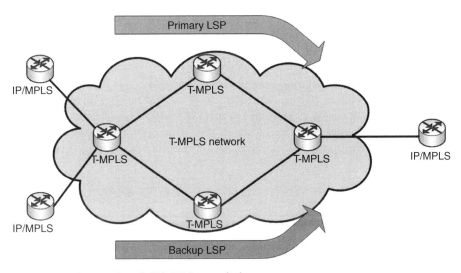

FIGURE 15.4 T-MPLS network with IP/MPLS router links.

T-MPLS in order to avoid confusion. IETF and ITU-T agree that MPLS and T-MPLS are disjoint networks. An LSP initiated from either network would encapsulate into Ethernet before transiting to the other network. Client support in T-MPLS is based on the IETF Pseudo wire model for layer-2 VPN. The two-layer architecture includes top layers as client Virtual Circuit LSPs and bottom layers for aggregating the VC LSPs into Trunk LSPs. The key differences between MPLS and T-MPLS include the use of bidirectional LSPs in T-MPLS and the absence of: PHP, merging of LSPs and ECMP routing in T-MPLS [14].

15.4 Voice Transport Mechanisms Over MPLS

Voice is carried on MPLS network either directly or with VoIP stack. VoIP is implemented with Real Time Protocol (RTP). Voice segments are packetized with RTP headers and placed in UDP segments for onward transmission through the IP/MPLS network. On the other hand, VoMPLS [15,16] removes the overhead associated with the protocol stack of VoIP and carries voice directly over MPLS. Figure 15.5 shows the comparison of protocol stacks in different schemes. VoATM can also be carried in networks where ATM is implemented over MPLS. In such cases, various ATM Adaptation Layer (AAL) services can be configured over the network. VoIP results in large headers but has the clear advantage of providing end-to-end connectivity with mature protocols. The MPLS user-to-network interface (UNI) definition for bringing users in direct contact with MPLS networks was stated recently [17–19]. Using MFA IA 1.0 and IA 5.0, voice can be directly encapsulated in MPLS packets. The other option is to use the ATM over MPLS definition to enable VoATM voice service.

Both VoIP and VoMPLS, require the voice to be converted into digital format using any of the several proposed codecs. Signaling is also needed to establish, maintain, and terminate a call. Transporting the digitized voice requires special protocols that can handle the real-time characteristics of the speech traffic. Next we discuss the VoIP over MPLS and VoMPLS designs and issues.

15.4.1 VoIPoMPLS and VoIP

VoIPoMPLS involves encapsulating the coded voice bytes into the RTP packets and transmitting over the UDP/IP/MPLS protocol stack. The combined count of header bytes for RTP, UDP, and IP is 40 bytes. Figure 15.6 shows the voice data together with headers of L3 and above. For MPLS VPN transporting voice, this overhead reaches 48 bytes and exceeds the payload size, making it very important to multiplex different voice packets and to compress the header. Voice packets that belong to the same connection can be multiplexed together in one MPLS frame. However, this option needs to be exercised carefully, given

VOIP	MFA IA 1.0	MFA IA 5.0	AAL1, AAL2, AAL3
RTP	MPLS	MPLS	ATM
UDP			MPLS
IP			
MPLS			

FIGURE 15.5   VoIPoMPLS and VoMPLS protocol stacks.

IP header (20 bytes)	UDP header (8 bytes)	RTP header (12 bytes) plus optional RTP mux	Voice data (40 bytes for G.711 at 64 kbps)

FIGURE 15.6 L3 and above uncompressed headers for VoIPoMPLS.

the delay budget of VoIP. Various header compression techniques have been proposed, including compressed RTP, enhanced compressed RTP, and robust header compression. Composite IP (CIP) and lightweight IP encapsulation (LIPE) also concatenate voice packets together [20]. With RFC 2508 [21] compression of RTP/UDP/IP headers, the header length reduces to 2–4 bytes. Other problems in VoIP include jitter removal, silence suppression, echo cancellation, serialization delay, and loss recovery.

15.4.2 VoMPLS

VoMPLS refers to the scenario in which the voice packets are directly transported over the MPLS without the use of the RTP/UDP/IP protocol stack. As seen earlier in Figure 15.5, the VoMPLS protocol stack is more compact as compared with VoIP. The main requirement for customer edge deployment of VoMPLS is the User Network Interface (UNI) between the MPLS network and the customer. The MPLS UNI definition [18] allows users at Customer edge (CE) to establish LSPs to the Provider Edge (PE) network. It greatly simplifies the provisioning of end-to-end QoS. The LSP may be a single link or a layer-2 network but is considered a single hop LSP. As there may be thousands of CE-to-CE connections, there arises a potential scalability issue. The MPLS proxy admission control [19] solves this problem by letting the PE ingress router search for existing tunnels that can satisfy the traffic parameters Type Length Value (TLV) in the label request message sent by the CE. Once a suitable LSP is found, its Resource Index Label (RIL) is returned to the CE, that then uses it to encapsulate the traffic for that tunnel. The PE may police the traffic for a given RIL and may discard the packets with wrong IP header information and excess traffic. One important point is that the QoS is guaranteed only within the PE-to-PE network. For CE-to-PE and remote PE-to-CE, customers have to deploy layer 2 QoS techniques for provisioning of resources.

15.4.3 MPLS UNI with Proxy Admission Control

In voice deployment over MPLS, the Voice Gateway (VGW) acts as the CE requesting guaranteed LSPs to the remote VGW. The MPLS proxy admission control allows the VGWs to request LSPs dynamically, to remote VGWs, and to share the multiplexed LSP-TE tunnels among themselves through the local PE acting as a proxy for admission control [22]. With the VGW as an example as in Figure 15.7, the following exchange of control messages may occur between the VGW and PE before a call can be established:

1. Phone 1 requests a voice call to Phone 2.
2. VGW sends a label request message to the PE containing Forwarding Equivalence Class (FEC) and Traffic Parameters TLV for the intended call.
3. PE acts as a proxy for the local VGW and searches existing TE tunnels to remote VGWs that satisfy the traffic parameters and FECs in the label request message.

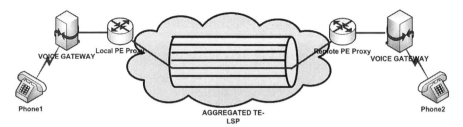

FIGURE 15.7 Voice calls through MPLS UNI and Proxy Admission Control.

4. If an existing tunnel is found that can satisfy the request, PE sends its RIL to the requesting VGW. If not, a new tunnel is created and its RIL is returned.

5. The requesting VGW can now initiate the call and release the resources once the call is terminated.

It is thus obvious that the MPLS UNI together with MPLS Proxy admission control provides a powerful mechanism to deploy thousands of voice calls through the PE network without control messaging overflow.

15.4.4 Bandwidth Management and Allocation

The network bandwidth management and allocation is of primary concern to providers who treat the bandwidth as a commodity to be managed efficiently to maximize profit. The DiffServ Aware Traffic Engineering (DS-TE) proposal [23] maps the DiffServ ToS values into Class Type (CT). A maximum of eight CTs are defined for a network, and they can be combined with preemption priority values to define TE classes. Preemption can be made possible across CTs, within CTs, and combined within-across CTs. Each LSP request is associated with a CT and subjected to bandwidth allocation constraints. Three bandwidth allocation models have been proposed:

- Under Maximum Allocation Model (MAM) [24], each CT is assigned a maximum bandwidth. LSPs requesting bandwidth under a specific CT are subject to the maximum bandwidth allocated to that CT. If the bandwidth of a CT is fully booked, LSPs may be denied admission even though other CTs may have bandwidth available.

- Under Russian Dolls Model (RDM) [25], an LSP requesting bandwidth can book bandwidth available in its own, as well as other, CTs. The booking of extra bandwidth beyond its own CT is subject to two conditions. First, the LSPs own CT bandwidth should be fully booked, and second, it should have access to only the bandwidth left over in the higher CTs. For example, in the absence of higher priority flows, a lower priority LSP can use all the available bandwidth in the higher CTs. Preemption is enforced to make sure higher priority flows can get their share of bandwidth whenever needed.

- Using Maximum Allocation With Reservation Model (MAR) [26], the MAM allocation is continued in all CTs until the total bandwidth booked reaches a preset threshold value, after which only the CTs that have not surpassed their limits can reserve further bandwidth.

15.4.5 VoMPLS Header Formats

In VoMPLS, various header formats are defined for different Payload Types. Payloads may include encoded audio information, dialed digits, silence descriptors, and control signaling. Two types of frames are defined in VoMPLS IA 1.0—primary and control frames. The VoMPLS primary frame header is shown [15] in Figure 15.8. Many primary VoMPLS frames may be multiplexed within a single MPLS packet.

VoMPLS primary frame header is 4 bytes long and includes five fields:

- The channel Identifier uniquely identifies the voice channel that is the source of the payload, so that a total of 248 different voice calls can be multiplexed into a single LSP.
- Payload Type identifies the encoding scheme being used, as well as silence removal/insertion descriptors. A value equal to, or above, 224 indicates a control payload (part of the control frame) that would allow dual tone multifrequency (DTMF) dialed digits as well as signaling for the channel to be carried [27].
- The revolving Counter field is set at the first sample or frame and keeps incrementing for each additional frame.
- Payload Length is read in conjunction with the pad length to keep the payload a multiple of 4 bytes.

Multiple primary frames may be multiplexed with the use of optional inner MPLS labels in addition to one mandatory outer MPLS label. The multiplexing arrangement is shown in Figure 15.9. Using CID, up to 248 voice calls can be aggregated with this arrangement. The payload of primary frames may consist of encoded audio data or Silence Insertion Descriptor (SID) parameters for support of Voice Activity Detection (VAD) and Comfort Noise Generation (CNG) during silence intervals.

Keeping the frames and header formats in perspective, the voice calls are established as follows [27].

1. Using MPLS signaling mechanism, a bidirectional LSP is created.
2. As voice call request arrives at the edge of the MPLS network, an existing CID is allocated to the new call within the LSP just created, or CID signaling is done to establish the new channel.

Channel identifier (8 bits)	Payload type (8 bits)	Counter (8 bits)	Payload length (6 bits)	Pad length (2 bits)

FIGURE 15.8 VoMPLS (MFA 1.0) primary frame header fields.

Outer label (mandatory)	Inner label optional)	Primary frame-1	Primary frame-2	----	Primary frame-N

FIGURE 15.9 VoMPLS multiplexing format.

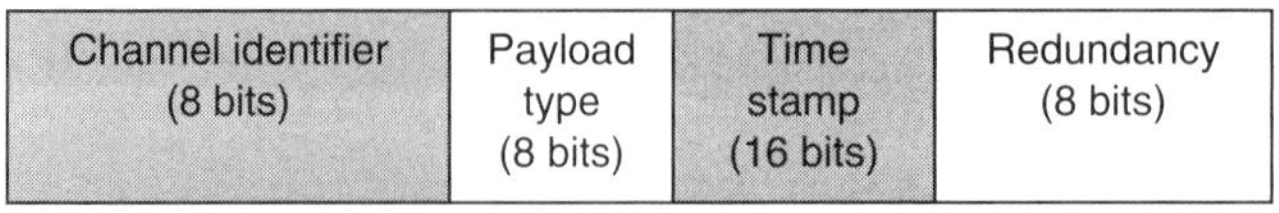

Channel identifier (8 bits)	Payload type (8 bits)	Time stamp (16 bits)	Redundancy (8 bits)

FIGURE 15.10 VoMPLS control frame header format.

3. Optionally, inner LSPs is created within the outer LSP, in which case the outer LSP label, inner LSP, label and CID uniquely identify the voice call.

The VoMPLS control frame header is also 4 bytes in length. The fields in the header are shown in Figure 15.10. Control fames cannot be multiplexed, and must be carried separately, however, the mandatory outer label and the optional inner label precede the control frame header fields in order to identify uniquely the voice call for which the control frame is sent.

The Payload Type field distinguishes between dialed digits and the channel related signaling. Time Stamp is relative to the first randomized time stamp. The Redundancy field is very important to ensure the receipt of the control frames. If the Redundancy field is set to 0, 1, or 2, the control packet is repeated the equivalent times.

Another solution for VoMPLS reuses ATM Adaptation Layer 2 (AAL2) components, but replaces ATM by MPLS [16,17], thus eliminating the ATM cell overhead. The MFA 5.0 implementation agreement proposes voice trunking over MPLS by directly encapsulating AAL2 common part sublayer (CPS) packets into MPLS (A2oMPLS). The gateway to the MPLS network should be able to function as an AAL2 switch. Multiple A2oMPLS connections may be multiplexed into a single LSP, and one MPLS frame may carry multiple CPS packets. The A2oMPLS subframes may be of different lengths, but the maximum CPS packet payload length is restricted to between 45 octets and 64 octets. Like the MFA IA1.0, the CID field in the CPS packet header allows up to 248 A2oMPLS connections to be multiplexed. Each A2oMPLS connection can be uniquely identified with the outer label of the LSP, optional inner label, and the CID value. When inner labels are used, the number of calls that are carried by a single LSP increases rapidly. The detailed header fields are shown in Figure 15.11.

The following fields are included in the A2oMPLS header:

• Reserved (10 bits) currently ignored.
• Length (6 bits) used for padding length. It is set to 0 if A2oMPLS packet length exceeds 64 bytes.
• Sequence Number (S No) (16 bits) used if guaranteed ordered packet delivery is required. Sequence number of 0 indicates otherwise.

The following fields are inside the CPS packet header; several CPS packets may be packed in one MPLS frame.

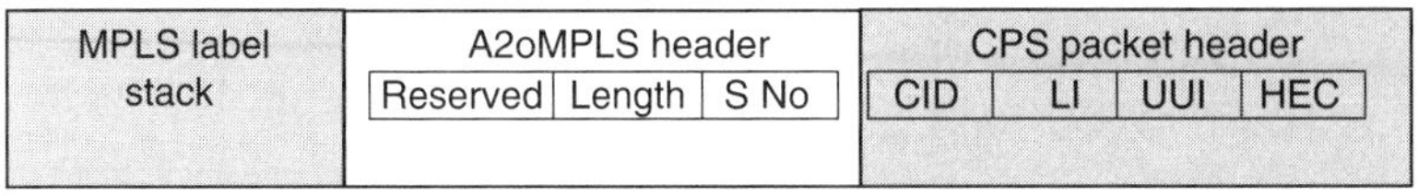

MPLS label stack	A2oMPLS header			CPS packet header			
	Reserved	Length	S No	CID	LI	UUI	HEC

FIGURE 15.11 A2oMPLS and CPS header in MPLS frame.

- THE CID (8 bits) identifies the A2oMPLS connection carried. Thus, a total of 248 connections (8 to 255) can be multiplexed into a single LSP.
- LI (6 bits) identifies the length of the CPS packet.
- UUI (5 bits) is used for user-to-user indication (0 to 27 for users, 30–31 for layer management, and above 31 reserved).
- Header error control (HEC) (5 bits) uses CRC checksum. However, this field may not be used in the MPLS environment.

The A2oMPLS transmitter operates in IDLE or PART states. In the idle state, the MPLS frame is empty, changing to PART as soon as a CPS packet is delivered for transmission, and the combined-use timer starts running. The MPLS frame is transmitted either on expiry of the timer or on reaching the maximum permissible length.

15.5 Conclusion

The use of packet based networks for transport of voice is increasing at a very fast pace. Considering the QoS nature of voice service, MPLS is the most preferred platform for the deployment of voice. MPLS can guarantee the bandwidth needed for a voice call and provide fast-protection switching using traffic engineering principles and methods. Voice can be deployed over MPLS using a variety of recently developed techniques, which. we have discussed here namely, VoIPoMPLS and VoMPLS including MFA IA 1.0, MAF IA 5.0.0, MFA IA 6.0.0, and MFA IA 7.0.0.

VoIPoMPLS deploys voice over the RTP/UDP/IP protocol stack, which uses MPLS tunnels. Since a large number of protocols is involved, the control information equals or exceeds the payload. With header compression, a reduction in bandwidth demand can be achieved. The most efficient method, however, is to run voice directly over MPLS. We have discussed VoMPLS in detail, including various recommended techniques to enable it in the network. VoMPLS is well suited for multiplexing in the core and carrying the bulk of voice calls through the MPLS domain. In fact, VoMPLS can be efficiently carried over the transport connections configured in the domain of T-MPLS, a newly developed ITU-T protocol. We have also explained the difference between MPLS and T-MPLS, and the advantages of using the latter.

The main obstacle in carrying voice directly over MPLS has been the lack of a UNI for MPLS domains. We have discussed the MPLS UNI proposal that was developed in 2004. Since the original proposal was not scalable, proxy admission control was introduced to pack new LSPs into existing tunnels. This has been explained in detail with an example.

The next-generation interactive telephony is based on MPLS. In future, many technologies will be integrated to allow exciting new features in VoP phones. For example, transparent connectivity through landline and mobile networks could let the user continue the conversation while changing the connection from landline network to mobile network, or vice versa. VoP phones will be flexible, allowing all the functionalities of current PSTN phones. The most significant new functionality will include multiway audio and text conference calling, videoconferencing, and several options for security, encryption, and authentication.

Glossary

IETF: Internet Engineering Task Force is an open volunteer organization that develops Internet protocols and standards.

ITU-T: International Telecommunications Union-Telecommunication Standardization Sector is a body that produces telecommunication standards for ITU.

IP/MPLS Forum (Formerly MFA Forum): An international association of telecom and networking companies advancing deployment of multiservice packet-based networks.

MPLS: Multiprotocol label switching is a mechanism to implement connection oriented and regulated services over the connectionless TCP/IP-based Internet. IETF and IP/MPLS Forum are responsible for advancing MPLS standards.

QoS: Quality of service is the probability of meeting the goals of reliability, timeliness, and delivery of packets in a network.

VoIP: Voice over IP is the routing of voice phone calls over the Internet fully or partly.

VoMPLS: Voice directly over MPLS is the routing of voice calls over MPLS instead of voice over IP over MPLS.

VoP: Voice over Packet is a general term that covers VoIP, VoIPoMPLS, and VoMPLS.

VPN: Virtual private network is a private network configured over the public Internet. It allows remote users to connect to their corporate network as if they were local users.

References

1. Teral, "Service provider and enterprise IP telephony markets," Infonetics Research, 30 August, 2006. http://www.infonetics.com/resources/purple.shtml?ms06.ngv.pbx.2q06.nr.shtml.
2. IP/MPLS 2-2008, White Paper, "Use of MPLS technology in mobile backhaul networks," February 2008.
3. IP/MPLS 1-2008, White Paper, "Addressing inter provider connections with MPLS-ICI," January 2008.
4. MFA, "MPLS ready to serve the enterprise," MPLS Frame Relay Alliance Forum White Paper, 2004, www.mfaforum.org/tech/superdemo_2004.pdf (accessed January 14, 2006).
5. E. Rosen et al., "Multiprotocol label switching architecture," IETF, RFC 3031, 2001.
6. L. Paulson, "Using MPLS to unify multiple network types," *IEEE Computer*, May 2004.
7. D. Awduche et al., "Requirements for traffic engineering over MPLS," IETF, RFC 2702, 1999.
8. Ramakrishnan et al., "The addition of explicit congestion notification (ECN) to IP," RFC 3168, 2001.
9. F. Le Faucheur et al., "MPLS support of differentiated services," RFC 3270, 2002.
10. V. Fineberg, "QoS support in MPLS networks," MFA Forum White Paper, 2003.
11. P. Lunk, "Traffic engineering for Ethernet: PBT vs. T-MPLS," *LightWave*, 2008, Online article, http://lw.pennnet.com/display_article/291600/13/ARTCL/none/none/Traffic-engineering-for-Ethernet:-PBT-vs-T-MPLS/ (accessed March 26, 2008).
12. D. Barry, "T-MPLS and PBT/PBB-TE offer connection-oriented packet transport," Lightwave, May 2007, http://lw.pennnet.com.
13. ITU-T, "Draft, New Amendment 1 to G.8110.1/Y.1370.1," November 2006.
14. Lum, "When networks collide: Putting the T into MPLS," 2006, FiberSystems.org, Online article, http://fibresystems.org/cws/article/news/25915 (accessed March 30, 2008).
15. MFA, "Voice over MPLS: Bearer transport implementation agreement (MFA IA 1.0)," MFA Forum, 2001.

16. MFA, "Voice trunking format over MPLS," MPLS/FR 5.0.0, I.366.2, MFA Forum, 2003.
17. ITU-T, Recommendation Y.1261, "Service requirements and architecture for voice services over MPLS," 2002.
18. D. Sinicrope and A. Malis, "MPLS PVC user to network interface," MFA, MPLS/FR Alliance 2.0.1, May 2003.
19. MFA, "MPLS proxy admission control protocol implementation agreement 7.0.0."
20. E. Vázquez, "Network convergence over MPLS," *Lecture Notes in Computer Science*, vol. 3079: High Speed Networks and Multimedia Communications, Springer Verlag, 2004.
21. S. Casner et al., "Compressing IP/UDP/RTP headers for low-speed serial links," IETF, RFC 2508, 1999.
22. V. Fineberg (with D. Sinicrope, T. Phelan, R. Sherwin, and D. Garbin), "The MPLS UNI and end-to-end QoS," *Business Communications Review*, pp. 27–32.
23. F. Faucheur and W. Lai, "Requirements for support of differentiated services," IETF, RFC 3564, July 2003.
24. Faucheur and Lai, "Maximum allocation bandwidth constraints model for Diffserv-aware MPLS traffic engineering," IETF, RFC 4125, 2005.
25. Faucheur, "Russian dolls bandwidth constraints model for Diffserv-aware MPLS traffic engineering," IETF, RCF 4127.
26. Ash, "Max allocation with reservation bandwidth constraints model for Diffserv-aware MPLS traffic engineering and performance comparisons," IETF, RCF 4126, 2005.
27. E. Fjellskål and S. Solberg, "Evaluation of Voice over MPLS (VoMPLS) compared to Voice over IP (VoIP)," Master's thesis, Agder University College, 2002.

Part III

Applications

16

Implementation of VoIP at the University of Colima

Pedro García Alcaraz and Raymundo Buenrosto Mariscal

CONTENTS

16.1 Introduction

This chapter describes a valuable experience at the University of Colima (a public academic institution), which arose from the need to provide the University's professors and researchers with telephone communication services, making full use of the existing infrastructure, while ensuring mobility and an adequate cost–benefit ratio.

16.1.1 Background

The Directorate of Telematic Services (DIGESET, in its Spanish acronym) of the University of Colima is responsible for implementing and managing telematic services within the University network. DIGESET was created in 1990 as the Network and Communications Department, within the Directorate of Academic Exchange and Library Development. It helped select, install, and maintain the first computer network at the University of Colima [1].

The significant progress made in the realm of telecommunications in recent years has shaped the way people communicate. The evolution of telephone networks into digital systems has permitted the emergence of a large number of services and facilities that enable institutions to work far more efficiently.

Voice over IP (VoIP) is a technology that makes it possible to transmit voice over IP networks in the form of data packets. IP telephony is an immediate application of this technology, in such a way that it makes it possible to make conventional telephone calls over IP networks with the use of a PC and standard telephones. In general terms, communication services are transmitted via IP networks, usually the Internet, instead of using the conventional telephone network.

16.1.2 Problem Statement

How could the voice communication service be improved at the University of Colima while offering an alternative communication scheme and enabling an IP telephony client to become a telephone extension?

Extending the services of the telephone network at the University, while simultaneously enabling users to make telephone calls with the use of an IP telephony device to dial any number of the conventional telephone network (public switched telephone network, PSTN) and vice versa, provides a better communication service at a lower cost with the implementation of an IP telephony scheme.

16.1.3 Current Situation

The University of Colima has had a major impact on society by achieving significant growth, extending its services, and building more facilities to meet the natural demands of the student community and staff. The University commissioned DIGESET so that it could remain in the forefront of telecommunication services.

The telephone network at the University of Colima has the equipment shown in Table 16.1 and Figure 16.1.

16.1.4 Justification

There is more than one factor leading us to the implementation and use of IP telephony. The main reason is the need to communicate in real time at a low cost.

The cost of using conventional telephone lines of telephone corporations, and of long-distance calls in particular, has resulted in the innovation of new technologies that provide these facilities more economically. In addition to the benefits of low cost, VoIP has important technical advantages over circuit commutation.

VoIP networks are based on open architecture. This means that VoIP services are more interchangeable and more modular. In addition, most institutions have an infrastructure of networks and telecommunications of which full use could be made.

TABLE 16.1

Equipment of the Telephone Network

Equipment	Location	Capacity Installed
Colima Campus		
M3 4400 Alcatel switchboard	Main switchboard of the entire telephone network. Located at DIGESET	
M2 Alcatel switchboard	Located at the Library of Science	866 extensions
WM1 Alcatel switchboard	Located at the Medical Library	
WM1 Alcatel switchboard	Located at the Library of Social Science	
WM1 Alcatel switchboard	Located at the University's Law School	25 extensions
Villa de Álvarez Campus		
WM1 Alcatel switchboard	Located at the Villa de Álvarez Library	92 extensions
Coquimatlán Campus		
M2 Alcatel switchboard	Located at an independent building	86 extensions
Manzanillo Campus		
M2 Alcatel switchboard		127 extensions
Common OXE hardware	Located at the external marketer	
Tecomán Campus		
M2 Alcatel switchboard	Located at the campus library	
WM1 Alcatel switchboard	Located at the Office of the Head of the Tecomán Campus	100 extensions
Common OXE hardware	Located at the University's High School 20	

The convergence of voice and data networks allows for the creation of new applications such as unified messaging, thus enabling users to communicate (fax, mail, telephone, etc.) with the use of the same user interface, videoconferencing, and teleteaching applications, which would be technically difficult to implement over separate networks.

This project provides the users in the University with an alternative communication solution and reduces costs by implementing an IP telephony scheme that uses the existing network platform.

16.1.5 General Objective

To analyze and implement an alternative telephone communication scheme by means of an IP technology that makes it possible to reduce installation and communication costs.

16.1.6 Specific Objectives

- To offer an economical alternative by using open standards and free source codes
- To reduce the cost of telephone calls by using the IP protocol
- To provide users of the telephone network with mobility
- To combine voice and data services in a network platform

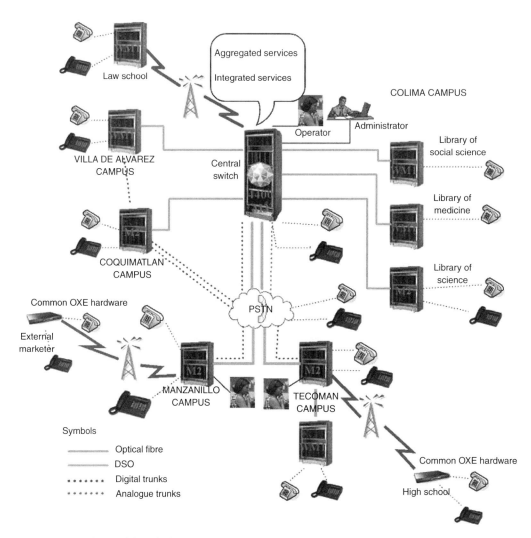

FIGURE 16.1   Scheme of the telephone network.

16.2 Life Cycle of IP Telephony

A number of processes are necessary to launch any project. In the case of the life cycle of IP telephony, there are four processes: planning and assessment, predeployment testing, ongoing operations, and optimization, as can be seen in Figure 16.2 [2]. The definitions of these stages are explained here, along with the experiences encountered at each stage.

16.2.1 Planning and Assessment

This stage determines whether the existing infrastructure of the IP network meets the requirements of the implementation of the VoIP service, one of which may be bandwidth.

Bad planning may increase the cost of launching the project. During this stage, the scope of the research is also known. Voice reliability and security must be examined.

Our planning involves the following:

- Implementation of IP telephony using the infrastructure of the existing IP network
- Basing the scheme of the project on open standards in order to reduce licensing and implementation costs.

16.2.2 Predeployment Testing and Implementation

If the tests performed at the stage of planning and assessment are accepted, we proceed to the implementation of the system. In the implementation of the project was used as a communication protocol, SIP (Session Initiation Protocol), as PBX VoIP server, the PBX Asterisk and as softphone, the SJPhone (made by SJ Lab, Inc.) installed on Personal Digital Assistant (PDAs) and Xlite (made by CounterPath Corporation) installed on computers. There is only a brief definition of the SIP protocol, because this chapter is not aimed at discussing it in depth. We also explain what the Asterisk server and the clients are.

a) **SIP** is a text-based protocol, similar to HTTP and SMTP, for initiating interactive communication sessions between users [3]. SIP has emerged as a key protocol with strong industry support for the deployment of IP-based telephony. In addition to the rich media session and information that it can convey, SIP offers the following additional benefits [4]:

- Converged network
- Mobility
- Enhanced audio quality
- Integrated presence

b) **The Asterisk server** is a PBX software (private branch exchange or private business exchange) that uses the free software concept (GPL, General Public License). Asterisk runs on a Linux platform and on other Unix platforms with or without hardware connected to the PSTN [5].

Asterisk allows for real-time connectivity between the PSTN and the VoIP networks. The benefits and advantages of Asterisk are listed here.

- Interoperability with conventional telephone systems and VoIP systems
- The possibility of increasing the number of functions, users, and lines without having to change the PBX
- The possibility of interconnecting several offices through the Internet
- Costs that are lower than those of a conventional PBX
- Constant evolution.

Asterisk has a large number of features [6], some of which are listed below.

Call Features

ADSI On-Screen Menu System	Call Detail Records
Alarm Receiver	Call Forward on Busy
Append Message	Call Forward on No Answer
Authentication	Call Forward Variable

Automated Attendant
Blacklists
Blind Transfer
Call Recording
Call Routing (DID & ANI)
Call Transfer
Caller ID
Caller ID on Call Waiting
Calling Cards
Database Integration
Direct Inward System Access
Distributed Universal Number Discovery (DUNDi™)
E911
Fax Transmit and Receive (3rd Party OSS Package)
Interactive Directory Listing
Local and Remote Call Agents
Music On Hold
Music On Transfer:
- Flexible MP3-based System
- Random or Linear Play
- Volume Control
Roaming Extensions
SMS Messaging
Voicemail:
- Visual Indicator for Message Waiting
- Stutter Dialtone for Message Waiting
- Voicemail to email
- Voicemail Groups
- Web Voicemail Interface
Streaming Media Access
Supervised Transfer

Call Monitoring
Call Parking
Call Queuing
Call Retrieval
Call Snooping
Call Waiting
Caller ID Blocking
Conference Bridging
Database Store / Retrieve
Dial by Name
Distinctive Ring
Do Not Disturb Predictive Dialer
ENUM
Flexible Extension Logic Privacy
Interactive Voice Response (IVR)
Macros
Open Settlement Protocol (OSP)
Overhead Paging
Protocol Conversion
Remote Call Pickup
Remote Office Support
Route by Caller ID
Time and Date
Transcoding
Spell/Say
Talk Detection
Three-way Calling
Text-to-Speech (via Festival)
VoIP Gateways
Trunking
Zapateller

Codecs

ADPCM
 G.711 (A-Law & μ-Law)
G.722
G.723.1 (pass through)
G.726
LPC-10

G.729 (through purchase of a commercial
 license)
GSM
iLBC
Linear
Speex

Protocols

IAX™ (Inter-Asterisk Exchange)
H.323

SIP (Session Initiation Protocol)
MGCP (Media Gateway Control Protocol)
SCCP (Cisco® Skinny®)

The 1.4.2 Asterisk version has been installed on the Centos 5.0 operating system on a 2333 GHz IBM Blade HS21 Xeon server with a 2 GB RAM.

Asterisk is controlled by configuration files in the directory /etc/asterisk. There are a considerable number of files. In this chapter, we show the configuration of two only.

sip.conf: an important file for the chan_sip channel to work. It enables Asterisk to interact with other SIP devices, including other Asterisk PBXs.

```
[general]                    ; General options
bindport = 5060              ; Port for the SIP server
Bindaddr = 0.0.0.0           ; The IP address to be used or all those existing
context = entrantes          ; Default context for all incoming calls
disallow = all               ; We first deny the use of all codecs.
```

```
allow = ulaw              ; We add the use of codecs in order of preference.
canreinvite = no
[alcatel]
type = peer
host = X.X.X.X
dtmfmode = rfc2833
disallow = all
allow = ulaw
canreinvite = no


#include /etc/asterisk/sip_usuarios.conf
```

The extensions of the SIP devices have been set up in the file sip_usuarios.conf.

```
[54001]
type = friend
dtmfmode = rfc2833
disallow = all
allow = gsm
host = dynamic
secret = 820
context = categoria5
callerid = "Pedro Garcia A." <54001>
canreinvite = no
```

extensions.conf: this is the file where call-routing decisions are made.
```
[entrantes]
include => extensiones
include => error-alcatel


[extensiones]
exten => 54000,1,Dial(SIP/54000,90,Ttr)
exten => 54001,1,Dial(SIP/54001,90,Ttr)
⋮
exten => 5400n,1,Dial(SIP/5400n,90,Ttr)
[error]
exten => _X.,1,Background(invalid)
exten => _X.,2,Hangup


[error-alcatel]
exten => _X.,1,Background(error-custom)
exten => _X.,2,waitexten(20)
exten => _X.,3,Hangup


[categoria1]
include => extensiones
include => exten-alcatel
include => error
```

```
[categoria7]
include => extensiones
include => exten-alcatel
include => locales
include => cel-local
include => cel-fuera
include => nacionales
include => usa-canada
include => resto-mundo
include => error

[exten-alcatel]
exten => _3XXXX,1,Dial(SIP/${EXTEN}@alcatel,90,Ttr)
exten => _4XXXX,1,Dial(SIP/${EXTEN}@alcatel,90,Ttr)
exten => _5XXXX,1,Dial(SIP/${EXTEN}@alcatel,90,Ttr)
exten => 9,1,Dial(SIP/9@alcatel,90,Ttr)

[locales]
exten => _3129267,1,Congestion()
exten => _3XXXXXX,1,Dial(SIP/0${EXTEN}@alcatel,90,Ttr)
exten => _13XXXXX,3,Dial(SIP/0${EXTEN}@alcatel,90,Ttr)
exten => _020,4,Dial(SIP/0${EXTEN}@alcatel,90,Ttr)
exten => _030,5,Dial(SIP/0${EXTEN}@alcatel,90,Ttr)
exten => _040,6,Dial(SIP/0${EXTEN}@alcatel,90,Ttr)
exten => _050,7,Dial(SIP/0${EXTEN}@alcatel,90,Ttr)
exten => _065,8,Dial(SIP/0${EXTEN}@alcatel,90,Ttr)
exten => _066,9,Dial(SIP/0${EXTEN}@alcatel,90,Ttr)
exten => _071,10,Dial(SIP/0${EXTEN}@alcatel,90,Ttr)
exten => _090,11,Dial(SIP/0${EXTEN}@alcatel,90,Ttr)
exten => _01800XXXXXXX,1,Dial(SIP/0${EXTEN}@alcatel,90,Ttr)

[cel-local]
exten => _044XXXXXXXXXX,1,Dial(SIP/0${EXTEN}@alcatel,90,Ttr)

[cel-fuera]
exten => _045XXXXXXXXXX,1,Dial(SIP/0${EXTEN}@alcatel,90,Ttr)

[nacionales]
exten => _01XXXXXXXXXX,1,Dial(SIP/0${EXTEN}@alcatel,90,Ttr)
exten => _019008494949,1,Dial(SIP/0${EXTEN}@alcatel,90,Ttr)
exten => _01900XXXXXXX,1,Congestion()

[usa-canada]
exten => _001XXXXXXXXXX,1,Dial(SIP/0${EXTEN}@alcatel,90,Ttr)

[resto-mundo]
exten => _00XXXXXXXXXXXXX,1,Dial(SIP/0${EXTEN}@alcatel,90,Ttr)
```

c) **The client** is the user of IP telephony. The client enjoys privileges depending on those of the server, or as a result of the features of the server. This can be either software or

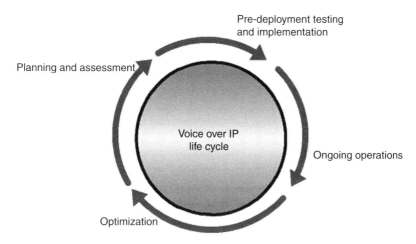

FIGURE 16.2 Life cycle of IP telephony.

hardware. The software clients are virtual telephones called softphones, such as xlite and sjphone, to name a few, that need a computer acting as a link. In contrast to softphones, we have hardware telephones called hardphones that are directly connected to the network, such as the VoIP USB phones or the SIP telephones. An analog telephone can also be used with an analog telephony adapter (ATA), as shown in Figure 16.3.

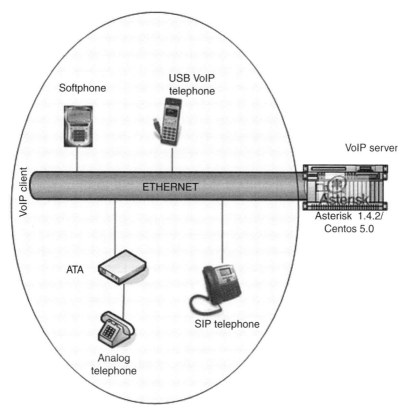

FIGURE 16.3 VoIP clients.

The IP telephony software must have the following features in order to meet the needs of the users of the University of Colima.

- List of contacts
- Quick search and easy-dial service
- User-friendly interface
- Call record (outgoing, incoming, and missed)
- Redial
- Support in English and Spanish
- Minimum hardware requirements
- Minimum requirements of software libraries
- Ability to work with different operating systems
- Call forwarding
- Call-divert service
- Volume adjustment of the microphone and the handset
- Selection of the audio codec.

Tests were run with different softphones, whose selection depended on their license type, kind of operating system, codecs supported, VoIP protocols supported, and number of simultaneous calls, among other considerations. The XLite Pro softphone [7] was finally selected and installed on the computers, while the SJPhone [8] was installed on the IPAQ.

Figure 16.4 shows the configuration of the XLite softphone, which is rather simple.

Tests were also conducted with Nokia N95 telephones [9]. These telephones include a wireless LAN connection and a SIP softphone, thus making it possible to have an SIP extension.

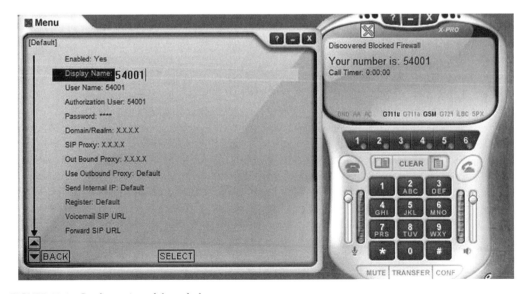

FIGURE 16.4 Configuration of the softphone.

At those departments of the University of Colima renting broadband Internet connection from telephone companies, we set up hardphones using a G729 audio codec. It should be noted that this codec was being used under a license, as it was bought from a VoIP communications company [10].

16.2.2.1 Dial Plan

With the aim of identifying and dialing the extensions that are installed under the IP telephony scheme, it is necessary to implement a dial plan that must adapt to the one already implemented, so as to ensure transparency in terms of dialing among extensions. A consideration that must never be forgotten when designing a dial plan is the future increase in the demand for IP telephony services.

At present, the University of Colima has a dial plan that is used to reference its analog and digital extensions throughout the departments constituting the University.

The extensions of the University comprise five digits, and the IP telephony extensions have been planned similarly, so as to ensure transparency for the user.

Table 16.2 shows the distribution of the extensions.

The aim has been to achieve transparency by making sure that user dialing is similar to that in a conventional telephone.

1. Dialing from a mobile device or a conventional telephone
Users will be able to dial directly one of the following:

- An SIP extension (54000...54009)
- A University extension, local numbers, Nextel numbers, local, national, and international cellular phones, without first dialing a code, depending on the user category, as shown in Table 16.3. The user may also have access to the following numbers: 020, 030, 040, 050, 065, 066, 071, 090, and 01800.

The process followed with this kind of dial plan is shown in Figure 16.5. The SIP application is registered on the Asterisk, so that the Asterisk can authenticate it and link it with the desired telephone extension.

2. Dialing the extension of a mobile device from a conventional telephone
Users will be able to dial the SIP extension (54000...54009) directly from an extension within the University of Colima.

If users call from an external line, they must call the switchboard and ask to be put through to the desired SIP extension.

TABLE 16.2

Extensions of the Telephone Network of the University of Colima

Range of the Extensions	Campuses
30000–49999	Colima
50000–50999	Villa de Álvarez
51000–51999	Coquimatlán
52000–52999	Tecomán
53000–53999	Manzanillo
54000–54999	Asterisk extensions

TABLE 16.3

Categories of the Dial Plan

Privilege Category	SIP Extensions	Alcatel Extensions	Local	Local Cell. (044)	National	External Cell. (045)	USA-Canada	Rest of the World
Category1	X	X						
Category2	X	X	X					
Category3	X	X	X	X				
Category4	X	X	X		X			
Category5	X	X	X	X	X	X		
Category6	X	X	X	X	X	X	X	
Category7	X	X	X	X	X	X	X	X

When a call is received on a softphone, it can be transferred to another extension by clicking the **Transfer** button and by entering the number to which the call will be transferred. Afterward, **Transfer** is clicked again. This process can be seen in Figure 16.6.

Users call the switchboard from a conventional extension or an external telephone line. From the switchboard, the caller is put through to the desired SIP extension. If the SIP extension is outside the network of the University, it may be necessary to use a virtual protocol network (VPN) to ensure better quality.

The codec that was used to call from the switchboard was Asterisk G.711, in view of the fact that it is the one that supports Alcatel's Release. SIP users may use codecs G711, G729, or GSM. Table 16.4 shows the codecs that are used, as well as the required bandwidth.

16.2.2.2 Billing or Call Record

It is necessary to know who has made telephone calls, which numbers have been dialed, and the duration of the calls. In order to gather this information, a record of the calls that

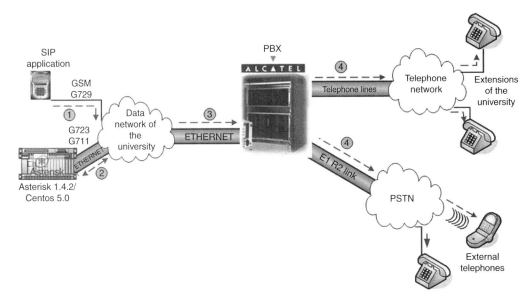

FIGURE 16.5 Dialing from a mobile device or a conventional telephone.

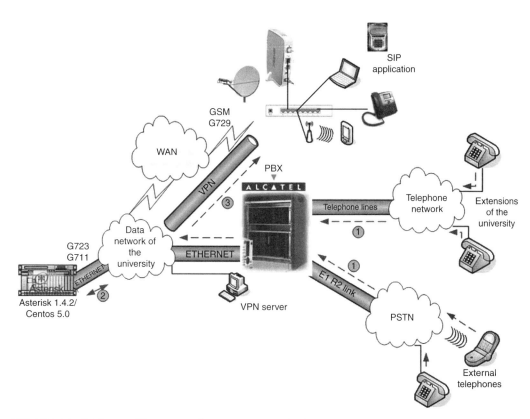

FIGURE 16.6 Dialing an SIP extension from a conventional telephone.

are made by the IP telephony users should be kept. These calls are always made through the Asterisk server, which makes it possible to gather, at the time of the calls, all data that is needed to monitor and determine the cost of the calls later. This is also known as call detailrecord (CDR).

The information provided by the CDR file is shown in Table 16.5. The types of database in which this information can be stored are SQLite, PostGreSQL, MySQL, or unixODBC. This time, for easier management, it was decided to store the information in a MySQL database.

The information of the CDR must be modified, as not all the fields are necessary. The only fields required are calldate, clid, src, dts, and duration. Detailed information about this process is shown in Figure 16.7.

TABLE 16.4

Codecs Used in this Project

Audio Codec	Bandwidth (kbps)	Mean Opinion Score (MOS)	Licensing Type
G.729	8	3.92	Patented
G.711	64	4.1	Open source
GSM	13	4.0	Proprietary

Source: J. Van Meggelen, J. Smith, and L. Madsen, *Asterisk: The Future of Telephony*, Editorial, 2005, O'Reilly Media, Inc., Sebastopal, CA, pp. 145–147.

TABLE 16.5

CDR Fields

Field	Type	Null	Default
calldate	datetime	NO	0000-00-00 00:00:00
clid	varchar(80)	NO	NULL
src	varchar(80)	NO	NULL
dst	varchar(80)	NO	NULL
dcontext	varchar(80)	NO	NULL
channel	varchar(80)	NO	NULL
dstchannel	varchar(80)	NO	NULL
lastapp	varchar(80)	NO	NULL
lastdata	varchar(80)	NO	NULL
duration	int(11)	NO	0
billsec	int(11)	NO	0
disposition	varchar(45)	NO	NULL
amaflags	int(11)	NO	0
accountcode	varchar(45)	NO	NULL
uniqueid	varchar(45)	NO	NULL
userfield	varchar(45)	NO	NULL

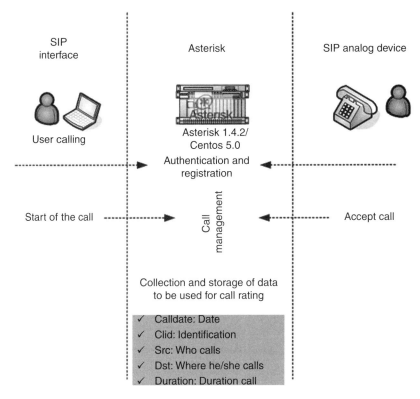

FIGURE 16.7   Process of information recording.

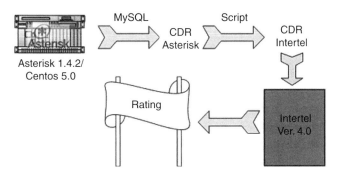

FIGURE 16.8 Dataflow and processes to be used for call rating.

TABLE 16.6

Data Provided by the CDR

2007-11-27 17:50:30	"Telefono IP" <54009>	54009
32038	Categoria5	SIP/54009-b00580f0
SIP/alcatel-b0022420	Dial	SIP/32038@alcatel | 90 | Ttr
56 42	ANSWERED	3

Once the CDR file is retrieved with the data that are required, it is necessary to create a script to design a rating system, as shown in Figure 16.8.

Table 16.6 gives detailed information as provided by the Asterisk CDR, Table 16.7 shows the data required, and Table 16.8 explains how the University's rating system can read the information.

TABLE 16.7

Data Required

Calldate	Clid	Src	Dst	Dcontext	Channel
2007-11-27 17:50:30	"Telefono IP" <54009>	54009	32038	categoria5	SIP/54009-b00580f0
Dstchannel	**Lastapp**	**Lastdata**	**Duration**	**Billsec**	**Disposition**
SIP/alcatel-b0022420	Dial	SIP/32038@ alcatel | 90 | Ttr	56	42	ANSWERED
amaflags	**accountcode**	**uniqueid**	**userfield**		
3					

TABLE 16.8

Data Ready to be Given to the University's Rating System

Abonado	Nombre	CCN	HoraFinLla	Duracion	Coste	VSACMFR	C	C	C	C
Tlf. Trf	Numero llamado	P	Codigo	PNI	NodTar	EnlNod	IdGE	IdEnl	C	C
54001	54001		0709242224	000:00:09	0.00	S	0			
	3142551	P		0	000001	000001	0	500	B	A

FIGURE 16.9　Scheme for the registration and authentication of users.

When an SIP user makes a call, it is necessary to make sure that there is a security system in place to prevent connections to alien users, and to ensure user mobility, which is one of the objectives of this project. The user is required to enter a login ID and password, as shown in Figure 16.9.

With the completion of this phase of the project, the IP telephony network is integrated into the conventional telephone network, as seen in Figure 16.10.

16.2.3　Ongoing Operations and Optimization

This phase is necessary in order to assess the new requirements for the launching of the project. It is necessary to examine the implementation process so as to determine the required changes. It is also during this phase that we establish the procedure by means of

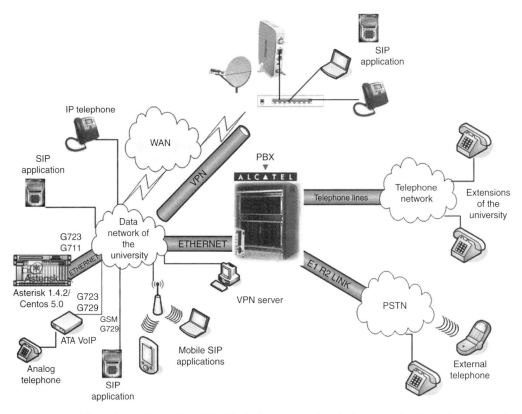

FIGURE 16.10　Scheme for the integration of the IP telephony network with the conventional telephone network.

TABLE 16.9

Assessment of the Quality of Service (QoS) of IP Telephony

Assessment of the quality of service of the Asterisk IP telephony at the University of Colima

SIP extension: _____

Date/Time	Number Dialed	Fault
		Noise
		Time delay
		Echo
		The call is interrupted

Comments:

which we assess the performance of the implemented IP telephony system. The results have been presented below.

Testing mechanisms were determined with the aim of assessing the performance of the system in a real-life environment. Later, decisions were made as to whether the resources that had been allocated were sufficient.

The assessment criteria that were established were the following:

1. User authentication tests
2. Proper rating of calls
3. Monitoring of network traffic caused by telephone calls
4. Assessment of the performance of the server components (memory, CPU, disk)
5. Jitter tests, as this is the main factor affecting the quality of voice.

Table 16.9 shows the form using which a fault is recorded when calls are made.

The Sniffer Clear Sight Analizer v6.0 was used to monitor jitter behavior during a call.

The monitoring of jitter behavior in Figure 16.11 shows that 83% of the packets that are transmitted from one campus to another within a single University network do not report a time variation, while 2% report a time variation of 45 ms or less. The rest of the packets report a variation no higher than 40%.

The jitter is mainly caused by network congestion, although a variation of 100 ms is acceptable. Figure 16.11 shows that the data network of the University of Colima is sufficiently suitable for the implementation of voice communication.

16.3 Conclusion

The development and implementation of this project makes it possible to establish an IP telephony system with the use of an open source on the VoIP server when using Asterisk, version 1.4.2, installed on the Centos 5.0 operating system. It is also possible to implement a dial plan that is transparent for the user. This is possible when using the same number of digits as in the dial plan of conventional telephony. The system of registration and authentication of users provides IP telephony users with mobility when they have access to softphones installed on a mobile device and registered on the Asterisk server, as

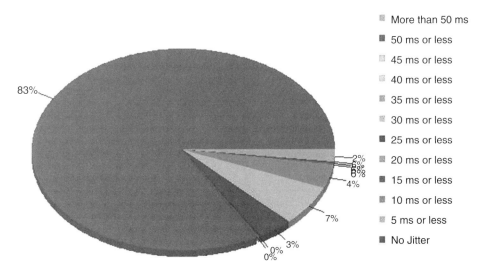

FIGURE 16.11   Jitter distribution in a different network segment.

shown in Table 16.10. For example, if a 10-minute call is made from Mexico City to the city of Colima from a conventional telephone to a cellular telephone, savings can be higher than 35%. In addition, if a call is made from a conventional telephone in Mexico City to another conventional telephone in the city of Colima, making it a local call, as shown in Figure 16.12, the savings that are achieved are higher than 88%.

At present, the University of Colima has 34 high schools, 14 of which have access to wireless Internet provided by the University. There are 11 high schools renting broadband Internet connections from a telephone company. Wireless Internet facility is available at six schools, and three high schools have access to a satellite network, whose cost is paid to an institution providing this service, as shown in Figure 16.13. VoIP was installed only in stocks with data network of broadband Internet (11) and satellite network (3). Those high schools paying for Internet service have conventional telephony whose cost is paid to a company providing this service. Of the 14 schools have been installed at six and remains installed at eight. Figure 16.14 shows that 43% of all progress has been made in terms of installation. Hardphones have been set up with the use of a G.729 audio codec, so as to reduce the use of bandwidth. However, the aim is to make sure that all the departments have access to IP telephony, so as to avoid the payment for telephone services.

The University's high schools that have satellite Internet have been facing the problem of seriously delayed packets, because of the company providing this service.

First, the Asterisk was installed on a personal computer, with a 1 GB RAM, Slaware operating system, version 11.0. The Asterisk is currently installed on an IBM server with a 2 GB RAM, as a result of an increase in the number of users.

TABLE 16.10

Cost Comparison Between Conventional Telephony and IP Telephony

	Conventional Telephony		IP Telephony	
	Cost Per Minute	**Total**	**Cost Per Minute**	**Total**
Cell phone	$2.85 min	$28.50	$1.85	$18.50
Home	$1.48 min	$14.80	$1.70	$1.70

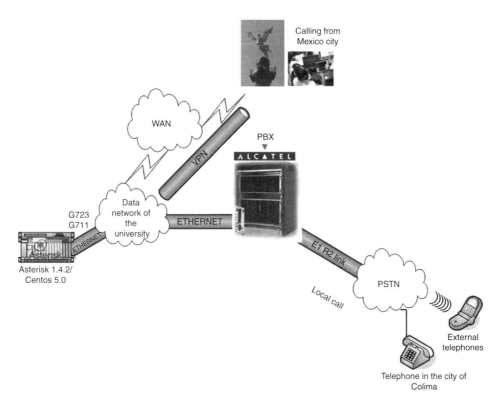

FIGURE 16.12 A long-distance call becomes local.

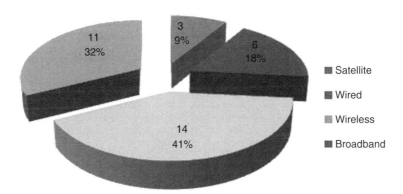

FIGURE 16.13 Type of data network at the University of Colima.

With the use of the ClearSight Networks software, we obtained the value of the VoIP index, which is 4.25, based on the mean opinion score (MOS). The other values are as follows:

    5 - "In-person" voice quality
    4 - Toll phone voice quality
    3 - Cellular phone voice quality
    <2 - Unacceptable voice quality

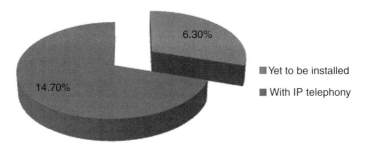

6.30%

14.70%

■ Yet to be installed
■ With IP telephony

FIGURE 16.14   Progress made on the installation of the IP telephony network.

TABLE 16.11

Performance Index QoS

	RTP Flow 1 148.213.4.85 → 192.168.1.65	
Packet loss (%)	0.00	0.35
Jitter (ms)	22.093	2.273
Out of sequence (%)	0.00	0.35
Latency (ms)	32.816	32.816

Table 16.11 shows the QoS values that were obtained with Clearsight. It can be seen that the values that were provided are within the allowed range.

This technology has been readily accepted in view of the advantages that it provides, including mobility and the low cost of calls. The users of the University's high schools can dial the 1296 telephone extensions of the University free of charge.

References

1. Digeset. 2004. Red de telefonía universitaria, Dirección General de Servicios Telemáticos. [online], http://www.ucol.mx/telefonia/ (2007, Diciembre, 04).
2. Cavanagh, J. 2006. *Successful deployment of VoIP and IP telephony.* Prognosis.
3. Ietf. 2008. The Internet Engineering Task Force [online], http://www.ietf.org/html.charters/sip-charter.html (2008, January, 01).
4. Intel. 2008. *Intel Tecnology Journal* [online], http://www.intel.com/technology/itj/2006/volume10issue01/art02_sip_evolution/p01_abstract.htm (2008, January, 03).
5. Gonçalves, F. 2006. Asterisk PBX Guía de la configuración.
6. Asterisk, 2008. Asterisk Digium [online], http://www.asterisk.org/support/features (2007, Diciembre, 08).
7. Counterpath. 2008. Build a winning VoIP strategy with our innovative softphones and SIP applications [online], http://www.counterpath.com/ (2008, Enero, 04.)
8. Sjlabs. 2008. The world leader in softphone production [online], http://www.sjlabs.com/ (2008, Enero, 04).
9. Nokia 2007. Nokia connecting people [online], http://www.nokia.es/ (2008, Enero, 10).
10. Digium. 2008. VoIP telephony communication solutions for business [online], http://digium.com/en/index.php (2008, Enero, 10).
11. J. Van Meggelen, J. Smith, and L. Madsen, *Asterisk: The Future of Telephony*, O'Reilly Media, Inc., Sebastopol, CA, 2005.

17

Multiparty Video Conferencing over Internet

Ahmet Uyar

CONTENTS

Multiparty video conferencing differs from point-to-point communications in that there are more than two parties in real-time audio/video sessions, which changes the architecture of the system significantly. In point-to-point real-time sessions, intermediary servers in the system for the delivery of audio and video streams are usually unnecessary. The only servers needed are those that are used to discover other parties and establish real-time sessions. Once a real-time session is established, the delivery of audio and video streams is performed directly between the two communicating end points. In the case of multiparty video conferencing sessions, it is extremely difficult for individual participants to send and receive audio/video streams directly to, and from, all other participants. Therefore, intermediary servers are often used in meetings to route audio/video streams to all participants more efficiently. It is also possible to organize participants in a session as a peer-to-peer network, and to deliver audio/video streams without any dedicated server, as has been done by End System Multicast [1] and Skype. In any case, multiparty video conferencing sessions require some type of efficient mechanism to route audio/video streams.

Another important difference between point-to-point communications and multiparty conferencing sessions is that the latter may require some media processing on intermediary servers. There are many cases in which this might be necessary: some participants in a meeting may not support all the codecs used in a session and intermediary media processing units are required to transcode some streams into desired audio and video formats. Another common media-processing task is to mix audio streams. This process combines many audio streams into one and reduces the bandwidth required to receive all audio streams during a session. Similarly, multiple video streams can be combined into one. This is also a common media processing function that provides more video streams to low-end participants with limited capability. Media processing is usually computing-intensive and may introduce extra delays to package delivery times. Therefore, they should be incorporated into the architecture of video conferencing systems with care.

Multiparty video conferencing sessions may also require floor control mechanisms of some kind [2]. Depending on the nature of the conferencing, participants might have various roles such as speaker, listener, or moderator. While a listener may need permission to talk, a speaker can talk any time in a session. Other resources in meetings such as chat, shared slides, whiteboards, and the like should also be managed in an orderly fashion to avoid uncertainties.

One of the most important characteristics of real-time audio and video communications over the Internet is the presence of package latencies from source to destinations. ITU recommends [3] that the mouth-to-ear delay of audio should be less than 300 ms to obtain communication that is of good quality. This latency includes not only the delivery times of packages but also the processing and buffering times at the speaker and listener machines. Therefore, all video conferencing architectures aim to minimize the latency of audio and video packages from sources to destinations by providing efficient media distribution mechanisms that can route media streams through the best possible paths. High bandwidth requirement is another prominent feature of real-time audio and video streams. Video streams, in particular, require higher bandwidths. In view of these requirements, video conferencing systems try to avoid sending multiple copies of a stream in the same link, and duplicate packages at intermediary nodes, an essential process during routing.

In this chapter, our aim is to outline the most important issues in multiparty video conferencing over Internet and to review the proposed solutions and standards. We then present some experimental performance results for a multiparty conferencing system that we developed at the Community Grids Labs at Indiana University. Starting with a brief history of remote video conferencing, we proceed to outline IP-Multicast media delivery systems and review two video conferencing standards. In the sections that follow, we provide an overview of two successful video conferencing solutions; one based on IP Multicast, and the other uses a reflector network for the delivery of real-time audio and video traffic. Next we present performance test results for real-time audio/video delivery using a software-based media delivery network. Lastly, we outline media processing architectures and present a few performance test results.

17.1 Brief History

From the time the television and telephony were invented, much effort has been made to provide teleconferencing services to the general public. The first notable example of this was Picturephone [4] in 1970 by Bell System. However, these efforts all failed until Internet provided an inexpensive and widespread delivery medium. In 1988, Steve Deering proposed IP Multicast as an efficient mechanism to distribute real-time audio and video streams over Internet to large groups. Multicast offers IP-level support for group formations and package routings, aiming to minimize package latency times and avoid unnecessary retransmission of bandwidth-intensive streams. However, these advantages were accompanied by increased complexity at IP routers and resulted in management and security problems [5]. Despite these shortcomings, Multicast is deployed in many networks and continues to be widely used, especially among academic institutions. The most successful video conferencing system built on IP-Multicast today is Access Grid [6]. It is a room-based group-to-group communications medium designed for high-end clients.

In the 1990s, some companies started to develop video conferencing systems, and their efforts were speeded up by the release of the H.323 [7] video conferencing protocol over IP networks by ITU in 1996. While some companies provided desktop-based video conferencing solutions, others developed hardware-based systems with cameras, monitors, and microphones designed for the purpose. These companies usually provided a hardware-based MCU that acted as a central unit to manage sessions and deliver audio and video streams. This central solution usually limited the number of supported participants, but it was adequate for many practical uses such as remote business meetings in which there were only a handful taking part. As a consequence, this area of application continued to grow, and today, there are many video conferencing companies in the market, among which the following three are the current market leaders: Polycom, Radvision, and Tandberg.

Another video conferencing protocol, sessions initiation protocol (SIP) [8], was developed by the Internet Community, and its first version was released in 1999. Both H.323 and SIP have their strong points. Today, many commercial products support both protocols, which continue to exist together.

Many efforts have been undertaken to design overlay networks to deliver real-time audio and video streams over the Internet. The main advantage of this type of solution is that, unlike IP-Multicast, it does not require any special IP-router level service support. Many projects have explored ways to implement multicast functionality at the application level. The End System Multicast [1] tried to implement multicast functionality at end points instead of IP-routers. In this approach, end points automatically formed an overlay network and streams were exchanged among them as if they were in a multicast-enabled network. Scalable Application Level Multicast [9] organized end points into a hierarchy and used it to route streams. Content addressable networks (CAN) organized endpoints in an n-dimensional logical coordinate space with respect to their neighbors. Each endpoint kept the state of its neighbors, and this structure was used to route packages [10]. Today, a successful video conferencing system that uses an overlay network to deliver real-time audio and video streams is the EVO collaboration network (formerly known as VRVS) [11], which deploys many reflectors around the world to route real-time media streams.

17.2 IP-Multicast

Multicast [12] is a set of transport-level protocols that provides group communications over IP networks. Some of these protocols are PIM-SM, PIM-DM, IGMP, MSDP, BGMP, and MADCAP. They are implemented in IP-routers and specify the group formations and management, package delivery mechanisms, inter-domain interactions, and so on. Similar to UDP, Multicast provides a best-effort package delivery service, but unlike UDP, it delivers packages to a group of destinations, instead of to one. The groups are formed dynamically and are transparent to the end user. Each group is identified by a virtual IP address and a port number. When a user sends a package to the group IP address and port number, all participants receive that package. It is not necessary for the sender to know anything about the receivers; routers are responsible for delivering the packages to registered destinations [13]. In a multicast meeting, intermediary servers are unnecessary. All participants in the meeting send their audio and video streams to the meeting address and routers deliver the streams to all.

IP was originally designed as a stateless transmission protocol in which intermediary routers did not keep any state information about the communicating endpoints and packages being transferred. In case of failures, endpoints were supposed to retransmit packages. This choice of design was made to keep routers simple and connect many types of networks flexibly [14]. More complex services were to be implemented in the upper layers of the IP; however, an exception was made to this with the introduction of IP-Multicast. IP-Multicast requires routers to form and manage Multicast groups, and to intelligently deliver all audio and video streams to all participants in a group globally. Since any user in a Multicast-enabled network can join any multicast group and send data, multicast networks are vulnerable to flooding attacks by malicious users. In addition, since multicast requires router-level support, it is comparatively difficult for an organization to join a multicast backbone. Today, many academic institutions support IP-Multicast, but commercial networks usually do not. Access Grid is the most successful Multicast-based video conferencing system that has been developed and is used mostly by academic institutions.

17.3 H.323

H.323 [7] is an ITU recommendation that defines a complete video conferencing system including audio and video transmission, data collaboration, and sessions management. It is used to initiate and manage video conferencing sessions. It was the first standard to provide video conferencing systems a recommended method to set up and conduct sessions and is still the most commonly used protocol for video conferencing over the Internet. H.323 supports two types of multiparty conferencing: decentralized and centralized [15]. In the decentralized model, there is no intermediary server, but every user sends/receives audio and video directly to/from all other participants. When implemented using UDP, this solution does not scale well. Every user should be connected to every other user in a

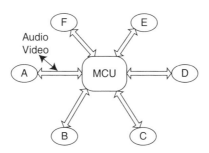

FIGURE 17.1 Centralized multipoint conferencing.

meeting, which requires a full mesh connection structure, signifying that this solution can only be used for very small meetings.

In a centralized model, the multipoint control unit (MCU) acts as the central server (Figure 17.1). Its main functions are to negotiate the link between the client and itself, receive and deliver audio and video data, and provide media-processing services such as audio mixing, video mixing, and audio and video transcoding. All participants connect to the MCU, through which they send and receive audio, video and data, and negotiate the type of media they will exchange. Today, this form of conferencing is much more common than the decentralized model. Some of the advantages of centralized conferencing are: not requiring multicast support, providing better control over the central component, easy for adding new services, and convenient for providing common functionalities such as audio and video mixing [16]. Currently there are many companies providing H.323-based video conferencing systems, a large number of them being hardware-based MCUs supporting small-scale meetings for 10–20 participants at a time.

17.4 Session Initiation Protocol (SIP)

SIP [8] is a session management protocol from the Internet Engineering Task Force (IETF) designed to discover other endpoints and establish real-time sessions by negotiating its characteristics, and is also used to implement video conferencing systems. Many existing H.323-compatible video conferencing systems also support SIP. SIP is a more flexible protocol than H.323, and one can easily implement a centralized or decentralized [17] video conferencing solution based on it. SIP does not propose any architecture to implement a video conferencing system, leaving it to the developer to provide one.

17.5 Access Grid

Access Grid [6] is a room-based video-conferencing system that provides group-to-group communications to institutions with high-end network connections. Each site has a room with multiple projectors that project many video streams from remote sites onto large screens. Multiple video cameras placed in various points send multiple video streams from unique perspectives. By providing many video streams from one site, and by projecting many video streams onto large screens, Access Grid imitates real-world face-to-face meetings. It tries to provide as much information as possible to create a sense of proximity and closeness. As of January 2008, there are 239 nodes listed globally at the Access Grid web site. Most of them are educational and research institutions.

Access Grid uses one multicast IP address and one port pair for video streams, and another multicast address and a port pair for audio streams. Everyone in a meeting sends their audio streams to the group audio address, and video streams to the group video address. Participating end points receive all audio and video streams in a meeting through these group addresses. The Access Grid 2.0 utilizes the Globus Toolkit mechanisms for authentication and user identification [18], encrypting RTP streams with a common key. A client who does not have this key can still obtain the streams but cannot decrypt them. This solution thus prevents unwanted users accessing meeting streams.

17.6 Enabling Virtual Organizations (EVO)

EVO (formerly known as VRVS) [11,19] is a mature video conferencing system from the Caltech CMS group for High Energy and Nuclear Physics (HENP) communities. The project started in 1995 and the first production version released in 1997. Since then, it has been used by thousands of researchers around the world. It provides audio and video conferencing capabilities for PC-based systems supporting both H.323 and SIP compatible end points. An endpoint can receive multiple audio and video streams in a meeting. Not being an open project we do not know the architectural details. However, as of 2003, they had 70 reflectors around the world [19], and users connect to one of these reflectors to join meetings. These reflectors form a middleware network to distribute real-time audio and video traffic. The organization provides a monitoring service in its web site showing the current status of the reflector network, the current configuration of the reflectors, and the links among them. It shows the amount of data being exchanged among those links along with package loss rates as well as the current number of active users and active meetings. However, there are no documents that describe the reflector network formation and package routing algorithms.

One of the advantages of the EVO collaboration system is its ability of the reflector to run behind firewalls, NATs, and proxy servers [20]. A user first connects to this reflector inside an organization, which establishes the connections with other participants in other parts of the network. EVO reflectors use only a single port to simplify the process of going through firewalls. In addition, they can use a TCP connection instead of a UDP connection to transfer media streams. Moreover, in each client a light VRVS proxy is installed to go through firewalls, NATs, and proxy severs, and to provide security services in the absence of a reflector in an organization.

17.7 Meeting Management for Video Conferencing

We can classify the tasks performed in a video conferencing system into three major categories: meeting management, audio/video stream delivery, and media processing. Meeting management includes starting and managing video conferencing sessions and specifying the mechanisms with which to discover and join meetings. On entering a meeting, endpoints negotiate the audio and video codecs to be used, and set up the required system resources for a particular meeting. For example, if audio and video mixers are used, these media processors must be initiated. During the meeting, access to shared resources such as microphones and whiteboards is also managed by the meeting management system.

IP-Multicast-based video conferencing sessions such as Access Grid meetings usually do not require meeting management. The meeting multicast address is usually advertised on a webpage, and participants join in at the time of the meeting. Media processing management in Access Grid meetings is unnecessary, as there are no media processing units. All participating endpoints are usually known to each other and floor control is not needed. Anyone is free to talk at any time and order is maintained in meetings by observing the general rules of courtesy.

In other video conferencing systems, the media stream delivery and media processing architecture are first designed, and the meeting management system is then built accordingly. Today most video conferencing systems use a web-based interface to initiate and schedule sessions. For example, the EVO system uses a web-based meeting management system. Users first register at their website, through which interface they find and join meetings.

17.8 Real-Time Stream Delivery

As we pointed out, real-time media stream delivery has two very important requirements: high bandwidth and low latency. These two limitations make real-time media stream delivery the most important part of any video conferencing architecture. All video conferencing architectures aim to minimize the distance from sources to destinations, so that the amount of time taken by packages to travel to their final destinations is minimized. In addition, they try to avoid sending multiple copies of a stream on the same link to save network bandwidth. IP-Multicast is implemented at the router level to deliver the packages to their final destinations in the minimum amount of time through the best possible routes, and to use the network resources most efficiently. However, IP-multicast is still not very widely supported, consequently overlay networks are frequently designed to deliver real-time media streams. The ever-increasing network bandwidth and computing power also helps these overlay systems to provide better performance. We developed a video conferencing system, Global-MMCS [21], for multiparty communications in the Community Grids Labs at Indiana University, and used a publish/subscribe message delivery network [22,23] as the media delivery middleware. In this section, we briefly introduce this network and provide a few performance test results for media delivery.

17.8.1 NaradaBrokering Middleware

NaradaBrokering [22,23] is a distributed publish/subscribe messaging system that organizes brokers into an hierarchical cluster-based architecture. The smallest unit of the messaging infrastructure is the broker, each of whom is responsible for routing messages to their next stops and handling subscriptions. A broker is part of a base cluster that is part of a super-cluster, which in turn is part of a super-super-cluster and so on. Clusters comprise strongly connected brokers having multiple links with brokers in other clusters, ensuring alternate communication routes, forming an organization scheme aiming to minimize the average communication path lengths among communicating endpoints.

Each client connects to a broker through which it exchanges messages and subscribes to the topics from which it wants to receive them. When a client publishes a message on a topic, it is delivered to all subscribing clients of that topic by the broker network.

Publish/subscribe systems can be classified into two major categories: topic-based and content-based. In topic-based systems, users exchange messages on shared channels. When a producer publishes a message on a topic, it is received by all subscribers of that topic. It is a many-to-many communication medium. In content-based systems, subscribers specify the kinds of messages in which they are interested, and the broker network checks the messages according to the user's specification, and delivers the message when there is a match. Topic-based messaging is better suited for audio and video distribution. As an audio or video stream is composed of many consecutive packages, it is more efficient to publish them on the same topic. It is also more convenient for receivers to subscribe to a topic to receive all the messages belonging to a media stream. Therefore, we used topics as the channels among the participants in a video conferencing session.

The NaradaBrokering publish/subscribe messaging system provided a scalable distributed architecture for content delivery, but did not provide any support for real-time traffic. We therefore added support for an unreliable transport protocol, UDP, and designed a compact message type to encapsulate media packages with a minimum increase in their size. Instead of using string-based topic names, we designed numerical topics of fixed length.

As a part of my Ph.D. thesis [24], we performed extensive performance tests for the delivery of audio and video streams, some of which are presented in this section to show the viability of using an overlay network to deliver real-time traffic.

We tested the capacity of a single broker for audio and video meetings with single and multiple concurrent meetings, on a Linux cluster with eight identical machines and a gigabit network switch among them. The machines had Double Intel Xeon 2.4 GHz CPUs, 2 GB of memory with Linux 2.4.22 kernel. Both NaradaBrokering software and our applications have been developed in Java.

17.8.2 Single Audio Meeting Tests on a Single Broker

We tested the capacity of a single broker for a single audio meeting by increasing the number of participants to the maximum that could be served during that session. There was only one speaker sending a ULAW audio stream for 2 min with no silence period, and transmitting one audio package every 30 ms, each 252 bytes in length. The broker was run on a dedicated machine. The results were collected from 12 measuring clients running in the same machine as the transmitter client. Other clients did not gather results, and were run in other machines, as shown in Figure 17.2. The speaker client recorded the package-sent times and the measuring receivers recorded the package-received times. Package latencies were thus calculated for every audio package.

Table 17.1 shows the experimental results. Each row in the table shows a test case with the given number of participants in the first column. The second shows the average latencies of the 12 participants from whom we gathered the results. The third column gives the average jitter values calculated for the average latencies of the 12 participants. Jitter values are calculated according to the formula given in the RTP specification [25]. The fourth column shows the percentages of packages that arrive later than 100 ms. The fifth presents the amount of data the broker receives from the speaker, and the last, the outgoing bandwidth from the broker and the amount of data sent out to all participants. As audio packages are sent periodically (a package every 30 ms), their routing does not affect the routing of the next one, until the broker is overloaded. This point is reached when the broker can no

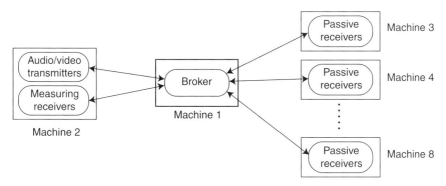

FIGURE 17.2 Single meeting test setting for one broker.

longer finish delivering an audio package to all participants in the meeting before the arrival of the next audio package. In such a case, each audio package delays the delivery of the next one, and the routing time of each increases continuously. In this test, the broker is overloaded when there are 1600 participants in the meeting.

These tests show that a broker can support up to 1400 participants in a single audio meeting, providing audio of very good quality and with no packages arriving late. The jitter values are also very small, as the routing of consecutive packages does not affect them.

17.8.3 Single Video Meeting Tests on a Single Broker

We performed single video meeting tests for a single broker in a similar fashion, with the same experimental setup as the audio meeting tests. One video meeting was in a single broker and one participant published a video stream for 2 min. Varying numbers of participants joined the meeting and 12 were used to measure the package latencies.

The publisher sent out an H.263 video stream with an average bandwidth of 280 kbps and a frame rate of 15 frames per second resulting in one frame sent every 66 ms. A frame had 3.1 packages on average but the video codec sent out one full picture update every 60 frames or every 4 s. These frames had much more packages than regular frames. The number of

TABLE 17.1

Test Results of Single Audio Meetings for One Broker

Number of Clients	Average Latency (ms)	Average Jitter (ms)	Average Late Arrivals (%)	In BW (kbps)	Out BW (Mbps)
12	0.6	0.18	0	64	0.76
100	1.4	0.15	0	64	6.4
400	4.2	0.21	0	64	25.6
800	8	0.18	0	64	51.2
1200	11.6	0.22	0	64	76.8
1400	13.5	0.26	0	64	89.6
1500	17.8	0.44	0.25	64	96.0
1600	2275	1.2	100	64	102.4

TABLE 17.2

Test Results of Single Video Meetings for One Broker

Number of Clients	Average Latency (ms)	Average Jitter (ms)	Average Late Arrivals (%)	In BW (kbps)	Out BW (Mbps)
12	1.2	0.44	0	280	3.3
100	4	2	0	280	27.8
300	13.2	7.8	0	280	83.4
400	17.3	10.1	0.05	280	111.2
500	23.4	13.2	2.6	280	139
700	36.8	18.1	7.6	280	194.6
900	102.7	23.8	38.2	280	250.2
1000	1609	27.8	98.9	280	278

packages for full updates fluctuated between 10 and 18 packages. The publisher transmitted 1800 frames over 2 min, which had 5610 packages in total. This video can be considered an average video stream with good quality for a video conferencing session.

Table 17.2 shows the experimental results for single-video meetings. The number of supported participants in a video meeting is much smaller than that supported in an audio meeting. In a video meeting, 400 participants can be supported with very low latency and jitter, and very few late-arriving packages. In the case of 500 participants, the number of late-arriving packages increases to 2.6%. The average jitter also goes up to 13 ms for the same test. Therefore, the quality of the stream delivery for 500 participants becomes unacceptable, although the broker becomes overloaded only when there are 1000 participants in the meeting. There are two reasons for this. The first is the higher number of packages in the video stream. Another, more important, is the uneven distribution of packages in the video stream throughout the transmission. The number of packages transmitted for each frame changes according to the actions in the picture. Moreover, the full picture updates sent by the video codec have many more video packages than the regular frames. Since the video codec sends all packages in a frame continually without delay, the later packages wait for the earlier ones to be routed at the broker. Therefore, some packages take more than 100 ms to arrive at their destination, long before the broker is overloaded. In summary, compared with single-audio meetings, single-video meetings utilize broker resources poorly, and supporting only 400 participants.

17.8.4 Audio and Video Combined Meeting Tests on a Single Broker

Contrary to the previous two cases, in this test there were two concurrent meetings, each having one media stream. On one hand, one meeting affects the performance of the other, while on the other hand, two meetings utilize broker resources more efficiently. When we performed some initial audio and video combined meeting tests, we observed that the delivery of video streams affected that of audio streams significantly. As audio communication is the fundamental part of a video conferencing session, we have given audio package routing at the broker priority over all other messages. This helped audio performance significantly while introducing only slight degradation in video transmission quality.

The latency values of combined audio and video meetings were comparatively higher than the average latency values of single-video meetings. However, the difference was not significant. When there were 400 participants in combined audio and video meetings, average latency was only 5 ms higher than in single-video meeting tests. There were no late-arriving packages for 300 participants and there were only 1.3% late-arriving packages for 400 participants. In summary, a broker supported up to 400 participants in combined audio and video meetings. This number is similar to single-video meeting tests. The main reason for this is the better utilization of broker resources when there are two concurrent meetings.

17.8.5 Multiple Meeting Tests on a Single Broker

We tested the performance and the scalability of a single broker for three types of multiple concurrent meetings: multiple audio meetings, multiple video meetings, and multiple combined audio and video meetings. We had one transmitter and 20 participants in all meetings. As multiple streams arrive randomly distributed in time, on average the packages wait much less at the broker than in single large-scale meetings. However, the processing of incoming packages uses some extra computing resources at the broker. Therefore, it provides a better quality of service to a smaller number of participants in multiple concurrent audio meetings. One broker supported 50 concurrent audio meetings with 1000 participants in total. In multiple concurrent video meeting tests, the broker provides a better quality of service to a higher number of total participants. It supports 35 concurrent meetings with 700 participants in total. This number is much better than the single-video meeting tests in which only 400 participants are supported. The main reason for this is the better utilization of broker resources in multiple concurrent video meetings. When audio and video meetings of smaller size were held together in a broker, it supported 20 audio and 20 video meetings concurrently, comprising 400 participants in audio meetings and 400 in video meetings. In summary, these tests demonstrate that a dedicated software-based server can easily serve hundreds of participants in video conferencing sessions, given enough network resources.

17.8.6 Performance Tests for Distributed Brokers

We have also conducted performance tests for multiple broker settings, including both single large-size meeting tests and multiple smaller-size meeting tests. Single large-size meeting tests demonstrated that the capacity of the broker network could be increased almost linearly by adding new brokers into the system. For example, while a single broker supported 400 participants in a video meeting, four brokers as a network supported 1600 participants.

The behavior of the broker network is more complex with multiple concurrent meetings than with a single meeting. Multiple meetings provide both opportunities and challenges. As we have seen in the single broker tests with multiple video meetings, conducting multiple concurrent meetings on a broker increases both the quality of the service and the number of supported users. This can also be achieved in multibroker settings as long as the size of the meetings and the distribution of clients among brokers are properly managed. If the size of the meetings are very small and the participating clients are scattered around the brokers, then the broker network could be poorly utilized. The best broker utilization is achieved when there are multiple incoming streams to all brokers, each of which is delivered to many receivers. If all brokers are fully utilized in this fashion, multibroker networks provide better services to a higher number of participants.

17.8.7 Performance Tests in Wide Area Networks

We conducted wide-area performance tests with four remote sites: Bloomington, IN; Syracuse, NY; Tallahassee, FL; and Cardiff, UK. The most important result of the wide-area tests was perhaps the fact that the networks that we tested provided communication of very high quality for audio/video streams. Even when transferring very high numbers of video streams, they provided excellent service for real-time vide conferencing applications. The loss rates were very small and negligible even for 200 video stream transfers. Similarly, the amount of latency and the jitter were very small. Passage through the Atlantic Ocean also did not pose a challenge. Therefore, the underlying network infrastructure was adequate for implementing a distributed brokering system above it to deliver audio/video streams.

In wide-area tests, the geographic distance among the sites was very important. It required approximately 55 ms for packages to travel from Bloomington, IN to Cardiff, UK. On the other hand, it required approximately 13 ms for packages to travel from Bloomington, IN to Syracuse, NY or Tallahassee, FL. These results indicate that it is very important to run brokers at geographically distant sites and minimize the transmission delays. For example, if there were no brokers in the UK, but there were multiple clients they would need to communicate through brokers in the US, which would add significant transmission delays, as the packages would need to cross the Atlantic Ocean twice. Having a broker in the UK with which clients can be connected eliminates this additional transmission overhead. Therefore, it is critical to run brokers at geographically distant sites when deploying a global media delivery network.

17.9 Media Processing

There might be many different kinds of media processing performed in video conferencing systems, the most common being audio mixing, video mixing, and stream transcoding. Although in a homogenous video conferencing setting where all users have high network bandwidth and computing power, media processing might not be necessary at the server, it is crucial in systems where users have varying network and device capacities. For example, Access Grid does not provide any media processing services at the server side, as each AG node can receive many audio streams and play them by mixing. They can also receive many video streams and display them concurrently. However, video conferencing systems that aim to support diverse sets of users with various network bandwidths and endpoint capabilities must provide media processing services to customize the streams according to the requirements of end users. For users who might have very limited network bandwidth, multiple audio and video streams should be mixed to save bandwidth, or few high-bandwidth streams should be transcoded to produce low bandwidth streams. Yet other users might have limited display or processing capacity, and for them, multiple video streams can be merged or larger-size video streams downsized.

Media processing usually requires high computing resources and real-time output. Therefore, the scalability of a video conferencing system should be limited when poorly implemented. More importantly, combined processing and distribution can affect the quality of audio and video distribution if the computing resources are shared. Therefore, the resources of media processing units should be independent of media distribution units to provide scalability.

Media processing usually introduces extra delays in media transmission when performed at the server side. Media streams are first transmitted to the server, processed by media processors, and transferred back to end points. This extra stop during transit from source to destination increases the transmission delays, particularly if the processors are placed at geographically distant sites that would reduce the quality of communication. Therefore, media processors should be located at central sites with advanced computing resources and high bandwidth.

In the GlobalMMCS video conferencing system, we developed three types of media processing services: audio mixing, video mixing, and image grabbing. Audio mixers combine multiple audio streams into one to support low-end clients with limited bandwidth, while video mixers merge four video streams. Image grabbers periodically record the snapshots of video streams as image files to provide users more information about them. All media processors receive media streams from the broker network, and publish the resulting streams back on the broker network. End points always receive media streams through the broker network. This architecture completely separates media distribution and media processing, thus providing a scalable architecture for both. New resources can be added to media delivery network or media processing units without affecting each other. In addition, it makes media processors independent of their location. They can be located anywhere as long as they are connected to the broker network. Here, we present an overview of audio mixing and provide experimental performance results as examples.

17.9.1 Audio Mixing

In video conferencing systems, it is best to avoid audio mixing at the server side to eliminate losses and prevent additional delays. The highest audio quality is achieved when all audio streams are delivered to all users and they are allowed to play those streams by mixing. Occasionally, however, a number of factors make audio mixing at the server unavoidable. Certain endpoints can receive only one audio stream, some do not have enough bandwidth, and some have no capability to process multiple streams. Therefore, audio mixing is necessary to provide a single mixed stream for such users.

The audio mixer receives all audio streams in a session and mixes them, and produces not one, but many, streams. It produces one stream for listeners and another separate one for every speaker. It works as shown in Figure 17.3. First, each received stream is decoded to a common audio format (8 kHz sampling rate, 8 bit per sample, mono). Next, repacketizers adjust the sizes of audio packages when necessary. Different audio codecs may use varying package sizes. After repacking, packages wait in a queue to be picked up by the audio mixer, which polls all queues regularly, and adds the values of the available data. A copy of the mixed data is passed to each subtracter, which then subtracts the data of that stream from the mixed data, if any, and passes it to the encoder. Encoders encode each stream according to the specific request of the user, and publish them on the broker network. In some cases, a speaker may send multiple audio streams into a session. Multicast groups in particular tend to send many streams, as they represent a group of participants instead of one endpoint. In such cases, the handling of streams is the same up to the mixer. After mixing, the subtracter subtracts the values of all streams belonging to that speaker from the mixed data.

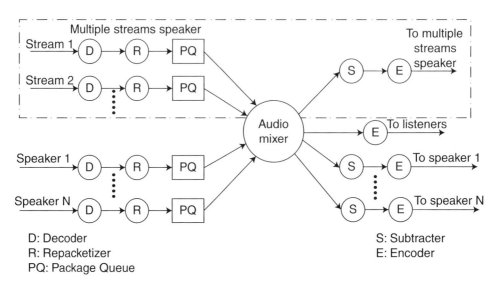

D: Decoder
R: Repacketizer
PQ: Package Queue

S: Subtracter
E: Encoder

FIGURE 17.3 Audio mixing.

17.9.2 Audio Mixing Performance Tests

While some audio codecs are computing intensive, others are not. Therefore the computing resources needed for audio mixing change accordingly. Audio mixers require prompt access to a CPU when they have to process received packages, or it could result in audio packages being dropped, leading to breaks in the communications. Therefore, the load on audio mixing servers should be kept at as low as possible.

We tested the performance of audio mixers for different numbers of mixers on a server. There were six speakers in each mixer, two of whom were continually talking, while the rest were silent. There was also one more audio stream constructed with the mixed stream of all speakers. Therefore, six streams entered the mixer and seven exited. All streams were 64 kbps ULAW. A WinXP machine with 512 MB memory and 2.5 GHz Intel Pentium 4 CPU hosted the mixer server.

Table 17.3 shows that this machine can support approximately 20 audio mixers. We should note, however, that in this test all streams are ULAW. This is not a computing intensive codec. When we conducted the same test with another more computing intensive codec, G.723, the same machine supported only five audio mixers.

TABLE 17.3

Audio Mixer Performance Tests

Number of Mixers	CPU Usage (%)	Memory Usage (MB)	Quality
5	12	36	No loss
10	24	55	No loss
15	34	73	No loss
20	46	93	Negligible loss

17.10 Conclusion

In this chapter, we have outlined the issues in multiparty video conferencing systems and given an overview of well-known video conferencing systems and standards. As IP-Multicast is not universally supported on the Internet, much effort has been undertaken to provide software-based overlay networks to deliver real-time audio and video traffic. The ever-increasing network bandwidth and computing power have helped these systems to provide better quality services. We also presented some performance test results from such a software-based media delivery network, which show that a single broker can easily deliver audio and video streams to a few hundred users, with very good quality. In addition, the capacity of the broker network can be increased by introducing additional brokers into the system. Therefore, software-based video conferencing solutions provide a strong alternative for the delivery of real-time traffic. In addition, we have presented some media processing test results, showing that media processing can also be performed at the server side on software-based units with good quality.

References

1. Y. H. Chu, S. G. Rao, and H. Zhang, "A case for end system multicast," *Proceedings ACM SIG-METRICS Conference (SIGMETRICS '00)*, June 2000.
2. H.-P. Dommel, and J. J. Garcia-Luna-Aceves, "Floor control for multimedia conferencing and collaboration," *Multimedia Systems*, 1997, pp. 23–38.
3. ITU-T Recommendation G.114, "One way transmission time," May 2003.
4. Picturephone. http://www.corp.att.com/attlabs/reputation/timeline/70picture.html (accessed January 2008).
5. K. Almeroth, "The evolution of multicast: From the MBone to inter-domain multicast to Internet2 deployment," *IEEE Network*, vol. 14, January 2000.
6. The Access Grid Project. http://www.accessgrid.org (accessed January 2000).
7. ITU-T Recommendation H.323, "Packet-based multimedia communication systems," Geneva, Switzerland, February 1998.
8. J. Rosenberg et al., "SIP: Session initiation protocol," Internet Engineering Task Force, RFC 3261, June 2002, http://www.ietf.org/rfc/rfc3261.txt.
9. S. Banerjee, B. Bhattacharjee, and C. Kommareddy, "Scalable application layer multicast," *Proceedings of ACM SIGCOMM*, ACM Press, New York, NY.
10. S. Ratnasamy, M. Handley, R. Karp, and S. Shenker, "Application-level multicast using content-addressable networks," *Proceedings of the Third International Workshop on Networked Group Communication*, 2001.
11. Evo Collaboration Network. http://evo.caltech.edu (accessed January 2008).
12. S. E. Deering, "Multicast routing in internetworks and extended LANs," *ACM SIGCOMM Computer Communication Review*, vol. 18 no. 4, August 1988, pp. 55–64.
13. M. Handley, J. Crowcroft, C. Bormann, and J. Ott, "Very large conferences on the Internet: the Internet multimedia conferencing architecture," *Computer Networks*, vol. 31, 1999, pp. 191–204.
14. D. D. Clark, "The design philosophy of the DARPA internet protocols," *ACM SIGCOMM Symposium*, Stanford, CA, 1998.
15. ITU-T Recommendation H.243, "Procedures for establishing communication between three or more audiovisual terminals using digital channels up to 1920 kbit/s," Geneva, Switzerland, February 2000.

16. Videoconferencing Cookbook Version 4.1, Video Development Initiative, Advanced Components and Management, http://www.vide.net/cookbook/cookbook.en, March 2005 (accessed in January 2008).

17. J. Rosenberg and H. Schulzrinne, "Models for multiparty conferencing in SIP," Internet Draft, Internet Engineering Task Force, November 2000 (Work in progress).

18. R. Olson, "Certificate Management in AG 2.0," http://fl-cvs.mcs.anl.gov/viewcvs/viewcvs.cgi/AccessGrid/doc, March 5, 2003 (accessed January 2008).

19. D. Adamczyk et al., "Global platform for rich media conferencing and collaboration," CHEP '03 La Jolla, California, March 24–28, 2003.

20. K. Wei, "Videoconferencing security in VRVS 3.0 and Future," *ViDe 5th Workshop*, March 25, 2003.

21. Global Multimedia Collaboration System. http://www.globalmmcs.org (accessed January 2008).

22. The NaradaBrokering Project. http://www.naradabrokering.org (accessed January 2008).

23. G. Fox and S. Pallickara, "NaradaBrokering: An event-based infrastructure for building scaleable durable peer-to-peer grids," Ch 22, *Grid Computing: Making the Global Infrastructure a Reality Grid*, West Sussex, Wiley, UK 2003.

24. A. Uyar "Scalable service oriented architecture for audio/video conferencing," PhD thesis, Syracuse University, 2005.

25. H. Schulzrinne, S. Casner, R. Frederick, and V. Jacobson, "RTP: A transport protocol for real-time applications," IETF, RFC 3550, July 2003, http://www.ietf.org/rfc/rfc3550.txt.

18

IMS Charging Management in Mobile Telecommunication Networks

Sok-Ian Sou

CONTENTS

With the decline in traditional voice services, telecom operators are eager to make profits in the data market. In particular, next generation networks are characterized by the integration and convergence of the wireless network, fixed Internet, and media content industries. Facilitating service deployment (including services from third-party providers) and providing billing using flexible business models are the keys to success. In this chapter, we use the Universal Mobile Telecommunications System (UMTS) as an example to illustrate the IP-based charging issue. UMTS provides high bandwidth data services to mobile users. To effectively integrate mobile technology with the Internet, 3GPP Release 5 introduces the IP Multimedia core network Subsystem (IMS) to control IP-based multimedia services [1]. The IMS architecture allows mobile operators and service/content providers to develop attractive and innovative services (such as VoIP, mobile gaming, and streaming services). In this all-IP architecture, two IP-based protocols defined by IETF are utilized: the session initiation protocol (SIP) adopted as the call session control and signaling protocol, and the diameter protocol adopted for authentication, authorization, and accounting (AAA) management.

Recently, IMS technology has significantly changed the existing billing model for traditional Internet services. An efficient IMS charging system is important for an operator to create revenue. For example, mobile users can send text messages, pictures, or videos while they are simultaneously engaged in Voice over Internet Protocol (VoIP) voice call sessions. This means that pre-event or mid-event authorization for online services should be performed. As the authors in [2] state: "Billing can now become a service in and of itself, as billing, in essence, will run like a service, with authorization, charging and rating all running as low-latency services within the IMS architecture." Therefore, the design and management of new sophisticated functions and protocols to support flexible and convergent charging mechanisms are essential. This chapter provides a comprehensive picture of the characteristics and practical aspects of online and offline charging based on the latest efforts undertaken by the 3GPP specifications. We describe the IMS architecture and the related charging components. Also, the offline and online charging solutions implemented for IMS calls will be discussed.

18.1 IP Multimedia Subsystem Architecture

This section reviews the IMS architecture for the UMTS network. As illustrated in Figure 18.1, the IMS architecture is an application-level architecture for networks that support a wide range of IP-based multimedia applications and services enabled by the flexibility of SIP protocol. It offers a unified architecture that facilitates communication across different networks, including fixed and mobile networks to support the market demand for multiple device types [3–6]. The IMS architecture divides the network into three layers: a bearer layer that transports the media IP packets, a SIP-based session-control layer, and an application layer that supports open interfaces. Figure 18.2 shows the IMS/UMTS network architecture, where the IMS connects the GPRS gateway support node (GGSN; Figure 18.2a) with the public-switched telephone network (PSTN; Figure 18.2b) and the external packet data network (PDN; Figure 18.2c). It can support both traditional telephony services and IP-based services such as instant messaging, push-to-talk, video streaming, mobile gaming, and so on. The other network nodes are described as follows:

- The home subscriber server (HSS; Figure 18.2k) stores the subscriber profiles and location information. It is equipped with Internet-based protocols for interaction with the IMS.
- The breakout gateway control function (BGCF; Figure 18.2g) selects the network in which the PSTN or circuit-switched (CS) domain breakout is to occur. If the BGCF determines that a breakout is to occur in the same network, it will select an media gateway control function (MGCF) that is responsible for interworking with the PSTN. On the other hand, if the breakout is to occur in another IMS network, the BGCF will forward the SIP request to another BGCF or an MGCF in the selected IMS network.
- The application server (AS; Figure 18.2i), which resides in the user's home network or in a third-party location, provides value-added multimedia services in the IMS.
- The media gateway (MGW; Figure 18.2e) transports the IMS user data traffic. The MGW provides user data transport between the UMTS core network and the PSTN/CS (including media conversion bearer control and payload processing).

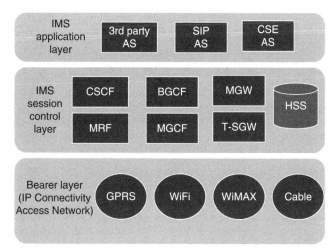

AS: Application Server
CSCF: Call Session Control Function
HSS: Home Subscriber Server
MGW: Media Gateway
T-SGW: Transport Signaling Gateway

BGCF: Breakout Gateway Control Function
CSE: CAMEL Service Evironment
MGCF: Media Gateway Control Function
MRF: Media Resource Function

FIGURE 18.1 IMS logical architecture.

- The MGCF (Figure 18.2f) controls the media channels in an MGW. It performs call control signaling conversion between the Integrated Services Digital Network (ISDN) user part (ISUP) for the PSTN/CS users and the SIP for the IMS users.
- The transport signaling gateway (T-SGW; Figure 18.2h) serves as the PSTN signaling termination point and provides the PSTN/legacy mobile network with IP

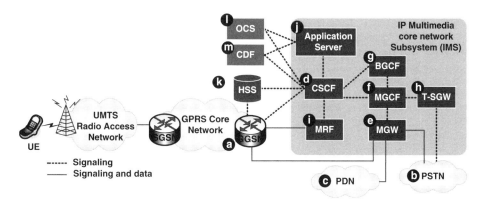

BGCF: Breakout Gateway Control Function
CSCF: Call Session Control Function
HSS: Home Subscriber Server
MGW: Media Gateway
OCS: Online Charging System
PSTN: Public Switched Telephone Network
T-SGW: Transport Signaling Gateway

CDF: Charging Data Function
GGSN: Gateway GPRS Support Node
MGCF: Media Gateway Control Function
MRF: Media Resource Function
PDN: Packet Data Network
SGSN: Serving GPRS Support Node
UE: User Equipment

FIGURE 18.2 The UMTS/IMS network architecture.

transport level address mapping. Specifically, it maps call-related signaling between the MGCF and the PSTN.

- The media resource function (MRF; Figure 18.2i) performs functions such as multiparty call, multimedia conferencing, and tone/announcement.
- The call session control function (CSCF; Figure 18.2d) is a SIP server responsible for call control and communications with the HSS.

IMS signaling is carried out by three kinds of CSCFs: the proxy CSCF (P-CSCF), the interrogating CSCF (I-CSCF), and the serving CSCF (S-CSCF). Figure 18.3 depicts an example of a scenario of an IMS call, where user equipment (UE)1 is the calling party and UE2 is the called party.

- P-CSCF is the first entry point of the IMS for the UE. All SIP signaling traffic to or from the UE goes via the P-CSCF.
- I-CSCF is the entry point within the operator's home IMS network. It hides the network topology from other networks.
- S-CSCF performs registration and session control for the UE. It is also responsible for routing SIP messages to the ASs.

When a UE gets attached to the GPRS/IMS network and performs packet data protocol (PDP) context activation, a P-CSCF is assigned to the UE. The P-CSCF contains limited address translation functions to forward the requests to the I-CSCF, which then selects an S-CSCF to serve the UE during the IMS registration procedure. This S-CSCF supports the signaling for call setup and supplementary services control. All incoming calls are routed to the destination UE through the S-CSCF.

In 3GPP Release 6, the online charging system (OCS; Figure 18.2l) performs online credit control with the CSCF and the AS. For offline charging, the CSCF and AS are responsible

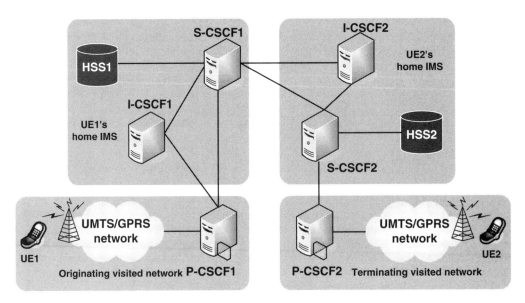

FIGURE 18.3   An example of a scenario for an IMS call.

for generating charging record for service delivery. The charging records are sent to the charging data function (CDF; Figure 18.2m). Details of the charging architecture will be discussed in the following section.

18.2 IMS Charging Architecture

As the IMS-based service control becomes a standard feature in fixed and mobile networks, there arises the challenge of keeping control of customer care and the charging/billing relationships. The success of IMS-based services relies greatly on charging systems that are capable of supporting their diversity while guaranteeing fine granularity.

In UMTS, the IMS and the OCS architectures are developed based on all-IP technologies [7,8]. Figure 18.4 illustrates the charging architecture and the information flow for offline and online charging in 3GPP Release 6. In this architecture, the charging trigger function (CTF) is a mandatory function integrated in all network nodes that provide charging functionality. The CTF provides metrics that identify the users and their consumption of network resources, and generates chargeable events from these metrics. The terminology for online and offline charging and the CTF are listed in Table 18.1. Details of the IMS offline/online charging are explained in the following sections.

18.3 IMS Offline Charging

In Figure 18.4, a major component in the IMS offline charging architecture is CDF (Figure 18.4a). The offline CTF of the IMS node (such as AS, MRFC, MGCF, S-CSCF, P-CSCF, I-CSCF, or BGCF shown in Figure 18.4) generates and forwards the offline charging information to the CDF through interfaces Rf, Wf, and Gz (implemented by the diameter accounting protocol [9,10]). The CDF processes the charging information and generates a well-defined CDR. The CDR is then transferred to the charging gateway function (CGF; Figure 18.4b) via the Ga interface (implemented by the GTP' protocol [11,12]). Capable of performing CDR preprocessing including filtering, validation, consolidation, reformatting, and error handling, the CGF acts as a gateway between the IMS/UMTS network and the billing system (Figure 18.4g). It passes the consolidated offline charging data to the billing system via the Bx interface, which supports interaction between CGF and the billing system for CDR transmissions by a common, standard file transfer protocol (e.g., FTP or SFTP) [13,14]. The CDR files may be transferred in either push or pull mode.

In IMS offline charging, the diameter protocol is utilized, where the CDF is a diameter server and the IMS nodes are diameter clients. An IMS node sends the diameter accounting request (ACR) message carrying the offline charging information to the CDF. The CDF then acknowledges the IMS node with the accounting answer (ACA) message. The offline charging of a session-based service is processed as follows: Upon receipt of a diameter ACR message with accounting-record-type "start," the CDF opens an IMS session CDR. During the IMS session, the accounting information can be updated when the IMS node sends the ACA message with accounting-record-type "interim" to the CDF, which closes the session CDR on receiving an ACR message with accounting-record-type "stop." For event-based

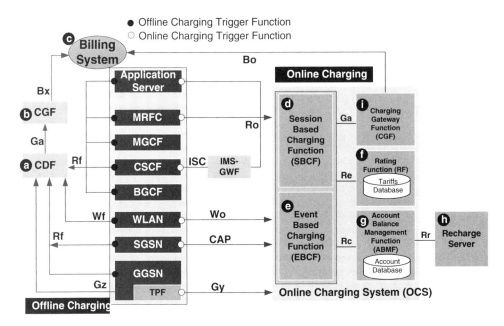

BGCF: Breakout Gateway Control Function
CDF: Charging Data Function
CSCF: Call Session Control Function
IMS-GWF: IMS Gateway Function
ISC: IMS Service Control
MRFC: Media Resource Function Controller
TPF: Traffic Plane Function

CAP: CAMEL Application Part
CGF: Charging Gateway Function
GSGN: Gateway GPRS Support Node
IMS: IP Multimedia Subsystem
MGCF: Media Gateway Control Function
SGSN: Serving GPRS Support Node

FIGURE 18.4 IMS charging architecture in UMTS.

procedures such as user registration or HSS interrogation, accounting information is transferred from an IMS node to the CDF using the ACR message with accounting-record-type "event." The CDF can be implemented as a network node or as a functionality residing in an IMS node.

TABLE 18.1

IMS Charging Architecture in UMTS

Term	Definition
Offline charging	A charging mechanism where charging information does not affect in real-time the service rendered
Online charging	A charging mechanism where charging information can affect, in real-time, the service rendered and therefore a direct interaction of the charging mechanism with session/service control is required
Offline CTF	A mandatory function integrated in the network element that provides offline charging functionality. It provides metrics that identify the user and the consumption of network resources, makes chargeable events from these metrics, and forward them to the CDF
Online CTF	A mandatory function similar to offline CTF that supports online charging with several enhancements. The online CTF must support requesting, granting, and managing resource

18.3.1 Offline Charging Record Correlation

Several charging correlations exist in an offline IMS session, including the correlation between CDRs generated by different IMS nodes, correlation between CDRs generated by the GPRS support nodes, and that between CDRs generated by different operators. These charging correlations are explained later. The CDRs generated in different IMS nodes are correlated by the IMS charging identifiers (ICIDs), which are globally unique across all 3GPP IMS networks for a period of at least one month [15]. In a CSCF CDR, the ICID is listed in the ICID field. A new ICID is generated for an IMS session at the first IMS node that processes the SIP INVITE message. The ICID is included in all subsequent SIP messages (such as 200 OK, UPDATE, BYE) for that IMS session until it is terminated. The SIP request includes the ICID specified in the P-Charging-Vector header [15]. This ICID is passed along the SIP signaling path to all involved IMS nodes. Each IMS node that processes the SIP request retrieves the ICID and includes it in the CDRs generated later. Therefore, CDRs from different IMS nodes within an IMS session can be correlated in the billing system through the ICID.

In an IMS session, the media (e.g., voice, video) packets are transferred through GPRS bearer 10 session. The GPRS charging information, such as the GGSN address and the GPRS charging identifier (GCID), is used for the GPRS (for the bearer level) and the IMS CDRs (for the application level) correlation. The GPRS charging information for the media session is included in the P-charging-vector header, and is passed from the GGSN to the P-CSCF. Note that the GPRS charging information for the originating network is used only within the originating network. Similarly, the GPRS charging information for the terminating network is used only within the terminating network. Within the SIP signaling path, the GGSN address, the GCID, and the ICID are passed from the P-CSCF to the S-CSCF and other IMS nodes. The S-CSCF also passes this information to the CDF. Charging correlation among different operators is achieved through the inter-operator identifier (IOI), a globally unique identifier shared between operator networks and service/content providers [16]. With IOIs, home IMS networks for both parties can settle account with each other. The originating IOI and the terminating IOI are exchanged by the SIP request and response messages. The originating S-CSCF includes the originating IOI in the P-charging-vector header of the SIP request. In a CSCF CDR, the IOIs are listed in the IOI field. Upon receipt of the SIP message, the terminating S-CSCF retrieves the originating IOI of the calling party's home IMS network. It then includes the terminating IOI in the P-charging-vector header of the SIP response message. Through the SIP this message, the originating S-CSCF retrieves the terminating IOI of the called party's home IMS network.

18.3.2 Offline IMS Call Procedure

Figure 18.5 depicts an offline charging example where the dashed lines represent signaling links, and the solid lines represent data and signaling links. The IMS call procedure is given in the following steps:

Step 1. By sending a SIP INVITE request, UE1 initiates an IMS call to UE2 [8]. The INVITE message contains the GPRS charging information (i.e., the GGSN address and the GCID). This message is first forwarded to the proxy CSCF (i.e., P-CSCF1) and then to the serving CSCF (i.e., S-CSCF1) of UE1. P-CSCF1 generates an ICID for this IMS call session.

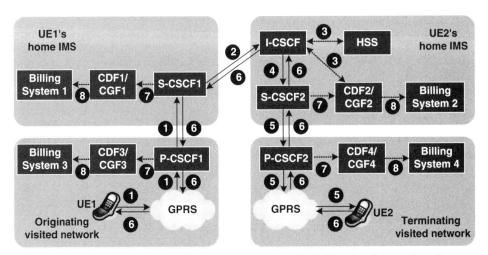

CDF: Charging Data Function
CSCF: Call Session Control Function
HSS: Home Subscriber Server
IMS: IP Multimedia Core Network Subsystem
S-CSCF: Serving CSCF

CGF: Charging Gateway Function
GPRS: General Packet Radio Service
I-CSCF: Interrogating CSCF
P-CSCF: Proxy CSCF
UE: User Equipment

FIGURE 18.5 An offline charging example of IMS call.

Step 2. S-CSCF1 sends the `INVITE` message to the interrogating CSCF (i.e., I-CSCF in UE2's home IMS). The `INVITE` message contains the ICID and the originating IOI of the IMS call.

Step 3. After interrogating the HSS, the I-CSCF sends the `ACR` message with accounting-record-type "event" to CDF2. Based on the event type, CDF2 creates an I-CSCF CDR (which can use for statistic purpose) and then replies the `ACA` message to the I-CSCF.

Step 4. In UE2's home IMS, the I-CSCF sends the `INVITE` request to the serving CSCF of UE2 (i.e., S-CSCF2).

Step 5. Through S-CSCF2 and the proxy CSCF of UE2 (i.e., P-CSCF2), the `INVITE` request is finally passed to UE2.

Step 6. If UE2 accepts the call, a SIP `200 OK` response will be sent back to UE1 through S-CSCF1 and P-CSCF1. The IMS nodes (S-CSCF1 and P-CSCF1) obtain the terminating IOI of the IMS call from the response message.

Step 7. On receiving the `200 OK` response, P-CSCF2 sends an ACR message to CDF4, S-CSCF2 sends an `ACR` message to CDF2, S-CSCF1 sends an `ACR` message to CDF1, and P-CSCF1 sends an `ACR` message to CDF3. These `ACR` messages include the accounting-record-type parameter "start" to indicate the opening of CDRs for an IMS call session. The CDFs create the related CDRs (i.e., S-CSCF CDRs and P-CSCF CDRs) and reply with the `ACA` messages. When the call is terminated, P-CSCF2, S-CSCF2, S-CSCF1, and P-CSCF1 send the `ACR` messages with accounting-record-type "stop" to the corresponding CDFs for closing the CDRs.

Step 8. Finally, these CDRs are transferred to the related billing systems through the corresponding CDF/CGF for offline processing. Note that in telecom industry, the operator may use a mediation device to collect CDRs before sending to the billing system. The mediation device performs filtering and consolidation on CDRs. The content contained in a CSCF CDR is listed in Table 18.2 [15].

18.4 IMS Online Charging

The 3GPP has recently devoted great effort to the IMS architecture support for online charging, the goal of which is to enable pre- and post-paid convergence. Such convergence is desired by operators who want to mitigate fraud and credit risks. When IMS supports real-time credit controls, operators can perform real-time credit balance checks, charges, discounts, and promotions. In addition, more personalized advice of charge and credit limit controls can be provided to the subscribers [18–20]. IMS supports online charging capabilities through the OCS, where an IMS node or an AS interacts with the OCS in real-time to process the user's account and controls the charges related to service usage. Details of the OCS and the IMS online charging are described in this section.

TABLE 18.2

Content in a CSCF CDR

Type	Description
Record type	The type of CSCF record (e.g., P-CSCF CDR or S-CSCF CDR)
Role of node	The role of the CSCF (either originating or terminating) [17]
Node address	The node providing the accounting information
Session ID	The call ID specified in the SIP message
Calling and the called party address	The addresses (e.g., SIP URL or TEL URL) of the calling party and the called party, respectively
Served party IP address	The IP address of either the calling or the called party
Service request time stamp	The time when the service is requested
Service delivery start time stamp	The time when the IMS session is started
Service delivery end time stamp	The time when the IMS session is released
Record opening time	The time when the CDF opens the CDR record
Record closure time	The time when the CDF closes the CDR record
Local record sequence number	A unique record number created by the IMS node
IOI	The globally unique identifiers of home IMS networks for MS1 and MS2, respectively
Cause for record closing	The reason for a CDR release. The reason can be "end of session," "service change," "CDF initiated record closure," and so on
IMS charging identifier (ICID)	A unique identifier generated by the IMS for the SIP session
List of SDP media components	The SIP request timestamp, SIP response timestamp, SDP media components (i.e., SDP media name, SDP media description, and GPRS charging ID (i.e., GCID)) and media initiator flag
GGSN address	The control plane IP address of GGSN context

In online charging, network resource usage is granted by the OCS based on the price or the tariff of the requested service and the balance in the subscriber's account. Figure 18.4 shows the OCS architecture defined in 3GPP 32.296 [7]. The OCS supports two types of online charging functions: session-based charging function (SBCF) and the event-based charging function (EBCF). The SBCF (Figure 18.4d) is responsible for network bearer and session-based services such as voice calls, GPRS sessions, or IMS sessions. Moreover, it is able to control session by allowing or denying a session establishment request after checking the subscriber's account. The EBCF (Figure 18.4e) is responsible for event-based services such as content downloading.

In the OCS, the rating function (RF; Figure 18.4f) determines the price/tariff of the requested network resource usage in real-time. Until now, the billing plan used in the Internet has been quite simple. Users are mainly billed with a flat rate, based on their subscription and access time. In mobile telecommunication networks, users are mainly billed based on their subscription and the call duration time, as well as a number of other parameters, such as the provision QoS, the location of the users, or the time segment.

The SBCF/EBCF furnishes the necessary information (obtained from the IMS/UMTS network nodes) to the RF and receives the rating output (monetary or nonmonetary credit units) via the Re interface. The RF handles a wide variety of ratable instances, including data volume, session/connection time, or service events (such as content downloading or messaging service). In addition, different charging rates can be adopted for different time segments (e.g., peak hours and off-peak hours).

The account balance management function (ABMF; Figure 18.4g) maintains user balances and other account data. When a user's credit depletes, the ABMF connects the recharge server (Figure 18.4h) to trigger the recharge account function. The SBCF interacts with the ABMF to query and update the user's account. The CDRs generated by the charging functions are transferred to the CGF (Figure 18.4i) in real-time. Note that both the CGF in Figure 18.4b and the CGF in Figure 18.4i collect CDRs through the Ga interface. The CGF passes the consolidated offline charging data and the online charging data to the billing system via the Bx and the Bo interfaces, respectively.

18.4.1 Diameter Credit Reservation Procedure

The online charging mechanism for circuit-switched services based on customized applications 16 for mobile network enhanced logic (CAMEL) technology [21,22] is not suitable for handling real-time rating for various IP-based multimedia services. Therefore, the 3GPP IMS adopts IP-based diameter protocol to achieve online charging capabilities [9,10]. In online charging services, the diameter credit control (DCC) protocol is used for communications between an IMS network node and the OCS. The IMS network node acts as a DCC client and the OCS acts as a DCC server. The OCS credit control is achieved by exchanging the credit control request (CCR) and the credit control answer (CCA) messages. A credit control message can be one of the following types:

- INITIAL_REQUEST initiates a credit control session.
- UPDATE_REQUEST contains update credit control information for an in-progress session. This request is sent when the credit units currently allocated for the session are completely consumed.
- TERMINATION_REQUEST terminates an in-progress credit control session.

- EVENT_REQUEST is used for one-time credit control, which can be DIRECT_DEBITING, CHECK_BALANCE or PRICE_ENQUIRY.

In the credit control process for event-based charging, the unit credits can be immediately deducted from the user account in one single transaction. The message flow is illustrated in Figure 18.6.

Step 1. When the network node (e.g., IMS AS) receives an event-based online service request, the debit units operation is performed prior to service delivery. The network node sends the CCR message with CC-request-type "EVENT_REQUEST" to the OCS. The requested-action field is set to "DIRECT_DEBITING."

Step 2. Having received the CCR message, the OCS determines the price of the requested service and then debits an equivalent amount of credit. The OCS acknowledges the network node with the CCA message to authorize the service delivery. This CCA message includes the granted credit units and the cost information of the requested service. Then the network node starts to deliver the event-based service.

Immediate event charging is adopted when the price for an event-based service is clearly defined. For immediate event charging, the network node must ensure that the requested service delivery is successful. If the service delivery fails, the network node should send another CCR message with CC-request-type "EVENT_REQUEST" and requested-action "REFUND_ACCOUNT" to the OCS, which then refunds the credit units that are previously debited [10].

Step 1. The network node (e.g., IMS AS) sends the CCR message with CC-request-type "INITIAL_REQUEST" to the OCS. This message indicates the amount of requested credit.

Step 2. Upon receipt of the CCR message, the OCS determines the price of the requested service and then reserves an equivalent amount of credit. Once the reservation has been made, the OCS replies to the CCA message to authorize the service delivery to the network node, which then starts to deliver the service.

Step 3. When the service delivery is complete, the network node sends the CCR message with CC-request-type "TERMINATE_REQUEST" to the OCS. This message terminates the credit control session and reports the amount of the consumed credit.

Step 4. The OCS debits the consumed credit units from the subscriber's account and releases the unused reserved credit units. The OCS acknowledges the network node with the CCA message. This message contains the cost information of the requested service.

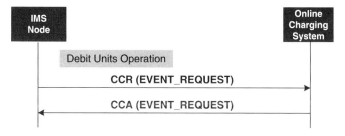

FIGURE 18.6 Diameter message flow for immediate event charging.

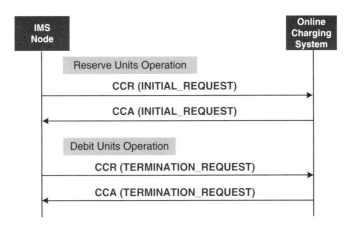

FIGURE 18.7   Diameter message flow for event charging with unit reservation.

The credit reservation procedure for session-based online charging includes three types of credit control operations: reserve units operation (Steps 1 and 2 in Figure 18.8), reserve units and debit units operation (Steps 3 and 4 in Figure 18.8), and debit units operation (Steps 5 and 6 in Figure 18.8).

Consider the scenario where a prepaid user requests an IMS session-based service from the AS. The following operations are execute:

Step 1. To start service delivery with credit reservation, the IMS AS sends the "INITIAL_REQUEST" CCR message to the OCS. This message indicates the amount of requested credit.

Step 2. Upon receipt of the CCR message, the OCS determines the price of the requested service and then reserves an amount of credit. After the reservation has

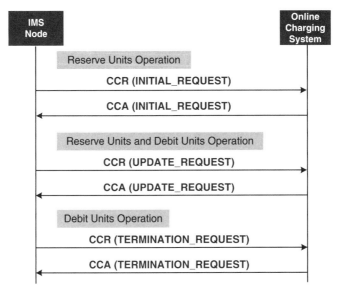

FIGURE 18.8   Diameter message flow for session charging.

been performed, the OCS acknowledges the IMS AS with the CCA message including credit reserving information.

Step 3. During the service session, the granted credit units may be depleted. If so, the IMS sends an "UPDATE_REQUEST" CCR message to the OCS. The IMS AS reports the amount of used credit and requests for additional credit units.

Step 4. When the OCS receives the CCR message, it deducts the used credit units and reserves extra credit units for the IMS AS. The OCS acknowledges the IMS AS with the CCA message with the amount of the reserved credit. Note that the reserve units and debit units operation (i.e., Steps 3 and 4) may repeat many times before a service session is complete.

Step 5. When the service session is complete, the IMS AS sends the "TERMINATE_REQUEST" CCR message to the OCS. This action terminates the session and reports the amount of the consumed credit.

Step 6. The OCS releases the unused reserved credit. The OCS acknowledges the IMS AS with the CCA message. This message may also contain the total cost of the service.

18.4.2 IMS Session-based Charging Scenario

Session charging with unit reservation is performed in credit control of IMS session-based services. Consider the online charging scenario shown in Figure 18.9, where UE1 makes an IMS call to UE2. The IMS message flow for session charging with unit reservation (see Figure 18.9) is described as follows:

Step 1. UE1 initiates a SIP INVITE request to S-CSCF1 through P-CSCF1.

Step 2. S-CSCF1 sends a CCR message with CC-request-type "INITIAL_REQUEST" to the SBCF of the OCS [4,23].

Step 3. On receiving the CCR message, the SBCF retrieves the account information and the subscribed profile from the ABMF. The SBCF then sends a Tariff Request message to the RF to determine the tariff of the IMS call. Based on the subscriber information, the RF replies to the SBCF with the Tariff Response message, which includes the billing plan and the tariff information for the IMS service.

Step 4. When the tariff information is received, the SBCF performs credit unit reservation with the ABMF. It then replies to S-CSCF1 with the CCA message containing the granted credit (e.g., specifying the number of minutes or bytes allowed).

Step 5. After obtaining the granted credit units, S-CSCF1 continues the call setup to UE2 (through P-CSCF2 and S-CSCF2). When UE2 accepts the call, a SIP 200 OK message is sent to UE1 through S-CSCF1 and P-CSCF1. At this point, the IMS session starts.

Step 6. During the service session, S-CSCF1 supervises the network resource consumption by deducting the granted credit units (quota). If the granted credit is depleted, the S-CSCF1 sends another CCR message with CC-request-type "UPDATE_REQUEST" to the SBCF. Through this message, S-CSCF1 also reports the amount of used credit and requests additional credit units for the remaining session. The SBCF deducts the consumed credit units and reserves extra credit units from the ABMF. The SBCF then acknowledges S-CSCF1 with the CCA message including the amount of the reserved credit. Note that this step may recur many times before a service session is complete.

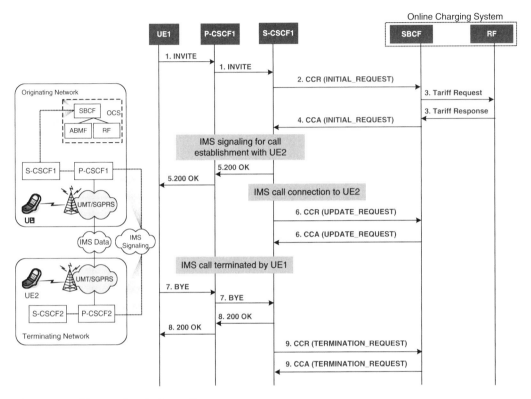

FIGURE 18.9　Message flow for ims call setup.

Steps 7–8. The IMS call is terminated by UE1 through sending a SIP BYE message. When the service session is complete, S-CSCF1 sends another CCR message with CC-request-type "TERMINATE_REQUEST" to the SBCF. The SBCF debits the consumed credit units from the subscriber's account with the ABMF. The SBCF then sends the CCA message to S-CSCF1.

18.5 Conclusions

IP Multimedia core network Subsystem offers a robust environment for developing and deploying multiple network services and data from a common network across different access networks. This chapter introduced the charging management for IMS services. Several 3GPP technical specifications describe online charging in IMS networks. For example, the 3GPP TS 32.260 and TS 32.296 standards describe an OCS with real-time charging functionality. In this chapter, we described the IMS architecture and the related charging components. Moreover, we demonstrated how the offline and online charging solutions are implemented for the IMS calls. Specifically, a convergent and flexible OCS used in 3GPP all-IP architecture is introduced. The OCS assists mobile telecom to incorporate data applications with real-time control and management. More details can be found in [24].

References

1. E. Larsson, "IMS: A key role in the future of telecoms," *TMCnet News*, April 2007.
2. S. Schwartz, "Standards watch: IMS could be the key to wireless–wireline convergence," Billing & OSS World, 2005.
3. IP Multimedia Services Charging, White Paper, HP, 2006.
4. IETF, "SIP: Session initiation protocol," IETF, RFC 3261, 2002.
5. Y.-B. Lin and A.-C. Pang, *Wireless and Mobile All-IP Networks*. Wiley, 2005.
6. H. Schulzrinne and J. Rosenberg, "The IETF internet telephony architecture and protocols," *IEEE Network*, Vol. 13, no. 3, 1999, pp. 18–23.
7. 3GPP. 3rd Generation Partnership Project; Technical Specification Group Service and System Aspects; Telecommunication management; Charging management; Online Charging System (OCS): Applications and interfaces (Release 6), 3G TS 32.296 version 6.3.0 (2006–09), 2006.
8. 3GPP. 3rd Generation Partnership Project; Technical Specification Group Core Network; IP Multimedia Subsystem (IMS); Stage 2 (Release 5), 3G TS 23.228 version 5.15.0 (2006–06), 2006.
9. 3GPP. 3rd Generation Partnership Project; Technical Specification Group Service and System Aspects; Telecommunication management; Charging management; Diameter charging applications (Release 6), 3G TS 32.299 version 6.10.0 (2007–03), 2007.
10. IETF, "Diameter credit-control application," IETF, RFC 4006, 2005.
11. 3GPP. 3rd Generation Partnership Project; Technical Specification Group Services and Systems Aspects; Telecommunication management; Charging management; Charging data description for the Packet Switched (PS) domain (Release 5), 3G TS 32.215 version 5.9.0 (2005–06), 2005.
12. 3GPP. 3rd Generation Partnership Project; Technical Specification Group Core Network; General Packet Radio Service (GPRS); GPRS Tunneling Protocol (GTP) across the Gn and Gp Interface (Release 6), 3G TS 29.060 version 6.15.0 (2006–12), 2006.
13. 3GPP. 3rd Generation Partnership Project; Technical Specification Group Service and System Aspects; Telecommunication management; Charging management; Charging Data Record (CDR) file format and transfer (Release 6), 3G TS 32.297 version 6.2.0 (2006–09), 2006.
14. IETF, "File transfer protocol (FTP)," IETF, RFC 959, 1985.
15. 3GPP. 3rd Generation Partnership Project; Technical Specification Group Services and Systems Aspects; Telecommunication Management; Charging Management; Charging data description for the IP Multimedia Subsystem (IMS) (Release 5), 3G TS 32.225 version 5.11.0 (2006–03), 2006.
16. 3GPP. 3rd Generation Partnership Project; Technical Specification Group Core Network and Terminals; IP Multimedia Call Control Protocol based on Session Initiation Protocol (SIP) and Session Description Protocol (SDP); Stage 3 (Release 5), 3G TS 24.229 version 5.18.0 (2006–09), 2006.
17. 3GPP. 3rd Generation Partnership Project; Technical Specification Group Service and System Aspects; Telecommunication management; Charging management; IP Multimedia Subsystem (IMS) charging (Release 6), 3G TS 32.260 version 6.7.0 (2006–09), 2006.
18. M. Koutsopoulou et al. "Charging, accounting and billing management schemes in mobile telecommunication networks and the internet," *IEEE Communications Surveys & Tutorials*, vol. 6, no. 1, first quarter 2004, pp. 50–58.
19. Z. Ezziane, "Charging and pricing challenges for 3G systems," *IEEE Communications Surveys & Tutorials*, vol. 7, no. 4, fourth quarter 2005, pp. 58–68.
20. S. Schwartz, "Diameter is growing up," Billing & OSS World, 2006.
21. Digital cellular telecommunications system (Phase 2+); Customised Applications for Mobile network Enhanced Logic (CAMEL); CAMEL Application Part (CAP) specification (Release 1997), GSM 09.78 version 6.5.0, (2000–07), 2000.

22. 3GPP. 3rd Generation Partnership Project; Technical Specification Group Services and System Aspects; Customized Applications for Mobile network Enhanced Logic (CAMEL); Service description; Stage 1 (Release 5), 3G TS22.078 version 5.15.0 (2005–03), 2005.

23. IETF, "SDP: Session description protocol," IETF, RFC 4566, 2006.

24. Y.-B. Lin and S.-I. Sou, *Charging for Mobile All-IP Telecommunications*, Wiley Publishing Co., 2008.

19

Commercial Interoperable VoIP IA Architecture

Bary Sweeney and Duminda Wijesekera

CONTENTS

19.1 Short History of the Old Order (TDM Voice Services)

The first commercial voice systems were manufactured in the late 1870s, but large-scale deployments did not occur until the early 1900s. The model selected to secure the systems that provided voice services remained relatively unchanged until the transition to VoIP in the early 1990s. Early voice systems were based on integrated hardware appliances that were located in the same physical location, typically contained in an equipment cabinet in a locked room designed to meet the power and air conditioning requirements of the appliances. Traditional systems were developed using proprietary operating systems and programming languages. Access to the system for configuration purposes usually required physical access to the system via the craftsman port and network management was provided via a closed out-of-band network.

The protocols used by legacy systems were not well understood outside of the voice vendor community and the standards associated with voice systems were expensive to obtain. Finally, signaling and bearer communications typically occurred along the same physical and logical path so it was easier to troubleshoot and secure.

The primary threats to legacy voice systems were toll fraud and tampering with subscriber privileges. Toll fraud is a criminal act that occurs when a hacker inappropriately accesses a managed telephone system and then probes the system for a weakness that will provide an outside telephone line. Once the outside line is obtained, the hacker uses the line to make calls anywhere in the world and toll fees are charged to the owner/operator of the system.

Tampering with subscriber privileges is a criminal act that occurs when a hacker inappropriately modifies the privileges purchased by a subscriber. An example would be the modification of a subscription profile to permit a subscriber to receive free roaming when their subscription charged for the service. The introduction of VoIP changed the fundamentals of securing voice systems.

19.2 The New World of VoIP

The emergence of Voice over Internet Protocol (VoIP) can be attributed to two developments in the mid-1990s. The first development was the acceptance of the Internet for commercial use. The second was the desire to bypass the PSTN for long-distance calls to avoid paying toll charges. As with the commercial IP model, VoIP developers adopted the distributed appliance concept for deploying VoIP systems. It is not uncommon to find the call control agent (CCA), which is the appliance that processes the signaling to complete a session, for the phones in one room in a facility with the media gateway for the connection to the PSTN in another room. Due to the smaller footprint of VoIP systems, it is possible to locate VoIP systems in locations that are not designed for them (e.g., the oft quoted "hallway broom closet"). In addition, due to the distributed nature of VoIP systems, it is no longer possible to locate all appliances associated with providing voice services in the same physical space. For example, the LAN components are typically distributed throughout the facility. Another change is that many of the VoIP operating systems offered by commercial providers use open source (i.e., LINUX) or well-known (i.e., Microsoft or Unix) operating systems and are developed using Java, Extensible Markup Language (XML), and C++ programming languages.

Unlike the model used in the development of legacy systems, the VoIP model consists of an open standards model with the Internet Engineering Task Force (IETF)[*] and the International Telecommunications Union-Technical (ITU-T)[†] taking the lead in the development of standards. Access to the systems for configuration purposes is usually accomplished via an Ethernet interface with standards-based protocols using both in-band and closed out-of-band networks. In addition to the traditional TDM voice threats, the migration to Internet Protocol (IP) has introduced a large number of new threats, which are summarized later in this chapter. In VoIP, typically the signaling and bearer portions of

[*] http://www.ietf.org/
[†] http://www.itu.int/ITU-T/

the communication take distinctly different paths with the end instruments often being the only appliances with the visibility of both portions. The change in the environment and technologies associated with VoIP has caused the information assurance (IA) community to develop new approaches to secure VoIP systems.

19.3 How Did We Get Here?

As stated earlier, the mechanisms used in voice networks for interfacing between different carriers and vendor equipment are essentially unchanged from 30 years ago. The mechanisms providing interoperability within current voice networks (both IP and TDM) are the traditional TDM interfaces such as primary rate interface (PRI),* common channel signaling system No. 7 (SS7),† or channel associated signaling (CAS). The reason for holding onto the past is the large investment (sunk costs) of carriers and businesses in TDM technologies. Approximately 60% of the cost of migrating from TDM to VoIP is associated with the replacement of TDM end instruments with VoIP end instruments.‡ As a result, the migration of voice to VoIP has been an inside-out approach with the carriers using IP for the long haul transport of voice services while the edge voice systems remain TDM. The migration has permitted the use of TDM technologies as the common interface for providing interoperability. However, within the next five years the evolution of VoIP systems within the enclave will result in the need for interoperable end-to-end VoIP solutions.

19.4 Interoperable VoIP Foundation

Information assurance as a standalone architecture provides limited or no benefit and is only useful as an overlay to secure an existing baseline architecture that provides some capability. This is also true for VoIP, which uses an interoperable VoIP foundation based on standards-based VoIP signaling and bearer protocols in combination with standards-based IP network management protocols. Before discussing the VoIP IA overlay, it is important to understand the protocols that are used for the VoIP foundation.

Industry has primarily chosen two standards-based protocols for providing interoperability signaling for VoIP systems. The first protocol is referred to as H.323¶ and is a collection of standards produced by the ITU-T. In the 1990s many vendors chose H.323 as the standards-based approach for implementing voice and video systems. Very few VoIP vendors implemented the protocol as specified by the ITU-T and instead implemented a proprietary version of H.323 in order to provide the vendor unique line-side features. As mentioned previously, most VoIP systems relied on conversions to TDM for

---

* Bellcore/Telcordia National ISDN1 (NI-1) and ISDN2 (NI-2) series of standards.
† ANSI T1.1XX-19XX series of standards.
‡ The costs referenced are focused on voice unique purchases and exclude any costs associated with upgrading the Local Area Network (LAN).
¶ http://www.itu.int/rec/T-REC-H.323/en

the long haul portion of the network and proprietary line-side protocols were a logical choice. However, the VoIP industry has found that H.323 has several technical limitations and the IA mechanisms to secure H.323 are somewhat limited and are not widely deployed. As a result, industry has turned to a protocol developed within the IETF called the session initiation protocol (SIP) for providing interoperable IP end-to-end signaling.

The requirements for implementing SIP are documented in the IETF's request for comment (RFC) 3261 and it is associated RFCs.* SIP was designed to provide IP end-to-end signaling support for VoIP and multimedia sessions by conveying session capabilities between session participants and allowing the session participants to negotiate compatible media types. One issue that will be discussed later is that RFC 3261 and many of the SIP RFCs describe the requirements and call flows associated with SIP proxies. However, most VoIP vendors are building SIP intermediaries (CCAs and session border controllers (SBCs)) that act as back-to-back user agents (B2BUAs) instead of SIP proxies. SIP proxies are transitive in nature, whereas B2BUAs typically anchor one SIP session on its input side and bridge that session to another SIP session on the outbound side. From an external view, the B2BUA will often appear to be two separate and distinct SIP sessions. Another issue is that SIP, by design, has limited IA capabilities and relies heavily on other protocols for its security. It is important to note that SIP and H.323 are only necessary for providing interoperability between multiple vendors. Within a vendor solution, proprietary signaling protocols are permitted and are typically used in order to permit the vendor to incorporate value-added features. The responsibility of the CCA is to convert between the proprietary signaling protocol on the line side and the standardized signaling protocol (SIP or H.323) on the trunk (WAN) side.

In comparison to SIP, industry has focused on a single protocol, the real-time transport protocol (RTP),† as the common protocol for providing the VoIP bearer transport. RTP is augmented by the real-time transport control protocol (RTCP), for monitoring RTP performance. Like SIP, both RTP and RTCP are independent of the underlying transport and network layers and have limited IA capabilities, which necessitate their reliance on other protocols for security.

Unlike the signaling and bearer protocols used to provide end-to-end IP interoperability, the story associated with VoIP network management protocols is quite complex. One constant within IP network management, regardless of whether the solution is voice or data related, is the use of the simple network management protocol (SNMP) for monitoring of the solution. Unfortunately, there are three versions of SNMP found within the VoIP industry with the most current version (version 3)‡ being the least deployed. Unlike SNMPv3, SNMPv1¶ and v2§ have limited IA mechanisms and by themselves are deemed to be insecure. After SNMP, the approaches used by VoIP vendors to provide network management vary and range from the use of the trivial file transport protocol (TFTP)** to the use of secure shell protocol (SSH)††. Due to the broad range of solutions taken, this chapter will focus on the protocols that should be used to secure the VoIP network management solution.

---

*   http://www.ietf.org/rfc/rfc3261.txt
†   http://www.rfc-editor.org/rfc/rfc3550.txt
‡   http://www.ietf.org/rfc/rfc3411.txt
¶   http://www.ietf.org/rfc/rfc1157.txt
§   http://www.ietf.org/rfc/rfc1902.txt
** http://www.ietf.org/rfc/rfc1350.txt
†† http://www.ietf.org/rfc/rfc4251.txt

19.5 VoIP Threat Summary

A complete discussion of the VoIP threats is beyond the scope of this chapter and other sources provide excellent discussions and are listed at the end. However, in order to understand the justifications for the commercial interoperable VoIP IA architecture it is important to provide a general discussion of the threats. Threats are typically segmented into one of six categories: authentication, authorization, nonrepudiation, confidentiality, integrity, and availability.

The first category is authentication and a common threat in this category is known as masquerading. Masquerading is often used by malicious users to obtain guarded information, deny services or resources, and/or alter the proper process for completing a session. Within VoIP, masquerading is typically used to improperly register/de-register end instruments (i.e., telephones) from the CCA, impersonating a legitimate subscriber (e.g., using caller-ID to pretend to be someone you are not), or as a means of accessing a CCA for malicious intent. A recent form of attack associated with masquerading is called "voice phishing," which uses voice mail as a mechanism for imitating email related "phishing attacks." The second category is authorization and is typically associated with a malicious user gaining access to data, services, or resources that they should not. In terms of VoIP, this threat is typically manifested in a malicious user gaining access to a provider's subscriber database, allowing the user to place calls that are not permitted within their subscription (e.g., international calls), and/or modifying the call detail records (CDRs) that are used for billing purposes. The third category is nonrepudiation and attacks against nonrepudiation are often used in combination with authentication and authorization attacks to obscure that an action has been taken. VoIP attacks associated with this threat often focus on obscuring configuration changes to the CCA or media gateway, such as the modification of the routing to ensure that sessions are routed to the malicious users network so that they can increase billing.

The fourth category is confidentiality and is primarily associated with eavesdropping on the signaling, bearer, and/or network management sessions. Eavesdropping on the signaling session is used to determine call patterns. Eavesdropping on the bearer session allows people to listen in on conversations, which is particularly onerous to the financial and medical communities. Eavesdropping on the network management sessions permits a malicious user to collect configuration information that facilitates other attacks.

The fifth category is integrity and within the VoIP network is typically focused on real-time sessions and databases. The real-time threats are typically concerned with the collection and replay of bearer, signaling, and network management sessions. For instance, a malicious user could capture the bearer stream, modify the conversation, and then replay the modified conversation to an unwitting party for malicious reasons. The database integrity threats are associated with the authentication and authorization threats and would focus on the corruption of VoIP databases.

The final category is availability and is typically associated with denial of service (DoS) and/or flooding attacks, which are similar to the attacks found within the data networks. A recent VoIP related flooding attack is called SPAM over Internet telephony (SPIT) and uses VoIP systems to inundate voice mail services in the same manner as SPAM inundates email servers. As mentioned previously, sometimes threats contain aspects of several categories. For example, a man-in-the-middle attack involves improper/lack of authentication and authorization, eavesdropping, and inappropriate modification of the information that may result in limited availability.

Although this section has identified numerous VoIP unique threats, many of the threats to VoIP system typically originate from the data systems that are connected to the converged LANs. Typically, a VoIP attacker compromises a data system on the same LAN and uses the compromised system as a staging point for the attack on the VoIP systems. Therefore, many of the threats associated with VoIP are the same threats as are found within the data network because successful attacks to the data network ultimately lead to compromised VoIP systems unless the VoIP systems are protected from the data systems, which is often not possible.

19.6 Secure VoIP Interoperable Protocol Architecture

As discussed in the previous two sections of this chapter, most protocols used within VoIP are inherently insecure and rely on other mechanisms for security. This section focuses on how to secure the most commonly used protocols within the VoIP environment. The first protocol that is discussed is RTP and the mechanisms used to secure the protocol. The following discussion is focused on the signaling protocols; H.323 and SIP. Finally, the protocol discussion concludes with the mechanisms used to secure the most common network management protocols.

The VoIP bearer path, which uses RTP as the base protocol, is secured using the secure real-time protocol (SRTP)* to provide confidentiality and integrity. SRTP is not a distinct protocol, but instead is a profile of RTP and is overlaid on RTP and RTCP to secure them. SRTP intercepts RTP and RTCP packets and then forwards an equivalent SRTP packet on behalf of the sending side, and intercepts SRTP packets and passes an equivalent RTP or RTCP packet up the stack on the receiving side. This is sometimes referred to as a "bump in the stack" implementation. One of the benefits of using SRTP is that it permits the use of end-to-end encryption and is amenable to network address translation (NAT) and network address port translation (NAPT), which is discussed later.

The primary utility provided by the SRTP RFC is the definition of cryptographic transforms that have been embraced by the VoIP community. In particular, the VoIP community has settled on the use of the advanced encryption standard (AES),† which is the default cipher, for providing confidentiality. In addition, SRTP defines hash message authentication code-secure hash algorithm 1(HMAC-SHA1)‡ as the default message authentication code for providing integrity. Due to the overhead associated with applying HMAC-SHA1 to the SRTP bearer packets, a truncated hash of 32 bits was chosen. This was deemed an acceptable tradeoff given the likelihood and usefulness of capturing and modifying the large number of encrypted bearer packets necessary to successfully conduct a replay attack.

One of the challenges associated with using SRTP is clipping that can occur at the start of a session while the key exchange is occurring between the two end instruments (i.e., phones). To avoid clipping, commercial VoIP vendors have chosen to use an approach called session descriptions (SDES)¶ when using SIP and H.235 when using H.323. SDES

---

* http://www.ietf.org/rfc/rfc3711.txt
† http://csrc.nist.gov/publications/fips/fips197/fips-197.pdf
‡ http://www.ietf.org/rfc/rfc2104.txt
¶ http://www.ietf.org/rfc/rfc4568.txt

```
INVITE sip:2222222;phone  -context=telecom.com@telecom.com;user  =phone SIP/2.0
Via: SIP/2.0/TLS clientA.telecom.com:5061;branch=z9hG4bK74bf9
Max-Forwards: 70
To: <2222222;phone -context=telecom.com@telecom.com;user  =phone>
From: <1111111;phone -context=telecom.com@telecom.com;user  =phone>;tag=9fxced76sl
Call -ID: a84b4c76e66710
CSeq: 314159 INVITE
Contact: <sip:1111111@10.10.10.1>
Contact-Type: application/sdp
Content-Length: 142
c=IN IP4 10.10.10.10
m=audio 49170 RTP/SAVP 0
a=crypto:1 AES_CM_128_HMAC_SHA1_32
    inline:NzB4d1BINUAvLEw6UzF3WSJ+PSdFcGdUJShpX1Zj|2^20|1:32
```

FIGURE 19.1 SIP INVITE example.

is an approach that embeds the encryption key in the session negotiation messages used to establish the VoIP session. SDES defines a new Session Description Protocol (SDP) attribute, called "crypto" for conveying the cryptographic session key. The H.235 protocol within H.323 provides a similar functionality using the h235Key field. Although the use of SDES reduces clipping, some clipping may occur if early media is transmitted during the session establishment process. Early media is associated with the caller transmitting bearer packets prior to the acceptance of the session by the receiving party and is associated with ringing tones and announcements. The impact of clipping is dependent on the vendor implementation. Because SIP and H.323 messages are encrypted and checked for integrity (discussed later), the inclusion of the SRTP encryption key in the signaling message provides a secure method for key exchange. Figure 19.1 shows how the cryptographic key is embedded within the SDP for a SIP INVITE.

A challenge for commercial VoIP providers is how to complete an end-to-end VoIP session where the intermediaries in the session may or may not be trusted. An obvious preference exists for end-to-end encryption, but the reality is that a hop-by-hop transitive trust security model is necessary due to the need for hierarchical signaling. Let's back up for a minute and discuss why the hierarchical signaling model is necessary.

End instruments (i.e., telephones) are inherently simple devices with limited processing and storage capacity. The routing and directory information associated with global voice networks is beyond the capacity of almost all end instruments. As a result, end instruments rely on CCAs for routing and directory information. In addition, the CCA is typically the source for configuration files and attachment (i.e., authentication and authorization) to the voice network. Even in the simplest of commercial end-to-end VoIP networks, the signaling is hierarchical with the CCA involved in the signaling path as shown in Figure 19.2.

As a result of the hierarchical signaling, an end-to-end trust model is not feasible resulting in the need to rely on a hop-by-hop security model. This requires that the appliances on both ends of each hop in the signaling path mutually authenticate to each other, to include the CCA and the end instrument. Based on the assumption that every hop is mutually authenticated, it is reasonable to assume that the end instruments are mutually authenticated via transitive trust. Vendors have considered mutually authenticating between end instruments after the session establishment, but the issues raised with early media in the preceding discussion preclude this as a viable option.

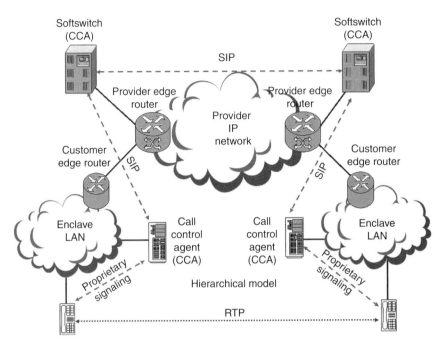

FIGURE 19.2   Hierarchical VoIP signaling.

Returning to the discussion at hand, with the assumption that a hop-by-hop transitive trust security model is necessary, the protocol used to establish the hop security is dependent on the signaling protocol chosen. Internet protocol security (IPsec),* in combination with H.235, is the ITU-T defined protocol for securing H.323 and is a viable alternative for securing SIP. However, industry has chosen the transport layer security (TLS)† protocol as the mechanism to secure SIP. First, let us discuss H.323 and its mechanisms for providing secure signaling. As mentioned previously, H.323 is a collection of standards from the ITU-T that are used by VoIP applications for signaling. The H.235.1 baseline security profile, which was developed by the International Multimedia Telecommunications Consortium (IMTC) and the European Telecommunications Standards Institute (ETSI), Telecommunications and Internet Protocol Harmonization Over Networks Project (TIPHON),‡ provides integrity to H.323 by defining mechanisms to provide a HMAC-SHA1 to the H.323 signaling packet.

To complete the package, IPsec is used to secure the entire H.225.0 and H.245 channel by providing confidentiality of the packets in transit over the IP network. The confidentiality is provided within IPsec using the AES encryption standard, which was also used to provide confidentiality within SRTP. IPsec is not an ITU-T standard like the rest of the H.323 protocols and is published by the IETF. At the time of this writing a new and improved version¶ of IPsec is published, but due to its recent publication, most vendors are

---

* http://www.ietf.org/rfc/rfc\2401.txt

† http://www.ietf.org/rfc/rfc2246.txt

‡ TIPHON is now named Telecoms & Internet converged Services & Protocols for Advanced Networks (TISPAN).

¶ http://www.ietf.org/rfc/rfc4301.txt

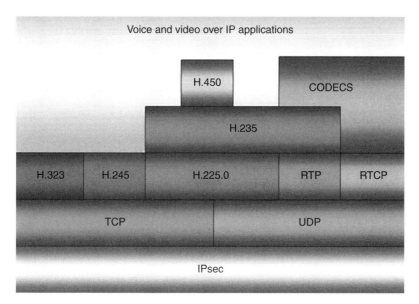

FIGURE 19.3 H.323 protocol stack.

still deploying the older (1998) version. IPsec was chosen over the other H.323 approaches because the encryption occurs at the IP layer and is capable of providing confidentiality for both the user datagram protocol (UDP) and transmission control protocol (TCP) portions of the H.323 protocol. As mentioned previously, H.323 is not defined in a single standard, but incorporates several different standards to work properly. The interrelation of the various standards is shown in Figure 19.3.

In contrast to H.323, industry has chosen TLS instead of IPsec for securing SIP. Like IPsec, at the time of this writing a new and improved version* of TLS is published, but due to its recent publication, most vendors are still deploying the older (1999) version. TLS was derived from the earlier versions of secure sockets layer (SSL) and for the purposes of this discussion, SSL is equivalent to TLS. Like IPsec and SRTP, the encryption algorithm and integrity mechanism used by TLS is AES and HMAC-SHA1. The advantage of industry selecting the same encryption and integrity mechanism for all three protocols is that it allows the standards community and the vendors to focus on monitoring, improving, and securing a limited subset of approaches. The disadvantage, of course, is that a compromise of either mechanism compromises all systems, but industry performed the cost-benefit analysis prior to making the decision. One advantage of using SIP with TLS is that it makes it easier to filter the signaling packets since the port used for SIP with TLS is limited to port 5061. Figure 19.4 shows how SIP and SRTP fit within the Open Systems Interconnection Basic Reference (OSI) model.

As stated earlier, the discussion of network management protocols is more complex. The network management foundation for VoIP systems is SNMP. Although SNMPv3[9] was published in 2002, most vendors are still implementing SNMPv1[10] and v2[11]. SNMPv3 has inherent IA capabilities, but for vendors that choose to use SNMPv1 and SNMPv2, they secure the protocol using IPsec. Due to the limitations of SNMP, many other protocols are

---

* http://www.ietf.org/rfc/rfc4346.txt

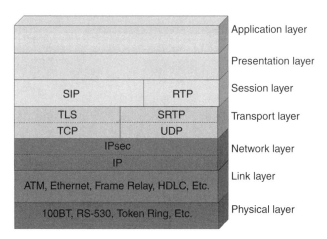

FIGURE 19.4 SIP and SRTP protocol stack.

used to provide network management. The protocol that is becoming increasingly popular is the hypertext transfer protocol over secure socket layer (HTTPS).* The reason is that the increasing configuration complexity of the VoIP systems require better configuration graphical user interfaces (GUIs), which are being manifested in web browser configuration interfaces. Since most commercial secure web browser approaches use HTTPS, it was a logical for the VoIP community to adopt HTTPS as the approach used to secure web-based applications.

Although HTTPS is becoming the second most commonly used secure protocol (SNMP is still number 1) due to the wide use of web browser-based interfaces, many systems still use command line interfaces (CLI) for network management. The typical protocol used to provide remote logon CLI capabilities in a secure manner is SSH,[13] which is commonly known as SSHv2. SSH like many other protocols within the VoIP environment relies on TLS as one of its three components. Like TLS and IPsec, SSH uses HMAC-SHA1 for integrity. However, commercial VoIP vendors typically use the Rivest, Shamir, and Adelman (RSA) algorithm as the default algorithm for providing confidentiality.

Now that we know how VoIP systems provide confidentiality and integrity, the next issue that should be discussed is the mechanisms to achieve interoperable authentication on a global scale in a secure manner. As of this writing (December 2007), it appears that the Public Key Infrastructure (PKI) is the only viable alternative for authenticating VoIP systems and VoIP network management personnel in a secure manner on a global basis. The installation of PKI certificates on VoIP systems allow VoIP systems to mutually authenticate each other using TLS and IPsec. The emergence of smart cards within industry for identifying individual users appear to have the most potential for mutually authenticating network management personnel and systems to the VoIP systems and provide a strong authentication capability that exceed the capabilities of the past. Smart cards are typically wallet size plastic cards that contain a picture on the outside and a PKI certificate internally. The PKI certificate is only transmitted if the user of the smart card enters a password thereby providing two factor authentication. Industry plans to use PKI to facilitate the hop-by-hop transitive trust security model needed for end-to-end VoIP.

---

* http://www.ietf.org/rfc/rfc2818.txt

19.7 Authorization Defense-in-Depth

As with every security architecture, a defense-in-depth approach is the best approach for VoIP.

In terms of authorization, the cornerstone of the defense-in-depth approach is the use of physical security to control access to the VoIP systems when feasible. Although it is not feasible to control access to the end instruments, it is often feasible to control access to the network infrastructure appliances (LAN switches & Routers) and signaling appliances (CCAs and media gateways). A simple locked door often provides adequate protection and providing multiple barriers, such as a lock on a cabinet housing a LAN switch in the open, is preferred to no barriers.

The next layer of the defense-in-depth strategy focuses on securing the appliances from unauthorized access. Securing access to a VoIP appliance uses the same approach as is taken for data appliances and involves ensuring user roles are defined and limited to the minimum set of actions required. Next, the operating system should be hardened by deleting unnecessary applications and installing the most current patches. Finally, it is essential to ensure that the system is appropriately audited and that the logs are maintained on a separate appliance in case of catastrophic failure. Additional defenses may be added depending on the specific threats associated with a system.

After securing the VoIP appliances and providing physical security, the next critical tenet of a defense-in-depth strategy is the separation of components (i.e., traffic, appliances, and users) and/or services from each other based on their characteristics. However, a converged network requires the opposite in that appliances within a converged network may service the voice, data, and video applications. As a result of this conflict, the interactions between the various component segments must be controlled to ensure that an attacker that gains access to one segment does not gain access to, nor can affect, the other segments. In addition, interaction control between various segments is also used to prevent configuration or user errors in one segment from impacting other segments. The actions of normal users of converged network services must not affect the other services, more specifically the voice service. The principal mechanisms that are used within this architecture for segmenting the network are virtual LANs (VLANs), segmented IP address space or subnets, and virtual private networks (VPNs) and are used in combination with filters, access control lists (ACLs), and stateful packet inspection firewalls (VoIP stateful firewalls) to control the flow of traffic between the VLANs and VPNs.

The 802.1Q VLAN tagging of the packet for the appropriate VLAN should occur at the end instrument (telephone), but may occur at the first LAN switch if the appliance is not capable of 802.1Q VLAN tagging. If the end instrument (telephone) supports a subtended data computer/workstation, it has several unique functions that it must perform. The first function is that it should be capable of VLAN tagging the data traffic with a different VLAN tag than the voice traffic. In addition, it must also be capable of routing the VoIP traffic and the data traffic to the appropriate VLAN. The objective of this function is to ensure that the data applications on the workstation do not have visibility to the voice-related packets.

Demilitarized zones are created for VoIP and data appliances that must service multiple segments of the converged network. An example is the CCA, which must service voice, video, videophone, and softphone appliances. Because access control between the various VLANs is significantly more complex, filtering is not adequate to achieve this goal and VoIP stateful firewalls are needed to ensure that only authorized packets are able to transit

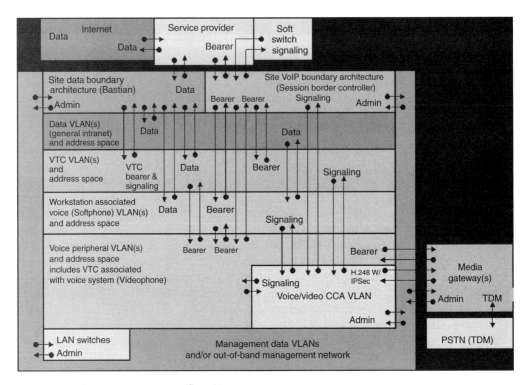

FIGURE 19.5   Component interaction flow diagram.

the VLAN boundaries. In addition, if the softphones are used in remote connectivity situations, such as for a long local for a mobile user, the system must be capable of supporting a VPN for VoIP traffic from the PC to the CCA. It is essential that the data and VoIP traffic be separated into the appropriate VLAN at the earliest point in the path. Figure 19.5 provides a depiction of the interactions between various VLANs within an enclave. The illustrations in this section are notional and address the scenario where ancillary services (i.e., DHCP, RADIUS, etc.) are internal to the system. The information is only provided as reference material and each enclave will need to determine their VLAN needs and boundaries based on their tailored requirements and security profile.

The last VoIP authorization defense-in-depth aspect that will be discussed in this chapter is requirement for a unique protection at the boundary of the enclave. VoIP suffers from certain difficulties with traditional perimeter defense mechanisms, such as NAT and data firewall behavior. Many protocols designed by the IETF used within the VoIP environment, such as SIP, are designed as end-to-end protocols. The end-to-end model is broken by the presence of firewalls and NAT appliances. In large deployments of VoIP, a specialized appliance is needed to facilitate the coexistence of VoIP and perimeter defense mechanisms. Industry has labeled this specialized appliance a session border controller (SBC) to distinguish it from traditional data firewalls. Before describing the requirements and functions associated with the SBC, it is important to explain the difficulties that VoIP protocols experience crossing network boundary devices and to explain the common types of solutions emerging from the standards communities and major vendors.

As mentioned previously, the use of NAT introduces several problems to an end-to-end protocol security model. The use of NAT is required at many enclave boundaries. In a traditional NAT employment, the NAT is conducted at the Network Layer (Layer 3) of the OSI model (NAPT is conducted at Layer 4). However, the VoIP environment requires that the SBC perform NAT at a higher layer in order to properly process the SIP messages. The common scenario provided for NAT is the use of private addressing in two remote enclaves, with a public address space in the interconnecting WAN.

During the initial SIP offer/answer exchange, both the originating and terminating SIP user agents (UAs) specify in the SDP payload the desired IP address and port combination for the caller and callee to receive the associated media stream and to properly direct the signaling stream. SIP UAs within the enclave may use private addressing for topology hiding reasons. A problem occurs when private addressing is used within the SDP payload since the private address is not resolvable from the WAN side of the NAT. A traditional NAT device will change the IP source address and/or port combination at packet header level, but not the IP address within the SDP payload. Consequently, the callee, or UA in the remote enclave, will not have the correct IP address to respond to from a signaling perspective and the call setup will fail. Even if the call is established from a signaling perspective, the bearer stream would be sent to the wrong address/port and the session would not be established.

An additional problem is that the signaling and bearer paths are different IP sessions, and NAT bindings in a traditional NAT appliance are finite in nature (not correlated). A scenario may result in that a VoIP session may continue for many minutes with active RTP streams, but with no signaling messages. Since there is no signaling, the signaling NAT binding could time out and the session would not be able to terminate properly. Difficulties experienced by SIP signaling and real-time transport protocol/real-time control protocol (RTP/RTCP) media protocols passing through firewalls stem mainly from the fact that bearer stream port numbers are selected dynamically for each call from a large pool of potential port numbers. Allowing this large range of potential port numbers to be open at the enclave boundary is unacceptable. The SBC needs to know which ports to open temporarily, when to open them and when to close them. Placing this functionality (essentially an application level gateway) into data firewalls has not happened yet and carries with it some disadvantages, for example, having the firewall perform actions it is not supposed to handle.

Solutions for these issues have evolved slowly within the standards bodies and major equipment vendors. A myriad of suggestions have been offered from both communities, yet none has attained widespread acceptance. In the interim, some vendors have developed limited solutions for their products to allow functionality, and several have acquired SBC functionality to enhance their product lines, however, no clear solution to achieve multivendor interoperability has come out of the major vendor arena. As a result, interoperability problems often occur between different vendor solutions due to the proprietary nature of the vendor unique solutions.

In response to a need for a solution to various problems with VoIP in the areas of firewall/NAT traversal, topology hiding, and lawful intercept, a large number of startup companies have produced SBCs. Topology hiding is accomplished by processing the SIP messages and performing the appropriate IP address and port translations within the IP header and the SDP payload. Figure 19.6, shows the typical end-to-end call sequence that establishes a SIP session. The diagram shows how the media is bi-directionally anchored by the SBC to ensure that topology hiding is provided.

Several issues are encountered by SBCs in attempting to meet the topology hiding requirements due to the VoIP signaling hierarchy and by the requirement to allow call forwarding and call transfer. The first issue is found in SBCs that front a Softswitch. Upon

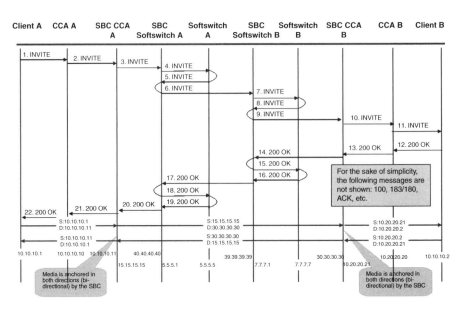

FIGURE 19.6   Typical end-to-end SIP call sequence.

receipt of an SIP INVITE, the SBC does not know whether that session terminates within the enclave or will be forwarded to another enclave (after processing by the Softswitch). Due to the uncertainty, the SBC must bi-directionally anchor the media stream. If the Softswitch returns the INVITE to the SBC for forwarding to the next hop, the SBC must restore the original IP address and port number so that it is no longer involved in the media stream associated with that session. The second issue associated with topology hiding is related to call forwarding and call transfer scenarios where the session no longer terminates within the enclave. Upon notification that a session is being forwarded or transferred, the SBC associated with the forwarding or transferring party must restore the original IP address and port number associated with that session to ensure that the media is not improperly anchored at the enclave since the session is no longer associated with the enclave. The next issue concerns the ability of the SBC to determine the appropriate next hop for signaling. For an SBC fronting a CCA, the SBC has only a primary and secondary TLS path in which to forward the SIP messages. The primary TLS path is to the SBC fronting the primary Softswitch for the CCA and the secondary path is to the SBC fronting the secondary Softswitch for the CCA. However, for an SBC fronting the Softswitch, it has numerous TLS sessions associated with the subtended CCAs of the Softswitch and with the other Primary and Secondary Softswitches for all the remote CCAs. Since the EBC does not have its own location service, it must rely on the Softswitch to inform it of the appropriate next hop. The Softswitch informs the SBC of the next hop using the ROUTE header construct defined in SIP. It is important to note that this issue is not relevant to INVITEs arriving from the WAN since the SBC is associated with one and only one SIP signaling system on the line side and always forwards INVITEs arriving from the WAN to its associated SIP signaling system (i.e., CCA or Softswitch). In addition to performing topology hiding, the SBC provides several other functions to include "pinholing."

"Pinholing" is accomplished by opening and closing "pinholes" that only allow approved sessions to transit the SBC based on the SIP messages. The coupling of the signaling and bearer stream requires that both streams must transit the SBC. If H.323 video is also transiting the SBC, the "pinholes" may also be associated with the H.323 messages. In addition, the SBC must have a timer associated with each "pinhole" to ensure that "pinholes" do not remain open indefinitely if a signaling message is not received to close the "pinhole." If the SBC closes the "pinhole," it must send a BYE message in each direction to notify the next hop signaling appliances that the session has been terminated.

Filtering is accomplished by allowing targeted IP flows to transit the SBC based on their "6 tuple" (source/destination IP address, source/destination port number, DSCP, and protocol identifier). Compared to "pinholes," filtering is a manual process and is not dynamic in nature. A type of traffic flow associated with filtering is a SNMPv3 session from the NMS to a CCA that is active all the time and is well defined. If the SBC detects an anomalous condition (DoS attack, abnormal number of invalid SIP requests, etc.) it must have the ability to notify the appropriate NMS of the event.

19.8 Softphones—The Way of the Future

Traditionally, end instruments (telephone) have taken the form of a dedicated hardware appliance with limited processing, memory, and storage capabilities. In VoIP we begin to see the emergence of the Softphone as the end instrument (telephone) of the future. There are numerous reasons for this emergence, but ultimately the reasons are distilled to money and functionality. Softphones are inherently cheaper than dedicated end instruments for the customer and vendor. This assumes that the cost of the host PC is a sunk cost. It is cheaper for the vendor since most CCA vendors no longer manufacture their own hard phones. As a result, the percentage of the profit and revenue generated by selling hard phones is passed to a third party company. In comparison, all vendors are developing their own softphones and any profit and revenue generated by purchase of a softphone is realized by the vendor.

From a functionality perspective, softphones provide extensive new benefits by allowing VoIP vendors to converge collaborative, instant messaging, voice, and video services within the same appliances. With the emergence of the Unified Communication offerings from every major vendor, it is the way of the future that traditional hard phone end instruments will become the rotary phone of the future.

Given that softphones are here to stay, how do we secure them? This is an issue that all of industry is currently grappling with as of this writing (December 2007). The first layer of defense is to segment them within their own VLAN away from the data and VoIP unique VLANs. If possible, the system should mark VoIP traffic differently than data traffic using the DiffServ Code Point (DSCP) or the VLAN tag. The second layer is to ensure that the platform is hardened to the extent possible and that all patches and upgrades are installed. Anti-virus, anti-spyware, and platform independent firewall protection is a must to protect the system from worms, viruses, and Trojans and should be the next layer of protection. If possible, application by application user profiles should be defined to limit the applications that can be compromised once an attacker gains access to the system. Finally, audit logs become critical to the successful investigation in the aftermath of an attack, which will inevitably occur. The deployment of softphones opens up a Pandora's box of threats associated with VoIP and should only occur after a thorough benefit versus security analysis is conducted.

19.9 Summary

Hopefully, I have convinced you that VoIP is a different beast from the traditional voice approach. As with alcoholics anonymous, acknowledging your problem is the first step in recovery. Fortunately, industry readily acknowledges that VoIP introduces unique challenges that must be overcome to provide reliable, secure, and quality voice services and are investing heavily in research and products to overcome the challenges. VoIP is here to stay and although additional security vulnerabilities will be discovered, it is expected that VoIP will be as secure as traditional voice technologies as it matures.

Bibliography

1. F. Andreasen, M. Baugher, and D. Wing, "Session description protocol (SDP) security descriptions for media streams," RFC 4568, July 2006, http://www.ietf.org/rfc/rfc4568.txt (accessed January 26, 2008).
2. M. Baugher, D. McGrew, M. Naslund, E. Carrara, and K. Norrman, "The secure real-time transport protocol (SRTP)," March 2004, http://www.ietf.org/rfc/rfc3711.txt (accessed January 26, 2008).
3. T. Dierks and E. Rescorla, "The transport layer security (TLS) protocol version 1.1 RFC 4346," April 2006, http://www.ietf.org/rfc/rfc4346.txt (accessed January 26, 2008).
4. D. Endler and M. Collier, *Hacking Exposed: VoIP*, McGraw-Hill, New York, 2007.
5. ETSI TS 102 165-1, Version 4.1.1. *Telecommunications and Internet Protocol Harmonization Over Networks (TIPHON) Release 4; Protocol Framework Definition; Methods and Protocols for Security; Part 1: Threat Analysis,* European Telecommunications Standards Institute (ETSI), 2003.
6. Federal Information Processing Standards Publication 197 "Advanced Encryption Standard," November 26, 2001, http://csrc.nist.gov/publications/fips/fips197/fips-197.pdf (accessed January 26, 2008).
7. D. Harrington, R. Presuhn, and B. Wijnen "An architecture for describing simple network management protocol (SNMP) management frameworks RFC 3411, December 2002, http://www.ietf.org/rfc/rfc3411.txt (accessed January 26, 2008).
8. ITU-T, "H.323—Packet-based multimedia communications systems," http://www.itu.int/rec/T-REC-H.323-200606-I/en (accessed January 26, 2008).
9. S. Kent and K. Seo, "Security architecture for the internet protocol RFC 4301," December 2005, http://www.ietf.org/rfc/rfc4301.txt (accessed January 26, 2008).
10. H. Krawczyk, M. Bellare, and R. Canetti, "HMAC: Keyed-hashing for message authentication. RFC 2104," February 1997, http://www.ietf.org/rfc/rfc2104.txt (accessed January 26, 2008).
11. D. R. Kuhn, T. Walsh, and S. Fries, *Security Considerations for Voice Over IP Systems: Recommendations of the National Institute of Standards and Technology,* Special Publication 800-58.
12. V. Kumar, M. Korpi, and S. Sengodan, "IP Telephony with H.323," John Wiley & Sons, Inc., New York, 2001.

13. D. Persky, *VoIP Security Vulnerabilities*, Fall 2007, http://www.sans.org/reading_room/whitepapers/voip/2036.php (accessed January 2008).
14. E. Rescorla, "HTTP over TLS," RFC 2818, May 2000, http://www.ietf.org/rfc/rfc2818.txt (accessed January 26, 2008).
15. E. Rescorla, *SSL and TLS*, Addison-Wesley, Indianapolis, 2001.
16. J. Rosenberg, H. Schulzrinne, G. Camarillo, A. Johnston, J. Peterson, R. Sparks, M. Handley, and E. Schooler, "SIP: Session initiation protocol," RFC 3261, June 2002, http://www.ietf.org/rfc/rfc3261.txt (accessed January 26, 2008).
17. H. Schulzrinne, S. Casner, R. Frederick, and V. Jacobson, "RTP: A transport protocol for real-time applications," RFC 3550, July 2003, http://www.ietf.org/rfc/rfc3550.txt (accessed January 26, 2008).
18. T. Ylonen and C. Lonvick, "The secure shell (SSH) protocol architecture," RFC 4251, January 2006, http://www.ietf.org/rfc/rfc4251.txt (accessed January 26, 2008).

Part IV

Reliability and Security

20

Security Issues of VoIP

Miguel Vargas Martin, Patrick C. K. Hung, and Adrienne Brown

CONTENTS

Recently, Voice over Internet Protocol (VoIP) technology has been increasingly attracting attention and interest in the industry. VoIP applications such as IP telephony systems involve sending voice transmissions as data packets over private or public IP networks as well as reassembling and decoding on the receiving side. Concern for security is the major barrier that prevents many businesses from adopting VoIP technologies.

IP telephony security is challenging as it involves many disciplines, from encryption to access control policies. In general, the engineering of a security policy starts with risk analysis and ends with a set of security assertions that is ready for integration into the security policy framework.

In this chapter, we first discuss security threats and vulnerabilities relating to VoIP protocols and technologies. We then discuss a security policy from the perspectives of communication and application security properties such as confidentiality, integrity, and availability for constructing a security framework for VoIP applications. Finally, we propose a security policy framework from the perspectives of network security for constructing a security framework for mobile VoIP applications in the future.

20.1 Introduction

This chapter begins with an analysis of why Voice over Internet Protocol (VoIP) is growing in popularity and how it faces security threats due to both its own protocol characteristics and the type of network it transmits data across. Security threats are classified and considered under the CIA model which emphasizes the need for confidentiality, integrity and availability. When analyzing each of these threats at an individual level, strategies already in use for similar problems may be applied to mitigate each threat. However, the collection of various protocols used by VoIP and the modularity of the VoIP system suggest that an end-to-end approach is needed to address the complexity of the VoIP security problem domain. More importantly, the IP network used to transmit VoIP calls is often the Internet, which is distributed across many countries and jurisdictions. Firewall policies, patch levels, and other mechanisms previously discussed would not be directly accessible to the administrator. We describe a model system for a VoIP security policy service designed for mobile VoIP systems. This model would allow the automatic query and processing of security policy requirements and security postures of VoIP applications and links along the call path, such that security could be verified and secured by a VoIP caller traversing the WAN.

20.1.1 Rationale for the Use of VoIP

Today, the Web is the main medium by which companies are carrying on their business. It is obvious that the Web has forced many companies to reorganize their business strategies and structures by using versatile Internet technologies in order to remain competitive in a business environment. VoIP technology is increasingly attracting attention and interest in the industry. A number of VoIP applications offer all the features available (e.g., caller id, call waiting, etc.) on traditional Private Branch eXchange (PBX) solutions.

A number of companies are moving to VoIP technology, because it allows the use of their current network infrastructure for transferring both voice and regular data traffic. This technology enables a lot of savings by reducing maintenance costs, long-distance fees, and other costs associated with a traditional telephone network.

20.1.2 Protocol Components of VoIP

In general, VoIP technologies employ a suite of protocols including signaling protocols such as session initiation protocol (SIP), and data control and transfer protocols such as transmission control protocol (TCP), real-time transport protocol (RTP), user datagram protocol (UDP), and Internet Protocol (IP). VoIP applications such as IP telephony systems involve sending voice transmissions as data packets over private or public IP networks as well as reassembling and decoding on the receiving side. However, concern for security is always the major barrier that prevents many businesses from employing VoIP technologies [1]. Security is considered as an essential and integral part of VoIP applications [2].

20.1.3 Intruders in a Traditional Telephone Network and a VoIP Network

Traditionally, intruders are known as persons who have gained unauthorized access to PBX or voice mail systems, and obtain free telephone calls by manipulating computer systems. Some studies even show that the intruding event occurs because of a perpetrator's need for revenge, sabotage, blackmail, or greed [3].

In VoIP applications, it is also conceivable that intruders can intercept incoming and outgoing phone numbers, break into someone's voice mail, or even listen in on confidential conversations over IP networks. One real case of VoIP application hacking happened in a company called Sunbelt Software in the United States. An intruder gained access to their VoIP application system through its remote access features. As a consequence, the company found itself facing a extended phone bill showing long-distance calls to international locations [4].

20.1.4 Attacks on Circuit-Switched and IP-Based Networks

There are a variety of attacks against circuit-switching telephone networks such as wiretapping [5,6], call masquerading, account fraud, disturbing calls, and denial of service (DoS). Many of these threats remain effective for packet-switching networks, including VoIP networks.

As VoIP is based on normal IP networks, VoIP applications inherit the known and unknown security weaknesses that are associated with the IP protocol. The resulting security problem domain is considerably larger compared to a traditional public switched telephone network (PSTN). Moreover, proprietary protocols, which many VoIP companies are using, have left VoIP customers vulnerable to virus attacks, security holes, abnormal port activities, etc.

VoIP is inherently vulnerable to network attacks, such as malicious code (i.e., worms, viruses, trojans), DoS, distributed DoS (DDoS), pharming, and (though nonmalicious) flash crowds. These attacks also damage infected systems by consuming resources, disrupting legitimate users, compromising confidential information, or by corrupting code and data [7]. While such damage affects infected systems, it also damages uninfected (or even nonvulnerable) ones. All systems connected to the Internet are sensitive to malicious code which tries to infect as many hosts as possible, causing congestion in the network infrastructure. Walsh et al. [8] analyze the challenges of securing VoIP, and provide a set of guidelines for adopters of this telephony technology.

20.1.5 VoIP Security Requires an End-to-End Security Model

Voice over IP security is facing more complex issues than those of traditional circuit-switching networks, because VoIP networks consist of multiple components. Each of those components has its own security issues and has also been deployed on current data networks, which impose some limitations on VoIP systems.

Therefore, the VoIP supported IP networks must be designed, implemented, and operated in a secure manner, providing end-to-end security between VoIP applications. Though many industry or research organizations are tackling this challenging topic in the context of secure real-time transport protocol (SRTP) and IPSec, there is still a lack of a unified security policy that specifies or regulates how VoIP applications should provide security mechanisms to protected resources in a single picture.

Considering the size of the Internet, management is best achieved through distributed methods, where each network is controlled according to its predefined local policies, and global deployment of a specific security policy is infeasible to enforce. VoIP technology traverses international boundaries, and hence it is very difficult to enforce regulations consistently across the world.

20.2 VoIP Protocol Vulnerabilities

20.2.1 Security Issues of SIP

Session initiation protocol is a real-time signaling protocol for IP voice that was developed by the Internet Engineering Task Force [9,10]. It is used for a bi-directional communication session initiation in which messages are exchanged between two or more nodes. SIP is responsible for the basic call control tasks such as communication session setup and tear down, or signaling for call initiation, dial tone, and termination. SIP is also responsible for the other signaling used in features such as caller ID, call transferring, and hold. It acts as the Signaling System 7 (SS7) protocol used in standard telephony and H.323 which is used in IP. It is an application-level protocol, and hence it can be used on top of different protocols. It can be carried by UDP, TCP, or SCTP [11]. If SIP is carried over UDP, speed and efficiency are increased.

Transmission control protocol is used when for security purpose, secure socket layer/transport layer security (SSL/TLS) is needed. Stream control transmission protocol (SCTP) offers increased resistance to DoS attacks through a four-way handshake method [11]. SCTP can use additional security services via "TLS over SCTP" or "SCTP over IPSec."

Session initiation protocol consists of different components including a user agent (UA), redirect server, registrar server, location server, and proxy server. UA software includes client and server components. Outgoing calls are made at the client side, whereas incoming calls are received at the server side. After some processing or translation is done, traffic forwarding is performed by a proxy server. Request authentication is handled by the registrar server, and the redirect server is responsible for resolving information for UA clients. UA clients send requests to UA servers for call initiation. The user notifies a registrar server of her current location to allow others to communicate with her.

As has been observed, the backbone of traditional phone systems (PBXs) is replaced by IP voice servers which usually run on Linux or Microsoft operating systems. Servers that deliver VoIP services and log call information are very sensitive to malicious software attacks and hacking activities.

Session initiation protocol uses a simple set of request messages such as INVITE, ACK, OPTIONS, CANCEL, BYE, and REGISTER. A UA client who wishes to initiate a session sends an INVITE message. This message is responded to with OK followed by an ACK message. To tear down a connection, a BYE message is sent. CANCEL cancels a pending invite. To query or change optional parameters of the session such as encryption, OPTIONS is used.

The VoIP-specific protocols are a major source of vulnerabilities. SIP is text encoded which makes it easier to analyze with standard parsing tools such as Perl or lex and yacc [11]. SIP traffic is plain text in its basic form. Thus voice traffic is vulnerable to packet sniffers (looking for caller IDs or passwords), and allows an attacker to forge packets for manipulating of device and call state. For example, this kind of manipulation could result in premature call termination, call redirection, or easily performed toll fraud. It is also relatively simple to intercept unencrypted VoIP calls. Hackers can download free software available on the Internet and intercept calls with ease. To protect caller IDs, account information, etc., SIP traffic needs to be encrypted. Although some efforts have been made for developing encrypted signaling, so far no widespread solution has been found to adopt.

There are specific attacks based on the vulnerabilities in the VoIP signaling protocols, including SIP. One of these attacks is known as a BYE attack. The goal of a BYE attack is to tear down an existing communication session prematurely, which can be considered as a DoS attack. Imagine three SIP UAs: Alice, Bob, and the attacker, and an ongoing dialog session between Alice and Bob. The attacker can send a fake BYE message to Alice. Alice believes that Bob is willing to tear down the connection. Alice stops sending RTP flow but Bob still is sending RTP packets to Alice because Bob is not aware of what has happened in between. Certain Intrusion Detection Systems (IDSs) are able to detect such attacks. By creating a rule, an IDS can detect flowing of RTP packets after a BYE message. In the previous scenario, if Bob is the real entity who is willing to stop the connection, Alice should not see RTP flow from Bob after she gets a BYE message from him. In the IDS, an alarm is activated if RTP flow is detected after receiving a BYE message.

The SIP protocol is not only used in VoIP call setup, but also in instant messaging. An attacker can also manipulate the header of an instant message and send a forged message to Alice and make her believe that the message has come from Bob. This attack is called fake instant messaging. There is also another signaling-based attack which is called call hijacking. In this attack, the attacker takes advantage of the REINVITE message which is used for call migrating. In the case of call hijacking attacks, one of the UA clients will experience continuous silence since the other part is not sending voice packets to it any more. Such an attack can also be considered as a DoS attack. This attack results in breach of confidentiality. For detecting such an attack, a similar approach as for the BYE attack can be used.

20.2.2 Security Issues of H.323

H.323 is an International Telecommunication Union (ITU) standard for video and audio communication over a packet zed network. H.323 includes several protocols such as H.225, H.245, etc. Each of these protocols plays a specific role in the call set up process [11]. An H.323 network normally includes a gateway, a gatekeeper, a multipoint control unit (MCU), and a back end service (BES). The gateway acts like a bridge between the H.323 network and external non-H323 networks such as SIP or PSTN networks.

The gateway also supports bandwidth control and address resolution. A gatekeeper is optional and is responsible for optimizing network tasks. In the case of the presence of a gatekeeper, a BES may exist for supporting functions such as maintaining of endpoint's data including permissions, services, and configuration. The MCU is also an optional component of H.323 networks. Multipoint conferencing and other types of communication between more than two endpoints are also facilitated by the MCU.

The H.323 protocol traffic is almost always routed through dynamic ports which is quite challenging for non-VoIP-specific firewalls which are not VoIP aware. Thus, it is necessary to use a VoIP-specific firewall. Another serious issue regarding H.323 networks is network address translation (NAT), because there is no match between external IP address and the port specified in H.323 headers and messages, actual IP address, and port numbers used internally. Therefore, the correct address and port numbers need to be sent to the endpoints for establishing a call connection. Compared with H.323, SIP is a more flexible solution, simpler and easier to implement. SIP is also more suitable for supporting intelligent user devices, as well as for implementing advanced features [12].

H.235 offers different security profiles. The variety of options offered by H.235 provides different possibilities for communication protection. In this chapter, we describe one of these defined possibilities—H.235v2.

H.235v2 supports elliptic curve cryptography (ECC) and advanced encryption standard (AES). To support product interoperability, it defines a variety of security profiles which are defined in annexes to H.235v2 (Annex D, Annex E, Annex F) [11]:

- Annex D: Shared secrets and keyed hashes
- Annex E: Digital signatures on every message
- Annex F: Digital signatures and shared secret establishment on first handshake, afterwards keyed hash usage

For instance, Annex D employs symmetric techniques. To provide authentication and message integrity, H.235v2 uses shared secrets. It supports H.245 tunneling and also secure fast connect.

20.2.3 Security Issues of RTP

Real-time transport protocol, a standard protocol which is designed for real-time data transmission, offers end-to-end data delivery services with real-time characteristics. RTP offers sequence numbering, timestamping, delivery monitoring, and payload type. VoIP traffic is transferred as RTP packets. RTP runs on top of UDP.

When signaling occurs, RTP is responsible for transporting the voice packets. It offers monitoring of its payload through sequencing and timestamping. It is appropriate for applications transmitting real-time data such as voice. Unfortunately, RTP does not guarantee quality of service (QoS) for real-time services; instead RTP relies on its lower layer services.

In addition to signaling-based attacks which threaten VoIP systems, there is an RTP-related attack that is based on the vulnerabilities in media transport. Attackers can send RTP packets with garbage contents, meaning that both the payload and header are replaced with random numbers and sent to the target. This attack can degrade the quality of voice or crash the victim's machine. One approach for detecting this attack is to check the sequence numbers in consecutive RTP packets. If this sequence is irregular, it can be considered as the presence of an attack.

20.3 Other VoIP Security Issues and Vulnerabilities

20.3.1 Convergence Issues

Although convergence of voice and data networks is one of the main advantages of VoIP systems (conveying the voice over a data network which results in efficiency and easy management), convergence is becoming another security concern of the VoIP system. The security threats against VoIP may affect convergence as well. VoIP is vulnerable to the same types of attack that threaten data networks. This includes well-known attacks such as DoS attacks, authentication attacks, etc. Voice is critical to business so it is desired that if something happens to the data network, the voice network can still function. If the voice and data networks are the same network, in case of any threat both of them will malfunction at the same time. Isolating the voice in its own network can be the best approach. It should be noted that the isolating approach depends on whether the business can accept the risk associated with malfunctioning of both data and voice networks or not. For those businesses that can accept this risk, current convergence is sufficient. However, there may be other businesses whose survival highly depends on voice; for them, the isolating process is the only solution. They may still be willing to have some features of convergence such as triggering of customer data displays by entering a caller ID. This requires the creation of architectures that can pull the required information from different networks. By the separation of data and voice, when a virus hits the data network it can never affect VoIP system. To reach this goal, most VoIP deployments use separate VLANs for data and voice. A virtual LAN sets asides a portion of bandwidth for voice traffic and separates voice and data traffic.

20.3.2 Man-in-the-Middle Attacks

Man-in-the-middle attacks refer to an intruder who is able to read, and modify at will, messages between two parties without either party knowing that the link between them has been compromised. The most common man-in-the-middle attack usually involves address resolution protocol (ARP), which can cause VoIP application to redirect its traffic to the attacking computer system. The attacking computer system can then gain complete control over that of VoIP application's sessions, which can be altered, dropped, or recorded. For example, an attacker can inject speech, noise, or delay (e.g., silent gaps) into a conversation [13]. In general, there are three types of man-in-the-middle vulnerabilities: (1) eavesdropping: unauthorized interception of voice data packets or RTP media stream and decoding of signaling messages; (2) packet spoofing: intercept a call by impersonating voice packets or transmitting information; and (3) replay: retransmit genuine sessions so that the VoIP applications will reprocess the information. To tackle all these vulnerabilities, VoIP applications can adopt the public key infrastructure (PKI), a security mechanism to ensure the confidentiality of all transmitted data, and to verify and authenticate the validity of each party in the context of public and private key. Without proper encryption, anyone can sniff any voice data packet transmitted over IP networks vulnerable to confidentiality and integrity threats [11]. In summary, man-in-the-middle attacks create security threats to confidentiality and integrity because this type of attack may release the voice data packets to unauthorized parties or modify the contents of the conversations.

Although using encryption can mitigate eavesdropping and also packet replay attacks, eavesdropping is still a real threat for VoIP systems. As mentioned earlier, separating data network and voice by using of VLANs is the main step for bringing security to the VoIP infrastructure, but attackers are getting smarter and always try to find a better way to break into the system.

There are a number of tools available on the Internet that help even non-expert attackers, without deep knowledge of security exploits, to be able to attack a VoIP system (see e.g., [13]). For instance, by using the DSNIFF tool, an attacker can capture the traffic and then encode the captured data to a playable file by tools such as Voice over Misconfigured Internet Telephones (VOMIT) which is an ideal eavesdropping tool for script kiddies. In traditional telephony systems, 64 kb of bandwidth is assigned to every voice call that leads to a possibility of 24 simultaneous calls over a leased line (T1 link). VoIP offers compression techniques which are capable of compressing the voice calls to 8 kb, which increases the capacity of leased lines. The advantage of bandwidth efficiency (saving) can be affected by the need of encryption even in IP telephones. There is a close relation between capability of end points to encrypting and decrypting in real time and QoS. There may not be any problems regarding encrypting a call by an end point, but increasing of the encrypted traffic can overwhelm media gateways. And also, as mentioned before, encryption can result in delay and jitter that affects QoS of the VoIP system.

Overall delay in VoIP system can be increased when encryption is performed. VoIP systems need to offer encryption for privacy protection and also for message authentication purposes. The encryption process can be done through stream cipher or block cipher. Compared to block cipher, stream cipher imposes less delay ("one bit" compared to a "block of bits" delay). Computing hash values which are used for authentication purposes can also introduce significant delays in the VoIP system because before processing, a full block of data needs to be hashed. Hash functions such as MD5, SHA are used in producing the hash values. For instance, MD5 produces a message authentication code (MAC) of size "128 bits," whereas SHA produces a MAC of size "160 bits."

20.3.3 Denial of Service, Distributed DoS

Denial of service attacks always refer to the prevention of access to a network service by bombarding servers, proxy servers, or voice-gateway servers with malicious packets. A DoS causes an incident in which a user is deprived of the services or resource they would normally expect to have. Intruders can launch the full spectrum of DoS attacks (e.g., unauthenticated call control packets) against a VoIP application's underlying networks and protocols just as they can against a traditional PBX. For example, voicemail and short messaging services in IP telephony systems can become the targets of message flooding attacks.

In general, there are three major types of DoS attacks:

1. Buffer overflow or Mcast—In this attack, the intruders usually send oversized Internet control message protocol (ICMP) messages, or more than the port can handle, to the targeted VoIP applications. To tackle this situation, VoIP applications can provide a security mechanism to: (a) turn off unnecessary ports; (b) enforce packet filtering outside the peer table; (c) turn off pinging and ICMP messages; and (d) verify received data packet size to avoid exceeding the bounds of the available memory stack.

2. SYN—In this attack, an attacker opens an overwhelming number of TCP connections in the victim's system in such a way that the ports are held until timeout. To tackle this situation, VoIP applications can provide a security mechanism to: (a) shutdown

unnecessary ports; (b) reduce timeouts in the TCP stack; and (c) increase connection buffer size.

3. Smurf—The intruders send an IP ping request to a receiving device, and the ping packet broadcasts to a number of devices within local network with a return address of the device under attack. To tackle this situation, VoIP applications can provide a security mechanism to turn off pinging and ICMP messages. In summary, DoS attacks create security threats to availability because DoS would render VoIP applications inoperable.

Denial of service and DDoS attacks may also target VoIP systems. These attacks consist of sending disruptive amounts of requests to a same server/telephone either from a single host (DoS) or from a number of them (DDoS). A number of network infrastructure countermeasures have been proposed such as fair bandwidth share and throttling techniques. For example, Demers et al. [14] propose to classify packets according to a combined criterion involving source host and packet sizes, as a measure to mitigate disruptive traffic. Mahajan et al. [15] introduce a "pushback" mechanism which consists of analyzing dropped packets in order to find signatures of aggregates responsible for congestion (without distinguishing malicious from nonmalicious packets) and subsequently filtering these signatures into a delay queue. Further, routers share signatures to upstream routers, which in turn will filter traffic matching these signatures. The pushback mechanism has been implemented and some variations have been proposed [16]. Wang et al. [17] propose a network IDS based on n-gram (an n-gram is a consecutive sequence of n characters or symbols in a text document) analysis, whereas Matrawy et al. [18] study a statistical classification method based on (p, n)-grams (a (p, n)-gram is a n-gram at position p). Vargas Martin [19] proposes a mitigation scheme where, upon congestion, packets are classified into equivalence classes. The classification of a packet is based on the number of times the packet has been forwarded. This scheme aims to alleviate congestion by classifying disruptive network traffic into only a few equivalence classes, while classifying nondisruptive traffic into many classes.

Qi et al. [20] propose a mitigation mechanism that would allow routers to classify and prioritize traffic such that disruptive network flows are not allowed to consume bandwidth needed by other users and services. This system keeps track of the number of times each packet is repeated based on a packet subset. To count packet subsets online, Bloom filters with counters are used. In summary, DoS attacks create a security threat to availability because DoS would render VoIP applications inoperable [21].

20.3.4 Botnet Threats

As mentioned earlier, a DoS attack is achieved by preventing the legitimate use of a service. DDoS attacks take advantage of multiple systems in the network to launch a DoS attack. To launch a DDoS attack, an attacker first looks for multiple vulnerable agent machines. This scanning process is normally performed automatically through the scanning of remote systems, looking for potential vulnerabilities. When a vulnerability is found, it is exploited to break into the machine. The attacker then sends remote control programs called "bots" to these selected machines. After planting the programs on the bot-infected machines (called zombies), those machines wait for commands coming from the attacker (bot-herder). Hence, an attacker can direct a large number of compromised systems against a target. A network of these bot-infected computers is called botnet. These compromised systems are normally selected among vulnerable computers which have high-bandwidth, always-on WAN connections. The real identity of an attacker is always hidden by attackers through IP spoofing in which the source address field of attack packets is spoofed. In a DoS attack, a server can be targeted with a flood of information requests which may bring the system down. Botnets are normally controlled through Internet relay chat (IRC) channels where bot-infected machines listen for instructions coming from the bot-herder. To catch the attacker, investigators monitor IRC channels and also can block traffic to the IRC channels which are used by hacked machines. Using thousands of these hacked machines to launch a DDoS attack and other attacks such as spamming, phishing, etc. against specific targets such as enterprises is becoming a common trend. VoIP is inherently vulnerable to network attacks including DoS and DDoS [22]. VoIP applications such as Vonage and Skype could provide better opportunities to attackers for controlling their bot-infected machines.

20.3.5 Spam over VoIP

Another latent problem is spam (i.e., unsolicited bunk messages). Just as with email systems, VoIP is susceptible to spam (also called Spam-over-Internet Telephony (SPIT)). SPIT can be considered as another threat of botnets, which could potentially disable the VoIP system. If a VoIP user gets a lot of calls from an audio spammer every day, it would make him reluctant to employ VoIP technology.

Spam over Internet Telephony is even worse than spam. If we receive our emails with even a few minutes delay it is not a grave problem, while SPIT is very noticeable to end users as it hits the gateways and degrades the quality of voice. SPIT can target any IP-based phone system. To address the SPIT issue, several companies are developing solutions for such threats. Techniques for combating SPIT include filtering, black/white lists, and caller's reputation [23–25], all of them becoming less effective as SPIT methods become cleverer.

20.3.6 Phone-Targeted Malcode

A virus is a piece of malicious code loaded onto the computer systems without your knowledge and runs against your wishes. As VoIP applications move beyond simply handling voice calls to running different applications, the virus risk is likely to increase because all VoIP applications have their own IP address like the computer systems on IP networks [26]. Thus, a virus attack could be very effective against the VoIP applications [27]. One of the common examples is when a virus injects a small replication code through a stack overflow to damage the VoIP applications or even bring down the IP networks. To tackle this scenario, VoIP applications should provide a security mechanism to verify received data packet size to avoid exceed bounds of available memory on the stack. In summary, virus attacks could generate security threats to integrity and availability.

Internet malicious code may also target VoIP phones directly. For example, the Cabir virus [28] infects cell phones (and other devices) that use Bluetooth (see http://www.bluetooth.com) wireless technology. Staniford et al. [29] present different classes of attacks and trends for future network attacks [23]. Another family of malicious logic is represented by the Santy worms [30], which target generic Web applications. The ability of Santy worms to infect Web applications makes them hard to prevent and detect, and they propagate more widely. The problem of detecting malicious logic on the Internet becomes harder when the malicious software is encrypted (i.e., the bulk of their payloads is encrypted using a different key per infection) or polymorphic (i.e., the malicious logic changes its representation on each new infection) [31,32].

Detecting and preventing malicious code attacks is a continuing battle, and a number of IDS have been proposed. Some IDSs rely on the fact that fast-propagating malicious codes typically perform a "large" number of similar actions within a "short" period of time [33,34].

20.3.7 Rogue Sets

Rogue sets attacks refer to performing a deception for the purpose of gaining access to someone else's resources. The intruders perform digital impersonation by adding a new set of VoIP applications to the attacked IP networks, then spoofing the identity of a targeted call participant. Then the malicious VoIP application can conduct any activities that may harm the attacked IP network. To tackle such an attack, VoIP applications can perform a network lock-down mechanism.

In a lock-down mechanism, only the network administrators can add new sets of VoIP applications to the network with administrative password, and logs are sent to the administrator when a new set is added. In addition, the VoIP application will be rejected if more than three entry passwords are attempted. In summary, rogue sets attacks create a security threat to confidentiality because the intruders get unauthorized access to an IP network.

20.3.8 Toll Fraud

Toll fraud attacks occur when an individual uses the equipment to place unauthorized calls. This is a type of traditional telephony attack in which a rogue device gains access to make long-distance calls. In this scenario, employees can use certain features of the telephony system to make long-distance calls such as return calls from voice mail (VM), trunk-to-trunk transfers, and call forwarding (CF) to external numbers. To tackle this attack, VoIP applications can enforce an authorization mechanism on an IP network. This means that rogue devices require authorization to gain access to the network. All devices are challenged on external calls. In addition, dialing rules are applied to groups of users and for certain times as defined by the administrator. All external calls have to satisfy the dialing rules prior to any external call.

20.3.9 Dynamic Host Configuration Protocol (DHCP)

Dynamic host configuration protocol (DHCP) attacks refer to a malicious computer (identity theft) on the network that can issue excessive requests to a DHCP server and force the server to issue all of its allocated IP addresses. The purpose is to spoof a DCHP server's responses. As there are many places in an IP network with dynamically configurable parameters, intruders have a wide array of vulnerable points to attack. Then, the DHCP server could not service the next legitimate request that comes in, thus stopping new sets from entering the network. Further, a malicious VoIP application can even reply to DHCP requests and provide incorrect information. This can cause a DoS, or enable a man-in-the-middle attack. To tackle this attack, VoIP applications can provide a security mechanism to revert to zero configurations for allocation of IP addresses and also verify the DHCP response to ensure that it is not being used. In an extreme scenario, we suggest assigning static IP addresses to each VoIP application for a completely secure network. In summary, DHCP attacks create a security threat to availability because such an attack may interrupt the normal operation of VoIP applications.

20.3.10 Pharming

Pharming attacks pose another potential DDoS problem to VoIP systems. Pharming is an evolution of a less-sophisticated attack called phishing, where a person is contacted (typically by email or instant messaging) with an apparently legal request to provide some information via a (fraudulent) Web page [35]. Pharming, on the other hand, consists of exploiting a domain name server (DNS) vulnerability to misdirect the communication between a client and a remote server. Thus, an attack on VoIP may consist of stealing sensitive information from a corporation's customers by having an impostor make the customers believe they are dealing with one of the corporation's representatives. Another pharming attack over VoIP consists of misdirecting large numbers of calls to a particular domain to perpetrate a DDoS.

20.3.11 Flash Crowds

Besides DoS and DDoS, another threat capable of congesting VoIP systems is imposed by flash crowds (i.e., sudden massive number of nonmalicious requests to the same server). A number of attempts to mitigate flash-crowds have been proposed. For example, Jung et al. [36] find characteristics of flash crowds. They use these characteristics to design a dynamic load-balancing algorithm for Web caches. Chen et al. [37] present a flash crowd mitigation system based on requests regulation for Web servers. The system is based on the observation that high-bandwidth applications (i.e., applications with request rates of more than a predefined threshold) are more sensitive to flash crowds than low-bandwidth ones. The idea is to monitor the response rate of high-bandwidth applications and the rate of request arrivals and compare these two against the corresponding long-term averages. A flash crowd signal is sent to a request regulator (which in turn, throttles the requests arrivals) if the response rate of the fast connections decreases (below a predefined threshold compared to the long-term average) and the arrival rate of requests increases (above a predefined threshold compared to the long-term average).

20.4 Security Policy for VoIP Applications

By convention, security threats are usually thought to come from outsiders. However, security threats arise in a well-controlled environment from authorized insiders. Intentional errors and malicious acts by disgruntled employees and insiders seeking revenge cause a considerable amount of damage and loss such as the theft, alteration, and corruption of information experienced in the telecommunications industry.

In the United States, FBI surveys reported that most security incidents are inside jobs perpetrated by employees, former employees, contractors, vendors, and others with inside knowledge, privileged access, or a trusted relationship with other insiders [38]. What are the fundamental security requirements of VoIP applications? The answer stated in [39] is as follows: "An important requirement of any information management system is to protect data and resources against unauthorized disclosure (confidentiality) and unauthorized or improper modification (integrity), while at the same time ensuring their availability to legitimate users." In order to circumvent security threats, three fundamental security requirements, namely confidentiality, integrity, and availability, have to be addressed.

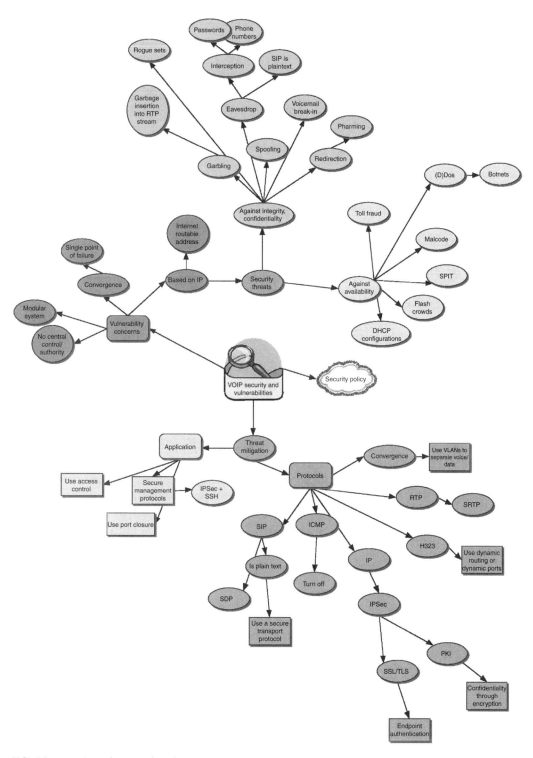

FIGURE 20.1 A roadmap to this chapter.

- **Confidentiality** It is the assurance that sensitive information is not disclosed. The confidentiality of VoIP applications is violated when unauthorized parties obtain protected information (voice data packets) over IP networks.

- **Integrity** This prevents the unauthorized modification of information. The integrity of VoIP applications is violated when the correctness and appropriateness of the content of the conversation (voice data packets) is modified, destroyed, deleted, or disclosed.

- **Availability** It refers to the notion that information and services are not available for use when needed. The availability of VoIP applications is violated when the system is brought down or malfunctioned by attackers or intruders.

A security policy is a set of rules and practices that specify or regulate how a system or organization provides security services to protected resources. A security assertion is typically scrutinized in the context of security policy. In general, the engineering of a security policy starts with risk analysis and ends with a set of security assertions that is ready for integration into the security architecture of a subject such as a service locator [22]. Risk analysis identifies security threats in a business process and forms a set of security assertions, which refer to rules and practices to regulate how sensitive or activity information is managed and protected within a loosely coupled execution environment. A security policy is often formalized or semiformalized in a security model that provides a basis for a formal analysis of security properties.

Figure 20.2 shows a security policy from the perspectives of communication and application security. Referring to the communication security, there are three layers: SIP, RTP, and IP. First, IPSec is used to encrypt the packets at the IP level. To enforce end-to-end security, the security policy adopts the public key infrastructure (PKI) a security mechanism to ensure confidentiality of all transmitted data, and to verify and authenticate the validity of each party in the context of public and private key.

Then, SSL/TLS is used to authenticate peer clients with digital signed certificates. These protocols provide endpoint authentication and communication privacy over the Internet using cryptography. The SSL is a commonly used protocol for managing the security of message transmission on the Internet. SSL has recently been succeeded by TLS, which is based on SSL. SSL uses the public-and-private key encryption system from RSA, which includes the use of a digital certificate. TLS and SSL are not interoperable. However, a message sent with TLS can be handled by a client that handles SSL, but not TLS. Without proper encryption, anyone can sniff any voice data packets transmitted over IP networks, which results in security threats to confidentiality and integrity [11].

In the RTP level, RTP provides end-to-end network transport functions suitable for IP telephony systems transmitting real-time voice data over IP network services [40]. The security policy adopts the SRTP using AES countermode to provide security measures shown as follows [11]:

1. confidentiality for RTP by encryption of the respective payloads;
2. integrity for the entire RTP packets, together with replay protection;
3. the possibility to refresh the session keys periodically, which limits the amount of cipher text produced by a fixed key;
4. an extensible framework that permits upgrading with new cryptographic algorithms;

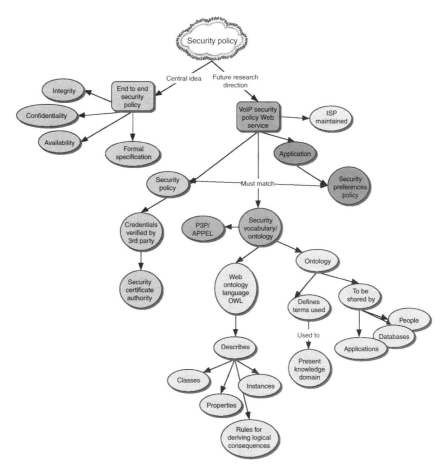

FIGURE 20.2 Security policy for VoIP applications.

5. a secure session key derivation with a pseudo-random function at both ends;

6. the usage of salting keys to protect against precomputation attacks; and

7. security for unicast and multicast RTP applications.

In the SIP level, the security policy adopts the key management extensions for session description protocol (SDP). The SIP is used to establish, modify, and terminate multimedia calls. During the setup process, communication details are negotiated between the endpoints using SDP, which contains fields for the codec used, caller's name, etc. Since the SIP message structure is a straight derivation from the HTTP request/response model, all security measures available for HTTP can also be applied to SIP sessions.

Back to the perspective of application security, the security policy uses IPSec and secure shell (SSH) for all remote management and auditing access control [11]. Access control is the process of limiting access to the resources of a system only to authorized users, programs, processes, or other systems. It is synonymous with controlled access and limited access. In general, access control is defined as the mechanism by which users are permitted access to resources according to their identities authentication and associated privileges authorization.

20.5 Security Policy Framework for Mobile VoIP Applications

With recent advances in mobile technologies and infrastructure, there are increasing demands for ubiquitous access to networked services. These services, generally known as mobile services, extend support from Web browsers on personal computers to VoIP applications, such as wireless IP phones, over the IP network. The capabilities and bandwidth of mobile devices are significantly inferior to those of desktop computers over wired connections. Mobile architectures are built on an insecure, unmonitored, and shared environment, which is open to events such as the security threats discussed in Sections 20.2 and 20.3. As is the case in many other applications, the information processed in mobile VoIP applications might be commercially sensitive such as voice messages and therefore it is important to protect it from security threats such as disclosure to unauthorized parties. As a result, each secure IP network should enforce the security policy stated by the organization.

Considering that mobile VoIP applications can work on laptops, IP phones, and wireless IP phones, one can imagine that these devices can move from one IP network to another. Referring to Figure 20.3, the mobile VoIP applications can be connected to a wireless network A (e.g., intra, inter, extra, ad hoc, etc.) at a company in the daytime. To maximize the communication security in such a wireless network, the mobile VoIP applications are adapting wired equivalent privacy (WEP) protocol to maintain the security properties of confidentiality and integrity in the communication. In addition, the communication between the mobile VoIP application and the VoIP security policy Web service is adapting SSL/TLS. With the trend of services computing, it is believed that security applications will adopt Web services technologies. A Web service is a software system designed to support interoperable application-to-application interaction over the Internet. It is well known that Web services are based on a set of XML standards, such as simple object access protocol (SOAP) [41,42]. The application (requestor) binds with the Web service via a SOAP message. SOAP is an XML-based messaging protocol that is independent of the underlying transport protocol (e.g., HTTP, SMTP, and FTP). SOAP messages are used both by services requestors to invoke Web services, and by Web services to answer to the requests. Therefore, the Web service receives the input SOAP message from the Web services requestor and generates an output SOAP message to the Web services requestor. For example, an employee at company A connects his laptop with a VoIP application to his company during the daytime. During the night, the employee may bring his laptop back home and continue to use the VoIP application for different purposes. As mentioned before, the VoIP application may contain confidential information such as voice messages. Thus, there exist security threats to the VoIP application in such a mobile environment. In particular, many

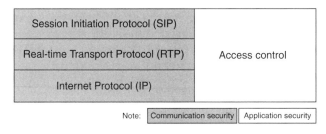

FIGURE 20.3   A security policy framework for mobile VoIP applications.

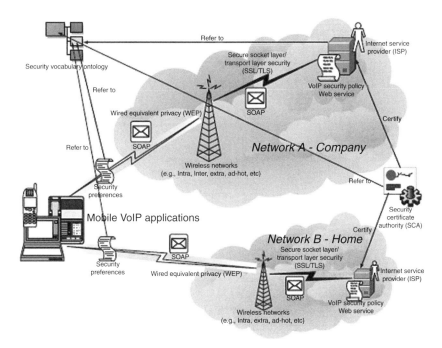

FIGURE 20.4   The conceptual model of generic security policy for VoIP applications.

IP parameters are established dynamically each time a VoIP application is added to a network. Therefore, intruders may have many vulnerable points to attack.

We propose a VoIP security policy Web service which contains a security policy like the one illustrated in Figure 20.4. When the mobile VoIP application connects to the network, the application checks for the security policy at the VoIP security policy Web service by comparing it to its security preferences policy. The VoIP security policy Web service is maintained and provided by the Internet service provider (ISP) of the network. Further, the credentials of such a security policy are verified by a third-party called the Security Certificate Authority (SCA). The SCA is the agency to certify that the claimed security mechanisms by the ISP are accurate and complete.

It is to be noted that SCA, security policy and security preferences policy all refer to a security vocabulary/ontology for a common language on different security terminologies. The semantic Web helps provide explicit meaning to information available on the Web for automatic processing and information integration based on the concept of an ontology. An ontology defines the terms, used to present a domain of knowledge, that are shared by people, databases, and applications. In particular, an ontology encodes knowledge possibly spanning different domains as well as describes the relationships among them. If the security mechanisms (protection) set in the privacy policy matches with the privacy preferences, the VoIP application will connect to the network. Referring to Figure 20.3, we define the security vocabulary/ontology by using the Web ontology language (OWL) [43]. OWL is an XML language proposed by the world wide web consortium (W3C) for defining a Web ontology. An OWL ontology includes descriptions of classes, properties, and their instances, as well as formal semantics for deriving logical consequences in entailments.

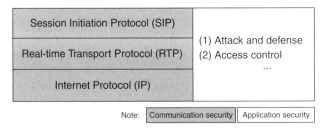

FIGURE 20.5　An illustrative security vocabulary in attack and defense.

Figure 20.5 shows a simplified OWL ontology that describes the vocabulary of security attack and defense. Next, the platform for privacy preferences project (P3P) working group at W3C developed the P3P specification for enabling Web sites to express their privacy practices [44]. P3P UAs allow users to automatically be informed of site practices and to automate decision-making based on the Web sites' privacy practices. Thus, P3P also provides a language called P3P Preference Exchange Language 1.0 (APPEL1.0), to be used to express user's preferences for making automated or semiautomated decisions regarding the acceptability of machine readable privacy policies from P3P-enabled Web sites [45]. Though P3P and APPEL were originally designed for tackling privacy enforcement in Web site and cookies in Web browsers, we revise the framework of P3P and APPEL to represent the security policy and security preferences (respectively) in the Appendix.

20.6 Conclusion

Before the advent of VoIP technology, Internet users were expecting the risks associated with sending data over the Internet, but they were expecting a confidential network for their voice calls. After the appearance of VoIP technology, and with the convergence of the voice and data networks, the data security issues became voice security issues as well. The fact that VoIP technology is using IP does not mean that it cannot be secured. To defeat VoIP security threats, a well-structured plan needs to be devised. The plan should include voice encryption, authentication, voice-specific firewalls, and separation of data and voice traffic. Redundancy in case of power loss should be considered in this plan as well [46]. It is also important that the voice servers and the other components of VoIP networks stay physically secure from intruders. To protect VoIP networks, these general recommendations should be considered: Using IDSs and intrusion prevention systems (IPSs), using VoIP specific firewalls, changing default passwords on all different components of the VoIP system, patching components against known threats, and also following vendor's checklists for securing all components after installation. We have summarized the basic concepts on VoIP, and outlined some of its most important threats. As VoIP gains terrain in the world of telecommunications, it seems that soon it will become one of the dominant technologies in telephony. VoIP has inherited a number of Internet vulnerabilities exploited by malicious code, DoS, DDoS, and pharming, as well as flash crowds, all of which pose a latent threat to network infrastructures. It is evident that security policies have to be put in place to avoid disasters.

20.7 Future Directions

Future work includes the analysis and design of secure mechanisms to effectively counter-act or mitigate these threats, including eavesdropping, call redirection, and spoofing. Future work also includes software attack prevention through the design and implementation of solid security policies. Also, it is imperative to design intrusion detection systems capable of coping with emerging encrypted and polymorphic malicious code. DoS and DDoS are a constant threat which a number of systems have not yet been able to contain. The continuing adoption of VoIP in corporations has made more evident the urgency of efficient defense systems against these attacks.

This chapter proposes a security framework for mobile VoIP applications in a dynamic environment integrated with various Web technologies (e.g., Web services, OWL, P3P, APPEL, SOAP) in a nutshell. We are currently building a full-scale model to support such a framework and designing simulation models to prove its correctness and completeness. This work is part of an undergoing project aimed to provide more secure networks.

Appendix

Referring to Figure 20.3, before the VoIP application connects to the network, the application must check with the security policy (Figure 20.6) posted on the VoIP security policy server with its security preferences (Figure 20.7) via a SOAP message (Figure 20.8). Figure 20.6 describes the security practices in the context of threat and defense into a P3P-like framework as applied to the security ontology described in Figure 20.9. On the other hand, Figure 20.10 describes the privacy preferences of the VoIP application into an APPEL-like framework. There are two statements (preferences) defined as follows:

1. For the security threat of malicious code, the VoIP application requires the network to have the security mechanism of signature or anomaly detection installed.

```
<owl:Ontology rdf:about="#securityVocabulary">
   <owl:versionInfo>v 1.00 2005/06/15 23:59:59</owl:versionInfo>
   <rdfs:comment>An    example    OWL    ontology    for    Security    Vocabu-
lary</rdfs:comment>... <-- evidence (abbreviated) --> ...
   <owl:Class rdf:ID="#securityAttack">
   <owl:unionOf rdf:parseType="Collection">
     <owl:Class rdf:about="#maliciousCode"/>
     <owl:Class rdf:about="#spit"/>
     ...
     <owl:Class rdf:about="#flashCrowds"/>
   ... <-- evidence (abbreviated) --> ...</owl:unionOf></owl:Class>
   <owl:Class rdf:ID="#securityDefense">
   <owl:unionOf rdf:parseType="Collection">
     <owl:Class rdf:about="#signature"/>
     <owl:Class rdf:about="#anomalyDetection"/>
     ...
     <owl:Class rdf:about="#serverLoadBalancing"/>
   ... <-- evidence (abbreviated) --> ... </owl:unionOf></owl:Class>
   ... <-- evidence (abbreviated) --> ... </owl:Ontology>
```

FIGURE 20.6 An illustrative security policy.

```
<POLICY> ... <-- evidence (abbreviated) --> ...
  <STATEMENT rdf:ID="#rule1">
    <ATTACK rdf:connective="or-exact">
      <THREAT ref="#spit"/></ATTACK>
    <DEFENSE rdf:connective="and-exact">
      <MECHANISM ref="#blackList"/></DEFENSE>
... <-- evidence (abbreviated) --> ...</POLICY>
```

FIGURE 20.7 Illustrative security preferences.

```
<appel:RULE behavior="securityPreferences">
  ... <-- evidence (abbreviated) --> ...
  <POLICY><STATEMENT rdf:ID="#preference1">
      <ATTACK>
        <THREAT ref="#spit"/></ATTACK>
      <DEFENSE rdf:connective=or-exact">
        <MECHANISM ref="#blackList"/></DEFENSE></POLICY>
  ... <-- evidence (abbreviated) --> ...</appel:RULE>
```

FIGURE 20.8 An illustrative SOAP message.

```
<owl:Ontology rdf:about="#securityVocabulary">
  <owl:versionInfo>v 1.00 2005/06/15 23:59:59</owl:versionInfo>
  <rdfs:comment>An    example    OWL    ontology    for    Security    Vocabu-
lary</rdfs:comment>... <-- evidence (abbreviated) --> ...
  <owl:Class rdf:ID="#securityAttack">
  <owl:unionOf rdf:parseType="Collection">
    <owl:Class rdf:about="#maliciousCode"/>
    <owl:Class rdf:about="#spit"/>
    ...
    <owl:Class rdf:about="#flashCrowds"/>
  ... <-- evidence (abbreviated) --> ...</owl:unionOf></owl:Class>
  <owl:Class rdf:ID="#securityDefense">
  <owl:unionOf rdf:parseType="Collection">
    <owl:Class rdf:about="#signature"/>
    <owl:Class rdf:about="#anomalyDetection"/>
    ...
    <owl:Class rdf:about="#serverLoadBalancing"/>
  ... <-- evidence (abbreviated) --> ... </owl:unionOf></owl:Class>
  ... <-- evidence (abbreviated) --> ... </owl:Ontology>
```

FIGURE 20.9 An illustrative security vocabulary in attack and defense.

```
<?xml version="1.0" encoding="UTF-8"?><env:Envelope
xmlns="http://intranet.com/2005/verifySecurityPolicy"
xmlns:env="http://www.w3.org/2003/05/soap-envelope"><env:Body>
  <POLICY><STATEMENT rdf:ID="#preference1">
      <ATTACK>
        <THREAT ref="#spit"/></ATTACK>
      <DEFENSE rdf:connective=or-exact">
        <MECHANISM ref="#blackList"/></DEFENSE></POLICY></env:Body>
</env:Envelope>
```

FIGURE 20.10 Privacy preferences.

2. For the security threat of DoS, the VoIP application requires the network to have the security mechanism of rate reduction installed.

Based on this example, it is obvious that the privacy policy at the VoIP security policy server matches with the rules set at the VoIP application's privacy preferences. Thus the VoIP application will connect to the network as illustrated in Figure 20.3.

References

1. P. C. K. Hung and M. Vargas Martin, "Through the looking glass: Security issues in VoIP applications," *Proceedings of the IADIS International Conference on Applied Computing*, San Sebastin, Spain, 2006.
2. VoIPSA, "Voice over IP security alliance," 2006, http://www.voipsa.org (accessed June 1, 2006).
3. B. H. Schell and J. L. Dodge, *The Hacking of America*, Quorum Books, Westport, CT, 2002.
4. L. Hensell, "The new security risk of VoIP," *E-CommerceTimes*, October 2, 2003, http://www.ecommercetimes.com/story/31731.html.
5. M. Sherr, E. Cronin, S. Clark, and M. Blaze, "Signaling vulnerabilities in wiretapping systems," *IEEE Security & Privacy*, vol. 3, no. 6, 2005, pp. 13–25.
6. S. Landau, "Security, wiretapping, and the internet," *IEEE Security & Privacy*, vol. 3, no. 6, 2005, pp. 26–33.
7. M. Bishop, *Computer Security: Art and Science*, Addison-Wesley, New Jersey, 2003.
8. T. J. Walsh and D. R. Kuhn, "Challenges in securing voice over IP," *IEEE Security & Privacy*, vol. 3, no. 3, 2005, pp. 44–49.
9. S. R. Ahuja and R. Ensor, VoIP: What is it good for? *Queue*, vol. 2, no. 6, 2004, pp. 48–55.
10. J. Peterson, "A privacy mechanism for the session initiation protocol (SIP)," IETF: Request for Comments 3323, 2002.
11. D. R. Kuhn, T. J. Walsh, and S. Fries, "Security considerations for voice over IP systems," *Recommendations of the National Institute of Standards and Technology (NIST)*, Special Publication 800-58, Technology Administration, U.S. Department of Commerce, 2004.
12. S. S. Gokhale and J. Lu, "Signaling performance of SIP based VoIP: a measurement-based approach," *Proceedings of the Global Telecommunications Conference (GLOBECOM)*, 2005.
13. D. Endler and M. Collier, *Hacking Exposed VoIP: Voice Over IP Security Secrets & Solutions*, McGraw-Hill, 2006.
14. A. Demers, S. Keshav, and S. Shenker, "Analysis and simulation of a fair queuing algorithm," *Proceedings of the Special Interest Group on Data Communication (SIGCOMM)*, Austin, USA, September 19–22, 1989.
15. R. Mahajan, S. M. Bellovin, S. Floyd, J. Ioannidis, V. Paxon, and S. Shenker, "Controlling high bandwidth aggregates in the network," *ACM SIGCOMM Computer Communication Review*, vol. 32, no. 3, 2002, pp. 62–73.
16. J. Ioannidis and S. M. Bellovin, "Router-based defense against DDoS attacks," *Proceedings of Network and Distributed System Security Symposium (NDSS)*, San Diego, USA, February 2002.
17. K. Wang and S. J. Stolfo, "Anomalous payload based network intrusion detection," Technical report, Columbia University, March 31, 2004.
18. A. Matrawy, P. van Oorschot, and A. Somayaji, "Mitigating network denial-of-service through diversity-based traffic management," *Proceedings of the 3rd Annual Conference on Applied Cryptography and Network Security (ACNS 2005), Lecture Notes in Computer Science*, Springer, New York, vol. 3531, 2005, pp. 104–121.
19. M. Vargas Martin, "A monitoring system for mitigating fast propagating worms in the network infrastructure," *Proceedings of the Canadian IEEE Electrical and Computing Engineering Conference*, 2005.
20. L. Qi, M. Zandi, and M. Vargas Martin, "A network mitigation system against distributed denial of service: a Linux-based prototype," *Proceedings of the IASTED European Conference on Internet and Multimedia Systems and Applications (EuroIMSA)*, Chamonix, France, March 14–16, 2007.
21. J. Larson, T. Dawson, M. Evans, and J. C. Straley, "Defending VoIP networks from distributed DoS (DDoS) attacks," *Proceedings of the Global Telecommunications Conference (GLOBECOM)*, 2004.
22. M. Vargas Martin, P. C. K. Hung, "Towards a security policy for VoIP applications," *Proceedings of IEEE Canadian Conference on Electrical and Computing Engineering*, Saskatoon, Canada, May 2005.

23. J. Rosenberg, C. Jennings, and J. Peterson, "The session initiation protocol (SIP) and spam," Internet Draft, 2005, http://www.jdrosen.net/papers/draft-ietf-sipping-spam-00.txt (accessed June 1, 2006).

24. H. Varian, F. Wallenberg, and G. Woroch, "The demographics of the do-not-call list," *IEEE Security & Privacy*, vol. 3, no. 1, 2005, pp. 34–39.

25. S. L. Pfleeger and G. Bloom, "Canning SPAM: Proposed solutions to unwanted email," *IEEE Security & Privacy*, vol. 3, no. 2, 2005, pp. 40–47.

26. J. Jung et al., "Flash crowds and denial of service attacks: Characterization and implications for CDNs and Web sites," *Proceedings of the 11th International World Wide Web Conference*, Honolulu, USA, 2002.

27. Defense Information Systems Agency (DISA), "Voice over internet protocol (VOIP)," Security Technical Implementation Guide, Version 1, Release 1, 2004, p. 13.

28. J. Blau, "Cabir worm wriggles into U.S. mobile phones," *PC World*, http://www.pcworld.com/news/article/0,aid,119763,00.asp.

29. S. Staniford, "How to own the Internet in your spare time," *Proceedings of the 11th USENIX Security Symposium*, San Francisco, USA, 2002.

30. E. Levi, "Worm propagation and generic attacks," *IEEE Security & Privacy*, vol. 3, no. 2, 2005, pp. 63–65.

31. C. Nachenberg, "Computer virus-antivirus coevolution," *Communications of the ACM*, vol. 40, no. 1, 1997, pp. 46–51, papers/draft-ietf-sipping-spam-00.txt.

32. C. Shannon and D. Moore, "The spread of the Witty worm," http://www.caida.org/analysis/security/witty/.

33. P. C. van Oorschot, J. Robert, and M. Vargas Martin, "A monitoring system for detecting repeated packets with applications to computer worms," *International Journal of Information Security*, vol. 5, no. 3, 2006, pp. 186–199.

34. D. Whyte, E. Kranakis, and P. van Oorschot, "DNS-based detection of scanning worms in an enterprise network," *Proceedings of Network and Distributed System Security Symposium (NDSS)*, San Diego, USA, 2005.

35. Anti-phishing working group, http://www.antiphishing.org.

36. J. Jung, B. Krishnamurthy, and M. Rabinovich, "Flash crowds and denial of service attacks: characterization and implications for CDNs and web sites," *Proceedings of the 11th International World Wide Web Conference*, Honolulu, USA, May 7–11, 2002.

37. X. Chen and J. Heidemann, "Flash crowd mitigation via adaptive admission control based on application-level measurement," Technical Report ISI-TR-557, University of Southern California, May 2002, http://www.isi.edu/~johnh/PAPERS/Chen02a.html.

38. P. H. Gregory, "Microsoft ignoring the biggest source of security threats?" *Computerworld*, February 25, 2004, URL: http://www.computerworld.com/securitytopics/security/story/0,10801,90466,00.html?SKC=security-90466.

39. P. Samarati and S. Vimercati, "Access control: policies, models, mechanisms," *Lecture Notes in Computer Science*, vol. 2171, no. 137, 2001.

40. Internet Engineering Task Force (IETF), http://www.ietf.org.

41. W3C, "SOAP Version 1.2 Part 1: Messaging Framework," World Wide Web Consortium (W3C) Proposed Recommendation, May 7, 2003, http://www.w3c.org/TR/2003/PR-soap12-part1-20030507/.

42. W3C, "SOAP Version 1.2 Part 0: Primer," World Wide Web Consortium (W3C) Proposed Recommendation, May 7, 2003, http://www.w3c.org/TR/2003/PR-soap12-part0-20030507/.

43. WebOnt. OWL Web Ontology Language. Web-Ontology (WebOnt) Working Group, World Wide Web Consortium (W3C), 2003, http://www.w3c.org/2001/sw/WebOnt/.

44. W3C, "The Platform for Privacy Preferences 1.0 (P3P1.0) Specification," W3C Recommendation, April 16, 2002, http://www.w3c.org/TR/P3P/.

45. W3C, "A P3P Preference Language Exchange 1.0 (APPEL1.0)," W3C Working Draft, April 15, 2002, http://www.w3c.org/TR/P3P-preferences/.

46. H. M. Chong and H. S. Matthews, "Comparative analysis of traditional telephone and voice-over-Internet protocol (VoIP) systems," *Proceedings of the IEEE International Symposium on Electronics and the Environment*, 2004, pp. 106–111.

21

VoWLAN Security Assessment through CVSS

Gianluigi Me and Piero Ruggiero

CONTENTS

21.1 Introduction

The Voice over Internet Protocol (VoIP) over WLAN (VoWLAN) is the transmission of voice information (VoIP) using (one or more of) the 802.11 Wi-Fi standards (a, b, g, n): this combined technology can be used to build autonomous WLAN VoIP networks (e.g., company internal wireless phones) or to integrate wired and wireless telephony in the same IP infrastructure, thus reducing calling charges and also addressing the problem of highly variable cell phone coverage inside buildings.

The VoWLAN technology could be widely adopted in the following scenarios:

- enterprises and healthcare infrastructures;
- the consumer space—as broadband service providers offer both VoIP services and wireless gateways bundled with a broadband connection;
- when prices of dual-mode Wi-Fi/cellular handsets decrease, enterprise users and consumers would roam across wireless home networks, the corporate wireless LAN and public Wi-Fi hotspots.

Since Wi-Fi (802.11) was designed to support IP data (not voice) connections within a limited area, network designers have to adapt the Wi-Fi technology in order to provide VoIP services. This has to be done by extending this area with multiple Access Points (APs) deployment, avoiding the following threats:

- Standard 802.11 APs have a reduced set of internetworking capabilities, as this task is out of their core purpose. Hence, an AP network is unable to load-balance user traffic among multiple APs, which would lead to stop to accepting requests for connections in case of overloaded AP.
- IP voice is a very time-sensitive application, and packet delivery delays will rapidly reduce the quality of service until the disruption.
- Only the Wi-Fi APs equipped with 802.11e can distinguish between IP voice packets and regular IP data packets: when 802.11e is not running, traffic data conflicts with voice data.

The misconfiguration (or bad design) errors can lead to unreliability of service due to unavailability in certain hours or locations. These, together with malicious (external and internal) attacks, pose the denial of service attacks as a major concern, such as kicking off other clients, consuming excessive bandwidth or spoofing access points. Furthermore, even an authorized client (internal security) may be able to sufficiently disrupt the service quality to make the network ineffective for legitimate clients. All these threats, related to the vulnerability of combined technologies are augmented by the threat posed by the unaware or malicious behavior of the user. In fact, recalling the Schneier statement, "Only amateurs attack machine, professionals target people," the same can be applied for example, to cellular network and to VoWLAN networks, representing a vehicle to infect different networks and, exactly for this reason, an optimal target for a professional (e.g., [1]).

The VoWLAN, considered as an instance of the Wireless Mesh Networks where the access nodes provide the interface for the users to communicate through this infrastructure, inherits the wireless mesh networks vulnerabilities related to the APs internet working. Other than mere connectivity, access nodes (VoWLAN APs) should provide support for resource reservation and security. These issues have not been completely and/or uniquely solved, together with the state-of-the art of WLAN and VoIP vulnerabilities and counter-measures. Our purpose is to propose a methodology to assess the security of a VoWLAN network, based on the First CVSS v 2.0.

21.2 Wireless Security

Most of the security properties in IEEE 802.11b standard are flawed, due to a lack of specification (e.g., key management, IV usage) and inappropriate choices (e.g., CRC32 as integrity algorithm). Further vulnerabilities rely on misconfigured AP, AP Denial of Service (DoS) and misuse of Simple Network Management Protocol (SNMP) over WLAN.

21.2.1 Authentication and Confidentiality

The WEP uses the RC4 stream cipher to encrypt data packets. This poses concerns about the choice due to the use of a synchronous stream cipher over the air [2]. Another concern about WEP is related to the Initialization Vector (IV) (24 bit) concatenation with the key, sent in the clear over the air: this reduces the key space from 64 to 40 bit or 128 to 104 bit, both considered unsafe [3]. These weaknesses are exploited in attacks of IV reuse [4], based on 2^{24} keystream collisions (made worse, in some implementation in PCMCIA 802.11 cards). Further attacks are the Known plaintext attack, which let the adversary to obtain the keystream (considering $RC4(v, k) = P \oplus C$, where C is eavesdropped), the Chosen Plaintext attack, and the Ciphertext only attack with the FMS attack [5] based on weak IVs in the Key Schedule Algorithm (KSA), in the form of $(A + 3, N - 1, X)$, releasing information on the static key. Further vulnerabilities rely on the lack of the secret key management recommendation in the 802.11 standard [4]. WPA adopts a Temporal Key Integrity Protocol (TKIP) for data confidentiality, integrity, and replay protection of 802.11 frames. It still uses RC4 for data encryption, but includes a key mixing function and an extended IV space (48 bits) to construct unrelated and fresh per packet keys derived from the four-way exchange. Since TKIP can be considered as a patch of WEP, it still has some of its weaknesses. As a long-term solution, IEEE 802.11i [6] defines a Counter-mode (CTR)/Cipher Block Chain (CBC) MAC Protocol [7] which provides strong confidentiality, integrity, and replay protection (using Additional Authentication Data (AAD)). Although the CTR mode can be considered secure as the capability to generate always fresh nonces, CTR is vulnerable to pre-computation attacks [8].

21.2.2 Integrity

CRC-32 is not an integrity check algorithm, but a correction code algorithm: for this reason WEP misuses it, since CRC-32 is a linear function of the message, enabling attacks as bit flipping. Furthermore, since it is not keyed, the checksum field can also be computed by the adversary who knows the message, enabling message injection from unauthorized users. This problem requires the substitution of integrity algorithm. The middle step was the introduction of a new Message Integrity Check (MIC), based on a lightweight algorithm called Michael, whose main vulnerability relies on ineffectiveness of its 29 bit security against active attacks. The ultimate 802.11i integrity check function is the CBC-MAC [7], which is computed over a sequence of blocks and contains AAD in computing the integrity checksum, and a nonce (unique per encryption key).

21.3 VoIP Security

During the last couple of years many vulnerabilities have been exploited [9,10]. Unfortunately, VoWLAN inherited vulnerabilities from VoIP and wireless. The VoIP protocols, defined in a plethora of standards and RFCs (you can find a collection of them in [11]) as Session Initiation Protocol (SIP) and Real-time Transfer Protocol (RTP), are more oriented towards functionality than security; this leads to the following non-exhaustive list of attacks.*

21.3.1 SIP Message as Clear Text

One of the main problems with SIP is related to its text-based nature: since it was created on HTTP and SMTP model, all the security issues that have previously affected HTTP web and SMTP mail communications are now threats to SIP-based communications. This choice was made for simplicity, ignoring that malicious users could use it to lead attacks against SIP simpler.

Registration Cracking   In [12] the reader can find how to authenticate all SIP devices in the same domain through MD5 digest. This process can be divided into four phases:

1. the client makes a request to the proxy;
2. the proxy returns a `407 Proxy Authentication Required`, together with some information to calculate digest;
3. the client uses these data and its secret parameters (generally a password) to calculate a new value of digest;
4. if the digest is right, the proxy authenticates the client.

This authentication mechanism is vulnerable to dictionary attacks: the free softwares in the Internet, for example, parse authentication requests previously sniffed to identify the information, calculate digest (i.e., fields `username`, `realm`, `nonce`, `uri` from the header `Proxy-Authorization`) and merge them with the passwords saved on a dictionary text file, checking if the MD5 output is the same as the genuine authentication request (i.e., `response` field of the header): so we have found a valid password for that particular username.

The best way to fight a cracking attack based on dictionary is to avoid using weak or default password: therefore, building the passwords following the well-known rules discourage dictionary and brute force attacks at the same time.

Registration Removal   A SIP device must register itself with the proxy server of the network during initialization, because the proxy must be able to forward correctly the calls for that device. This registration must be renewed after a default amount of time (typically 3600 seconds). A malicious user can send to the proxy `REGISTER_REQUEST` messages with the data of the target device and the fields `Contact` and `Expires` set respectively

---

* For a more complete attack list, refer to the previous chapter.

to * and 0: By doing so, the proxy will remove all the registrations for that device and it can no more receive the incoming calls.

Registration Addition   When users have more than one telephone in different places, they can register all these devices on a single contact: when the user receives a call, all his devices ring and the communication is forwarded to the first one which goes off hook. This feature can be exploited by an attacker who sends multiple registration requests for a contact (including one request for his own device), trying to answer to an incoming call sooner than the legitimate user.

Session Teardown   The communication parts of a call can explicitly terminate the session by sending a BYE message. This message must be concerning an established dialogue and must be followed by the right ToTag, FromTag, and CallID tags, to identify the voice session. Thus it is very simple to cause the interruption of a call, sniffing its instauration phase to gain the necessary parameters and then forging ad hoc BYE requests.

Reboot   SIP endpoints can subscribe and receive notification for asynchronous events: the device sends a SUBSCRIBE _ REQUEST to the proxy of its network for the interested events. The proxy sends the requested information through NOTIFY messages. Many SIP devices process NOTIFY messages even if they had no subscriptions. This vulnerability can be exploited by a malicious user: By sending NOTIFY with the tag Event set to check-sync, it is possible to cause the reboot of the target device.

Registration Hijacking   Through this attack, an intruder can hijack a call from the rightful client to another, often the hacker himself. In this way it is possible to receive all the inbound calls of a user, pretending to be him; sometimes it is better to launch a Man in the Middle (MiM) attack. The call happens between the correct users, but the hacker can hear and modify or save the voice stream.

The hacker needs the credentials for authentication to launch this attack, which can be completed together with the cracking of the password of the target user.

Audio Insertion   We know the importance of the Quality of Service (QoS) of communication during VoIP call and so the manipulation of RTP stream can be very annoying for the users. On reaching a privileged position through MiM attack, the hacker can sniff the information about the call and he inject an audio file in format *.wav previously saved. Mixing the current stream with the false one, produces plain confusion into the targets of the attack.

21.3.2 Limited Resource Devices

Very often the devices joining the VoWLAN network are portable devices (e.g., smart-phones and PDAs). Since their calculation resources are very limited, many concerns arise due to the attacks based on consumption of resources.

Flooding proxy   In a VoIP network the main component is the proxy server which provides the processing of the requests of the devices of the net. Therefore, attacking the proxy means causing the crash of the whole network. In order to reach this goal, it is possible to launch a flooding attack against the proxy. By sending thousands of packets or requests, attackers can compromise the response time of the server or cause its crash. For example, two main popular flooding attacks are

- **UDP packets flood:** the target of the attack is the UDP port used by the server, typically the 5060;
- **INVITE _ REQUEST flood:** generating INVITE requests with different Via, From, and Call-ID tags, the proxy is flooded with requests of instauration of new connections, leading to the partial or total interruption of the service.

Both attacks presented can be launched against a phone, too. If an attacker needs high performance hardware to attack a proxy with flood and if his target is a soft phone, the attack is very simple. As mentioned in the Introduction, the VoIP clients are often portable devices with limited resources: memory, CPU, and battery. Therefore, it is very easy to consume resources and cause delay or the crash of the device.

Fuzzing   Fuzzing is closely related to the discovery of bugs or vulnerabilities in an application, an operating system, or a device using a specific protocol: for example, an attacker can send forged packets to violate the rules of the attacked protocol. There are different class of vulnerability tested with fuzzing:

- **Buffer Overflow**   If the length of the data in input is not checked, this data could overwrite internal variables of an application or the stack, causing unpredictable results.

- **Format String**   It is a vulnerability regarding the poor program practices in the use of some functions of I/O; through a simple `printf()` used without filtering input strings, a hacker can read the values of variables on the stack or write data in arbitrary locations of memory. Consequences can be the crash of the system or the execution of malicious code.

- **Integer Overflow**   It can happen after an arithmetic operation that produces a result, is too big to be represented in the established location of memory. This kind of vulnerability can lead to a buffer overflow.

- **Logic Errors**   Malformed input can lead to memory leaks, high consumption of CPU, and to the crash of the system.

21.3.3 Network Vulnerabilities

This section includes attacks based on network vulnerabilities not closely related to VoIP. Prior to starting invasive and specific attacks, an intruder has to investigate about network topology, number of hosts, kind of active services, etc. The most common network attacks are

- scanning: this attack provides to create a list of devices that belong to the network;

- enumeration: once the active ports of the system to attack are found, we can obtain detailed information to launch invasive attacks afterwards;

- eavesdropping: once the access to the network is gained, it is very simple to sniffing voice streams to play and save the calls into many audio formats.

Practical examples of exploits of this vulnerability can be found in [13].

21.4 Guidelines for CVSS Assessment

In this section we examine the guidelines for the administration of VoWLAN, on the strength of their context of application. We can distinguish between five possible scenarios:

- SoHo: Small Office and Home Office;
- SME: Small Medium Enterprises;

- Corporate: enterprise;
- Health Infrastructures: hospitals and clinics;
- Public Hotspots: free network in high frequentation places.

The security requirements are very different depending on the context of application. However, since layered security can be applied to VoWLAN scenario, it is possible to assume a common top-down strategy, to provide wireless security and then the VoIP one in the order of priority. This is because most of the VoIP attacks can be launched only if the hacker is associated with the WLAN, or outside the network if the VoWLAN has an external point of access to Internet.

21.4.1 SoHo

The main requirement in SoHo scenario is to grant the privacy of communications between the parts, in a cost-effective manner: data exchanged in this typology of network are quite sensible, but the figure of the network administrator is often taken by an employee with low security skills. Very often, one of the goals of an attacker is to access the network to use it to launch attacks toward other targets. For these reasons, the network administrator must grant the defense of the access, through typical solutions such as WPA-PSK, which represents a simple and effective mechanism. For the VoIP side of the net, the application of best practice is sufficient to assure an acceptable service.

21.4.2 SMEs

The small medium enterprises (SMEs) are nowadays the most common and count from dozens to few hundreds of employees and it is not unusual that they have a specific department dedicated to IT. Regarding this typology of VoWLAN, data exchanged are more sensitive than SoHo and the communications represent a key issue for the business. Generally, the purpose of the administrator is to avoid the use of his network as a bridge to launch other attacks, as in the SoHo typology. This is the reason why a network administrator must adopt security schemes as WPA or WPA2 and other mechanisms of authentication based on 802.1X standard. Regarding the VoIP, the administrator of the network has to divide, where possible, the traffic of data from the voice through the use of Virtual LAN, to assure a simplified management and a safer network, decoupling different services. We also want to emphasize the risk from DoS attacks that could cause the crash of the net and so could be very dangerous for the business of the company. Therefore, it is necessary to use specific Intrusion Protection System (IPS) and Intrusion Detection System (IDS) against wireless and VoIP attacks.

21.4.3 Corporate

A corporate represents a big company with lots of employees, divided into departments and with a complex IT infrastructure. In this background the main goal on which the administrator of the network must focus is to grant the confidentiality of the data exchanged, that can be subject to industrial spying. For this reason, the administrator must use a security mechanism as WPA2 for the wireless side, plus the standard 802.1X for authentication. In particular, the introduction in the network of a server RADIUS is an excellent solution. For the requirements for confidentiality, it is used as a mechanism of coding such as Virtual Private Network (VPN) and then IPsec; for the VoIP side it is possible to use specific protocols against eavesdropping, such as Secure Real-Time Protocol (SRTP).

Another relevant problem is the availability of the service: a DoS attack could trigger the breakdown of communications with huge loss of money for the company. For this reason, it is fundamental to use IPS and IDS against wireless and VoIP attack, as well as specific sensors against jamming attacks.

21.4.4 Health Infrastructures

One of the applications more helpful and important in VoWLAN technology is in the health infrastructure scope such as hospitals and clinics. Through portable VoIP device, it is possible to contact all the care workers and the service operators in a simple and discrete way. The Health Insurance Portability and Accountability Act (HIPAA) protects the information systems housing Protected Health Information (PHI) from intrusions. It is stated [14] that when information flows over open networks, some form of encryption must be utilized. If closed systems/networks are utilized, existing access controls are considered sufficient and encryption is optional. However, the importance of information exchanged makes this scenery very critical. It is necessary to assure the availability and the quality of the service. This is why, in addition to a generic protection of access based on WPA, it is fundamental to use tools for monitoring the network, such as Intrusion Detection System, against wireless DoS. For the VoIP side it is important that the division of the network in Virtual LAN avoid DoS attacks and, at the same time, protect the communications that could concern the patients of the clinics. Moreover, solutions for coding the network traffic are not applicable, not to cause a degradation of quality performance of the service.

21.4.5 Public Hotspots

Public hotspots become more frequent everyday in transit places such as train stations and airports, where hundreds of persons have to wait for a long time, or crowded places such as shopping centers, where the free service of WLAN helps to charm more shoppers. When accessing this wireless network with portable device, it is possible to use some services based on VoIP. For this reason, it is necessary to set some rules to assure security of communications. The first priority is the availability of service, because if there is no network, people can not be connected. Hence, we have to avoid the possible DoS attacks monitoring the network with systems such as IDS and IPS. Moreover, the hotspot is a free service with free access: it is impossible to imagine mechanisms for access control. It is also unlikely that those providing this service "pro bono" are interested to spend further money for complex and expensive security systems. Indeed they may prefer to provide a service with basic functionality to grant a sufficient quality in a cost-effective manner. All these features make this scenario very insecure and it is often suggested to avoid transmitting sensible data during this kind of connection.

Table 21.1 underlines the most dangerous typology of attacks for the different scenarios and help the administrator to fix their risks; we use this categorization of attacks, according to the analysis made in [15].

21.5 Evaluation of Attacks and Countermeasures

In order to properly evaluate the vulnerabilities of the system regarding an attack, it is necessary to set some fixed parameters characterizing the attack and define some

TABLE 21.1

Summary of Countermeasures with the Changes in Application Background of VoWLAN

Countermeasures	SoHo	SME	Corporate	Health Infrastructure	Hotspot
Base best practice	✓	✓	✓	✓	
WPA	✓	✓		✓	
WPA2		✓	✓		
802.1X Authentication		✓	✓		
VLAN		✓	✓	✓	
VPN			✓		
IPS & IDS		✓	✓	✓	✓

variables that depends on the typology of the target system. Another key factor is the effort that the administrator has to spend on the system to grant its security. In order to facilitate the comprehension of examined cases study, we have decided to quantify numerically all the parameters and to use a well-known methodology, the CVSS.

21.5.1 Attacks Tree

Let us now introduce all the attacks studied and tested for this work, through two attack tree (Figures 21.1 and 21.2). The structure followed for their representation is explained in [16]. Each leaf of the tree is characterized with

- NSE: no special equipment is necessary to complete the attack;
- SE: the hacker needs special equipment (hardware or software) to launch the attack; they are listed according to the legend;
- $: it is the cost of equipment; obviously, it is equal to 0 for NSE attacks.

21.5.2 Characterization of Attacks

During our study, we have analyzed and tested in a laboratory network 27 different attacks for wireless side (most of them can be found in [17]) and 13 for VoIP. These tests helped us to investigate the risks of attack and on the way in which they are performed. After this experimental step, we were able to provide a common characterization of attacks and the consequent scoring. We based the approach on the Common Vulnerability Scoring System (CVSS, [18]) which is an open framework which converts vulnerability data into actionable information for IT managers. It is composed of three metric groups, Base, Temporal, and Environmental, each consisting of a set of metrics:

- **Base** represents the intrinsic and fundamental characteristics of a vulnerability which remain constant over time and user environments;
- **Temporal** represents the characteristics of a vulnerability changing over time but not among user environments;
- **Environmental** represents the characteristics of a vulnerability that are relevant and unique to a particular user's environment.

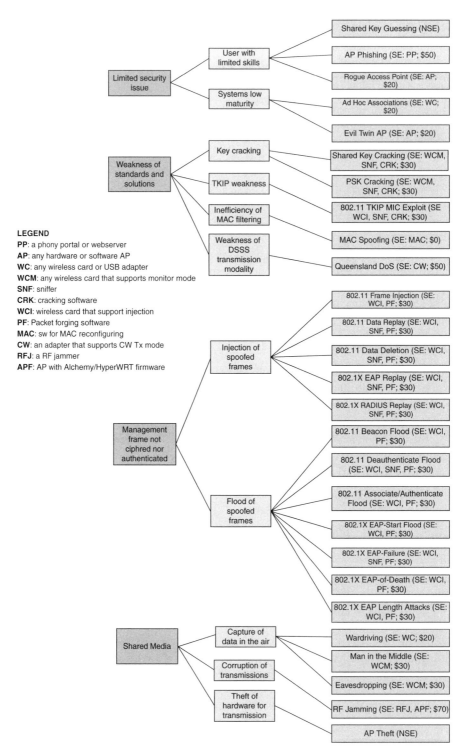

FIGURE 21.1   Wireless attacks tree.

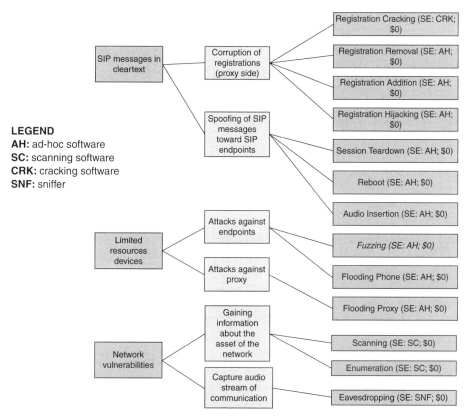

FIGURE 21.2 VoIP attacks tree.

The first set of metrics is always applicable and provide absolute values, suitable for every system. The others are optional and concern with specific features of the system, or with the period in which the attack is launched.

Base metrics consist of:

- **Access Vector (AV)** measures how the vulnerability is exploited;
- **Access Complexity (AC)** depending on the complexity required, it is used to exploit the vulnerability, once the attacker gains access to the system;
- **Authentication (AU)** represents the number of times the attacker has to authenticate to the system in order to launch the attack;
- **Confidentiality Impact (C)** measures the impact of attack on confidential information present on the target system;
- **Integrity Impact (I)** represents the impact on the genuinity of information provided from the system after an attack;
- **Availability Impact (A)** measures the availability of system resources after an attack.

Temporal metrics include:

- **Exploitability (E)** measures the current state of exploit techniques or code availability;
- **Remediation Level (RL)** represents the possibility to found a vendor solution to the vulnerability;
- **Report Confidence (RC)** measures the degree of confidence in the existence of the vulnerability.

Environmental metrics include:

- **Collateral Damage Potential (CDP)** measures the potential for loss of life or physical assets through damage or theft of property or equipment.
- **Target Distribution (TD)** measures the proportion of vulnerable systems;
- **Security Requirements (CR, IR, AR)** helps the analyst to tune the CVSS score depending on the importance of the affected IT asset to a users organization, measured in terms of confidentiality, integrity, and availability.

After the scoring of each vulnerability, using the standard equations we calculate a score ranging from 0 to 10, and a vector is created for each set of metric. It is a text string that contains the values assigned to each metric, and it is used to communicate exactly how the score for each vulnerability is derived.

Base and temporal score are not much useful if considered alone: we think that the third set of metrics, the Environmental, is the most meaningful in our context for its completeness, because before calculating this score we need to compute Adjusted Impact, Adjusted Base, and Adjusted Temporal values. This is why Environmental score is the most representative method to evaluate a vulnerability. In the rest of this document, we refer to this score with $Score_{ABS}$.

Let now analyze an example of scoring. The first attack we have studied is "Shared key guessing," an attack that relies on the attempts of an hacker to guess a WEP password to access a wireless network, wishing that the owner of the network has chosen weak or factory passwords. Obviously, this is an attack launched from the outside of the network, because the hacker is trying to join the WLAN through it. So we can assign the value Network for the Access vector value of CVSS framework. Access complexity can be considered as high because conditions for this exploit are quite unusual; the hacker does not need to authenticate to the network, so we can consider the value of authentication as none. There are no direct impacts for confidentiality, integrity, and availability in this attack, so their value are set to nil. CVSS guidelines recommends to score the vulnerabilities independently, not taking into account any interaction with other vulnerabilities. After compiling all Base metrics, we can examine the temporal ones: Exploitability can be set to high because no exploit is required (manual trigger) and details are widely available, whereas Remediation Level is unavailable because there is no solution available for this attack. Widespread exploitation lead us to assign to Report Confidence the value Confirmed. Regarding environmental metrics, Collateral Damage Potential can be valued as low-medium because this attack may result in moderate property damage, that is the possibility to launch other attacks as eavesdropping. Target distribution is low because there are few WLAN vulnerable to this attack. Then we must assign to the various environment we have considered the values of the requirements of Confidential, Integrity, and Availability: for example, regarding

SoHo scenario, confidential and integrity requirements are low, whereas availability can be considered medium (according to characterization of scenarios presented in Section 21.4.1). Finally, we can present the vector for this attack:

$$Base : AV : [N]/AC : [H]/AU : [N]/C : [N]/I : [N]/A : [N]$$

$$Temporal : E : [H]/RL : [U D]/RC : [C]$$

$$Environmental : CDP : [LM]/T D : [L]/CR : [L]/IR : [L]/AR : [M]$$

and concerning values:

$$Base : [1; 0, 35; 0, 704; 0; 0; 0] = 0$$

$$Temporal : [1; 1; 1] = 0$$

$$Environmental : [0, 3; 0, 25; 0, 5; 0, 5; 1] = 0, 8$$

However, the benefits of this process of scoring is quite limited if it is not applied to the system we want to protect: this is why it is necessary as a further step to transform this absolute score into a relative one:

$$Score_{REL} = Score_{ABS} \times \frac{COV}{100} \tag{21.1}$$

where **COV** is a factor of **coverage** of the system for that particular attack. This parameter is represented through a percentage and must be estimated by an expert in security, evaluating how much the system is presently vulnerable. The range of values of this factor can reset to zero the relative risk if the system presents a full coverage or preserve it unchanged if the architecture under consideration does not have any kind of defense.

21.5.3 Characterization of Countermeasures

Another important topic tied to the analysis of vulnerability is obviously the possibility of finding countermeasures to patch the unsafe system and to grant a higher degree of resistance to attacks. This analysis on countermeasures has been led to focus on two key points:

- **Cost of countermeasures (COST)** represents the cost of achievable countermeasures as human and technological resources. This index can assume values in the range from 1 to 5.
- **Efficacy of the countermeasures (EFF)** measures how much the achievable countermeasures are effective to protect the system from attack; its value is expressed in percentage.

Now it is possible to derive the final formula of **total effort (EFT)** that is:

$$EFF = COST \times \left(2 - \frac{EFF}{100}\right) \tag{21.2}$$

The result of this operation is a value from 0 to 10, to be easily understandable.

21.5.4 Attacks and Countermeasures

Our goal is to summarize remarkable issues of wireless and VoIP attacks and countermeasures, using the characterization presented in the previous sections. In the attacks table we have emphasized the most meaningful columns: Base score, Temporal score, and Environmental score. For Environmental metrics, there are five different scores because we applied the set of metrics to the scenarios of use of VoWLAN analyzed in Section 21.4. This methodology can help the network administrator to better understand the security requirements of his network.

21.5.5 Data Analysis

This quantitative characterization of attacks gives to us the possibility to analyze them numerically. This shows some trends that can help the system administrator in the hardening phase.

With regard to the risk of the attacks, wireless and VoIP reflect the same trends. Authentication and access control attacks have the lowest scores and this means that the impact of these attacks on the system is quite limited. Data confidentiality seems to represent an important issue for VoWLAN technology, more than data integrity. This can be true because the sniffing of a voice communication is very simple to put into practice, whereas it is more difficult to forge a fake conversation. Finally, the DoS attacks typology and represents the main problem for any network administrator because these kinds of attacks can lead to the total blackout of the network.

Wireless countermeasures are more effective and less expensive for authentication, access control, and data confidentiality attacks, whereas for data integrity and DoS they are slightly effective; VoIP technology presents best countermeasures for all the attacks tested in this chapter, even if they are quite expensive.

We can notice from the means of vulnerabilities that in Corporate and Health care scenarios the values of score are very high for both wireless and VoIP attacks, whereas in SoHo and SME are less incisive. Moreover, Public Hotspot environment can be placed in the middle of the two previous sets. These results comes from the analysis made in Section 21.4, where we have illustrated the main requirements for the scenarios and the tied security issues.

Intuitively, the environments where a network administrator needs more skill to plan a successful hardening strategy and where it is necessary to spend much money to enhance network security are those more sensitive to attacks (Table 21.2). This is not always true because we have to consider the relative score of the attack (and the tied coverage factor), and not only the absolute one.

TABLE 21.2

Contextualization of Risks with the Changes in Application Background of VoWLAN

Typology of Attacks	SoHo	SME	Corporate	Health Infrastrcture	Hotspot
Authentication	✓	✓	✓		
Access control	✓	✓	✓		
Data confidentiality			✓		
Data integrity			✓	✓	
DoS		✓	✓	✓	✓

After this analysis, we can introduce the **priority index** factor (**PR**) that represents the ratio of the risk of the attack and the effort for its countermeasures:

$$PR = \frac{Score_{ABS}}{EFT} \tag{21.3}$$

Through this factor it is possible to provide an order of priority to the actions to perform on the system: a high value of risk and low effort suggests to the network administrator the most dangerous attacks that can be easily fixed.

		Vulnerabilities							Countermeasures		
CVSS		BASE SCORE	TEMPORAL SCORE	ENVIRONM. SCORE SOHO	ENVIRONM. SCORE SME	ENVIRONM. SCORE CORPORATE	ENVIRONM. SCORE HEALTH	ENVIRONM. SCORE HOTSPOT	COST	EFFICACY	TOTAL EFFORT
Authentication attacks	Shared Key Guessing	0	0	0,8	0,8	0,8	0,8	0,8	1	40	1,6
	Shared Key Cracking	0	0	2,3	2,3	2,3	2,3	2,3	3	80	3,6
	PSK Cracking	0	0	3	3	3	3	3	1	100	1
Access Control attacks	Wardriving	0	0	1	1	1	1	1	2	60	2,8
	Rogue Access Points	5,8	5,2	1,1	1,3	1,7	1,6	1,1	3	80	3,6
	Ad Hoc Associations	5,8	5,2	1,1	1,3	1,7	1,6	1,1	2	70	2,6
	MAC Spoofing	0	0	1	1	1	1	1	2	10	3,8
Data Confidentiality attacks	Eavesdropping	5,5	5,2	5,7	7,1	8,4	7,1	5,7	3	80	3,6
	Evil Twin AP	7,1	7,1	1,7	1,9	2,1	2,1	1,7	5	80	6
	AP Phishing	7,1	7,1	1,8	2	2,2	2,2	1,8	5	80	6
	Man in the Middle	5,8	5,5	4,4	5,3	6,2	5,7	4,4	4	80	4,8
Data Integrity attacks	802.11 Frame Injection	7,1	7,1	6,3	6,3	9,5	9,5	6,3	3	30	5,1
	802.11 Data Replay	7,1	7,1	6,3	6,3	9,5	9,5	6,3	3	30	5,1
	802.11 Data Deletion	5,4	5,1	5	5	8	8	5	3	30	5,1
	802.1X EAP Replay	5,4	4,6	1,2	1,2	1,9	1,9	1,2	3	30	5,1
	802.1X RADIUS Replay	5,4	4,6	1,2	1,2	1,9	1,9	1,2	3	30	5,1
Denial of Service attacks	AP Theft	7,1	7,1	6,4	6,4	7,2	7,2	7,2	4	80	4,8
	RF Jamming	7,1	7,1	8,6	8,6	9,7	9,7	9,7	3	90	3,3
	Queensland Dos	5,4	4,9	1,9	1,9	2,1	2,1	2,1	2	100	2
	802.11 Beacon Flood	7,1	7,1	8,6	8,6	9,7	9,7	9,7	2	0	4
	802.11 Deauthenticate Flood	7,1	7,1	8,6	8,6	9,7	9,7	9,7	3	30	5,1
	802.11 Associate/ Authenticate Flood	7,1	7,1	8,6	8,6	9,7	9,7	9,7	3	30	5,1
	802.11 TKIP MIC Exploit	3,1	2,8	6,4	6,4	7,4	7,4	7,4	4	30	5,1
	802.1X EAP-Start Flood	5,4	4,6	1,8	1,8	2,1	2,1	2,1	3	30	5,1
	802.1X EAP-Failure	5,4	4,6	1,8	1,8	2,1	2,1	2,1	3	30	5,1
	802.1X EAP-of-Death	5,4	4,6	1,8	1,8	2,1	2,1	2,1	3	30	5,1
	802.1X EAP Length	5,4	4,6	1,8	1,8	2,1	2,1	2,1	3	30	5,1
	MEAN	4,9	4,6	3,7	3,8	4,6	4,6	4	2,9	51	4,2

		BASE SCORE	TEMPORAL SCORE	ENVIRONM. SCORE SOHO	ENVIRONM. SCORE SME	ENVIRONM. SCORE CORPORATE	ENVIRONM. SCORE HEALTH	ENVIRONM. SCORE HOTSPOT	COST	EFFICACY	TOTAL EFFORT
Authentication attacks	Registration Cracking	0	0	2,3	2,3	2,3	2,3	2,3	2	90	2,2
Access Control attacks	Scanning	0	0	1	1	1	1	1	4	80	4,8
	Enumeration	0	0	1	1	1	1	1	3	60	4,2
Data Confidentiality attacks	Eavesdropping	6,1	5,5	6	7,3	8,5	7,3	6	3	90	3,3
	Registration Hijacking	5,2	4,9	4,2	5,2	6,2	5,2	4,2	4	80	4,8
Data Integrity attacks	Audio insertion	5,3	4,5	4,4	4,4	5,6	5,6	4,6	3	80	3,6
Denial of Service attacks	Fuzzing	7,8	7,8	8,7	8,7	9,8	9,8	9,8	2	40	3,2
	Flooding Proxy	7,8	7,8	8,7	8,7	10	10	10	4	90	4,4
	Flooding Phone	6,1	6,1	7,7	7,7	9	9	9	4	90	4,4
	Registration Removal	5,7	5,1	5,7	5,7	6,4	6,4	6,4	4	80	4,8
	Registration Addition	6,4	5,8	5,4	5,6	6,2	6,2	6,2	4	80	4,8
	Session Teardown	5,7	5,4	5,8	5,8	6,6	6,6	6,6	4	90	4,4
	Reboot	6,1	5,8	5,9	5,9	6,7	6,7	6,7	4	90	4,4
	MEAN	4,8	4,5	4,2	5,3	6,1	6	5,8	3,4	80	4,1

FIGURE 21.3 Matrix of attacks and countermeasures.

TABLE 21.3

Priority Indexes for the Fixing Process

Typology of Attacks	$Score_{ABS}$	EFT	PR
Authentication	2.2	2.1	1
Access control	1.1	3.9	0.3
Data confidentiality	5	4.6	1.1
Data integrity	4.9	4.4	1.1
DoS	6.5	4.4	1.5

In Table 21.3 we present the convenience factors for the attacks, formerly grouped for typology. For this analysis we used the means of all the environmental scores to calculate $Score_{ABS}$ values.

Assume now a network administrator of a VoWLAN used in a clinic for communications between the doctors and a control room that manages the emergencies. This scenario fits the health infrastructure environment presented in Section 21.4.4; hence the administrator can refer to the hints shown in Table 21.1 to start an hardening strategy, that is, use of WPA for access control, use of virtual LAN to protect confidentiality of sensible communications, and IDS and IPS to avoid the possibility of DoS attacks that could lead to the blackout of the net. After this phase of initiation, he needs to calculate the coverage rates of his own network for the attacks presented in Figure 21.3 and then compute the $Score_{REL}$ values from the $Score_{ABS}$ listed in the same picture. For example, we assume that data integrity and DoS attacks have the highest values among all other attacks. Consulting Table 21.3, the administrator can decide to assign more priority to the fixing of DoS attacks than the integrity ones.

21.6 Conclusions

As we have analyzed in the previous section, attacks represent a real risk depending on the typology of the target VoWLAN: different kinds of network need different precautions. The network administrator should focus on placing a strategy for hardening in the right context. For this reason, we present the steps to which a good VoWLAN administrator should refer to protect his network from the attacks discussed in this chapter:

- choose from the listed environments, the one that better fits his network;
- apply the security issues suggested in Table 21.1 for a first hardening strategy;
- referring to the values of Environmental score $Score_{ABS}$ listed in Figure 21.3 and calculate the $Score_{REL}$ values, according to the coverage rates of his VoWLAN;
- plan a long-term hardening strategy through the knowledge of the most dangerous attacks for his network and most convenient strategies for defense: use Table 21.3 to identify a priority scale of vulnerabilities to fix.

References

1. C. Fleizach, M. Liljenstam, P. Johansson, G. M. Voelker, and A. Mehes, "Can you infect me now? Malware propagation in mobile phone network," www-cse.ucsd.edu/voelker/pubs/cellworm-worm07.pdf, November 2007.

2. P. Chandra, *Bulletproof Wireless Security*, Elsevier, 2005, p. 170.

3. J. Walker, *Unsafe at Any Key Size: An Analysis of the WEP Encapsulation*, IEEE Press, 2000, http://www.netsys.com/library/papers/walker-2000-10-27.pdf.

4. N. Borisov et al., "Intercepting mobile communications: The insecurity of 802.11," *Proceedings of 7th Annual International Conference on Mobile Computing and Networking*, ACM Press, 2001, pp. 180–188.

5. S. Fluhrer, I. Mantin, and A. Shamir, "Weaknesses in the key scheduling algorithm of RC4," Revised Papers, *8th Annual International Workshop on Selected Areas in Cryptography*, 2001, pp. 1–24.

6. IEEE Std 802.11i-2004, Amendment to IEEE Std 802.11, 1999 Edition, Amendment 6: Medium Access Control (MAC) Security Enhancements Part 11. IEEE Press, April 2004.

7. D. Whiting, R. Housley, and N. Ferguson, "Counter with CBC-MAC (CCM)," Request for Comments 3610, IETF, September 2003.

8. D. McGrew, "Counter mode security: Analysis and recommendations," http://www.mindspring.com/dmcgrew/ctr-security.pdf, 2002.

9. S. McGann and D. C. Sicke, "An analysis of security threats and tools in SIP-based VoIP systems," http://www.colorado.edu/policylab/Papers/Univ_Colorado_VoIP_Vulner.pdf, 2005.

10. G. Me and D. Verdone, "An overview of some techniques to exploit VoIP over WLAN," *Proceedings of the International Conference on Digital Communications*, Digital Telecommunications, 2006, DOI 10.1109/ICDT.2006.17.

11. SIP Standards, http://www.packetizer.com/ipmc/sip/standards.html.

12. Network Working Group, "SIP: Session initiation protocol," Request for Comments 3261, June 2002.

13. D. Endler and M. Collier, *Hacking Exposed VoIP: Voice Over IP Security Secrets and Solutions*, McGraw-Hill/Osborne, 2007.

14. Department of Health and Human Service, "Health Insurance Reform: Security Standards; Final Rule," Federal Register, February 2003.

15. Network Working Group—R. Shirey, *Internet Security Glossary*, Request for Comments 2828, May 2000.

16. B. Schneier, "Attack trees—Modeling security threats," *Dr. Dobb's Journal*, December 1999.

17. L. Phifer, "Wireless attacks," *A to Z*, SearchSecurity.com, May 2006.

18. P. Mell, K. Scarafone, and S. Romanosky, *CVSS v2—A Complete Guide to the Common Vulnerability Scoring System*, FIRST.org, 2007.

22

Flash Crowds and Distributed Denial of Service Attacks

Hemant Sengar

CONTENTS

The distributed denial-of-service (DDoS) attacks and flash crowds result in a large number of service requests on a server. Under such overloading conditions, the target server becomes sluggish and even unresponsive either due to its resource limitation or exhaustion of the path linking to the victim. In flash crowds, a large number of legitimate users try to access the server's resources simultaneously, whereas DDoS attacks contain malicious requests whose only goal is to subvert the normal operations of the server. Jamjoom et al. [1] and Park et al. [2] showed that the Internet is vulnerable to overloading caused by flash crowds and DoS attacks.

Internet Protocol (IP) telephony, being an Internet-based service, is equally vulnerable to overloading caused by flash crowds or flooding DoS attacks. In the past, there were many instances when landline and wireless telephone systems experienced flash crowd events. In 1996, there was an instance of televoting, in which within two hours, 2.5 million people called two telephone numbers to say "*yes*" or "*no*" to ITV's television program *Monarchy, Yes or No.* The "human repeated attempt" pattern was observed, in which the engaged number was repeatedly tried until either it responded or the customers hung up. Similarly, New York's SS7-based landline and wireless telephone systems experienced a flash crowd event during the September 11, 2001, terrorist attack and on August 14, 2003, the North America blackout. Thus, it is reasonable to predict that IP telephone systems will face similar flash crowd events. Further, due to the shift of intelligence toward the end devices (such as IP telephones and PC-based softphones), Voice over Internet Protocol (VoIP) is more likely to encounter overloading attacks than traditional telephone system [known as public switched telephone network (PSTN)].

22.1 INVITE Flooding Attack and Flash Crowd

Session initiation protocol (SIP) stacks can be found on PCs, laptops, VoIP phones, mobile phones, and wireless devices. To facilitate interconnection between these end devices, SIP uses *SIP Gateways* and *SIP Servers*. This gives an attacker a wide range of target devices, starting from end devices to routers, including switches, signaling gateways, media gateways, and SIP proxies. SIP-based entities are susceptible to two types of flooding attacks. In the first, bogus traffic is directed toward the SIP entity with the aim of exhausting CPU resources of the system and the resources lying in the path connecting to SIP entity. In the second case, the attacker exploits the vulnerabilities in the SIP protocol itself. SIP is a transactional protocol, in which each transaction consists of a request and its corresponding responses. SIP cores or *transaction users* (TUs) at the SIP entity (except stateless proxy) maintain transaction state for some time. INVITE transaction is particularly susceptible to flooding attack because it may take several seconds to complete. For example, a SIP proxy has the option to maintain INVITE transaction state up to 3 minutes. Similarly, when an user agent server (UAS) (i.e., SIP-based IP phone) accepts the INVITE request, TU generates a 2XX response and waits for an ACK while maintaining the transaction state. If UAS retransmits the 2XX response for $64*T1$ seconds (typical value of timer $T1$ is 500 milliseconds) without receiving an ACK, the session should be terminated by sending a BYE message. In all these and other examples, TU's finite capacity of maintaining state for the transaction could easily be exploited by flooding with spurious requests.

A SIP proxy server can operate in either a stateful or stateless mode. A stateless proxy server acts solely upon the current received SIP requests or responses (based on message content) without storing any message information. A stateful proxy server keeps track of SIP requests and responses (i.e., dialog information) received in the past. With the aid of the history information, current SIP requests and responses are processed. The stateless proxy server is less susceptible to DoS attacks compared to the stateful proxy server.

Under the flash crowd events, IP telephones or SIP proxies will notice a dramatic increase in the number of SIP INVITE requests similar to that when it is under INVITE flooding attack. By just looking into the number of requests or its message content, it is hard to predict the true cause of INVITE surge. Nevertheless, both the events overload the SIP proxy server to the point where it becomes sluggish and even unresponsive, consequently, causing a DoS. The main difference between two events lies in the nature of `INVITE` requests. In the case of flooding DoS attack, the `INVITE` requests are spurious, whereas a flash crowd event generates legitimate `INVITE` requests. This difference in the nature of `INVITE` requests results into two different observable protocol behavior. The call setup phase of a SIP-based IP telephony is completed by a three-way handshake of `INVITE`/`200 OK`/`ACK` messages. Under the `INVITE` flooding attack, within a short span of time, there will be a significant difference between {`INVITE`, `ACK`} SIP protocol attributes observed in the VoIP signaling traffic. Whereas, in a flash crowd event, difference in the same protocol attributes behavior rises slowly with time in comparison to flooding DoS attack, though the growth rate largely depends upon the severity of the flash crowd. The ability to detect and distinguish both types of events is important so that a differential treatment can be imparted to flooding attack requests by dropping malicious requests but entertain those requests arising due to a flash crowd event.

22.2 Signaling Traffic Behavior

22.2.1 Experiment Setup

To study VoIP signaling traffic behavior, we built a testbed consisting of a SIP proxy server and IP-based softphones. The testbed consisted of eight PCs equipped with the Linux operating system acting as SIP clients (i.e., UAs), SIP servers, a router with wide area network emulator, and an attacker with `INVITE` flood traffic generator. Figure 22.1 shows the layout of the testbed used to generate VoIP traffic and to evaluate the performance of the proposed detection and protection mechanisms. Our SIP proxy server was based on

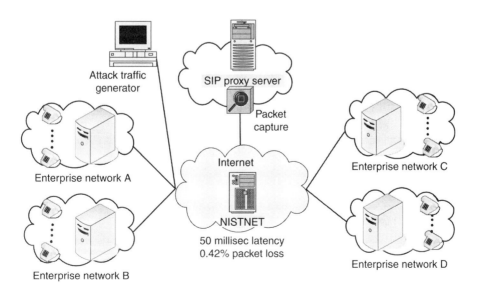

FIGURE 22.1 Simulated network topology.

SIP Express Router (SER) [3], an open source software project that provides call control, routing, and other operations service support and features. This system was scalable from a single box in a laboratory to a carrier-grade network supporting thousands of telephone subscribers. We installed and deployed SER SIP server onto a single PC (2 GHz Intel Pentium IV with 256 Mbytes of RAM). Our SIP server was provisioned with 10,000 telephone subscribers. Enterprise networks A, B, C, and D were simulated using (866 MHz Pentium III PC with 256 Mbytes of RAM) PCs equipped with SIP traffic generators that behaved as multiple UACs calling to UASs in other enterprise networks. The talk time (i.e., call duration) between any two subscribers is exponentially distributed with the mean value of 120 seconds. The wide area network emulator ("NISTNet" [4]) connected enterprise networks and SIP server with each other using 100 Mbps Ethernet links. The NIST package ran on a Linux router (633 MHz Pentium III PC with 128 Mbytes of RAM) in which packet delay distributions, congestion, loss, and bandwidth limitation, etc., were configurable. We set the Internet delay to 50 milliseconds and the packet loss rate to 0.42% in our experiments [5]. We used Network Time Protocol to synchronize the time of clients with that of the NISTNET server.

22.2.2 Normal Behavior of SIP Attributes

Figure 22.2 plots the normal SIP traffic behavior as observed near the SIP proxy server. We monitored both to-and-fro signaling traffic between subscribers and the SIP proxy server and vice versa (e.g., see Figure 22.1). SIP signaling messages are carried by UDP and SIP clients (i.e., UAs) use the default $T1$ timer value of 500 ms. At the end of the session, after sending a BYE message, each UA maintains the transaction state for the next 4 seconds. The stochastic nature of phone call arrivals varies with the change of time, and cannot be easily modeled by a random process with a deterministic time-varying arrival rate. Still, in order to emulate a near-realistic and normal call behavior, we assume Poisson distributed phone call arrivals. In our experiments, we use three independent Poisson processes $N_1(t)$, $N_2(t)$, and $N_3(t)$ with mean arrival rates of 30, 30, and 15 calls per second (CPS), respectively. In an effort to mimic diurnal call behavior as shown in Figure 22.2a, during an hour of experiment run, for the initial and last 25 minutes, the day-time peak call rate is generated by all three individual processes while during the in-between 10 minutes representing off-peak hours (such as night time or during the week-end), only N_3 was awake and N_1, N_2 slept.

As shown in Figure 22.2a, the SIP protocol attributes INVITE, (100 Trying), (200 OK), ACK and BYE curves show near-ideal normal behavior. It should be noted that in this chapter, the (200 OK) messages correspond to INVITEs only. During an hour of experimental run, we observe 975 and 11760 INVITE retransmissions from the clients and SIP proxy server, respectively. There are 2044 (200 OK), 4006 ACK, and 12490 BYE retransmissions from the SIP proxy server. Because of the Internet network conditions and processing delays at the server, the strict one-to-one relationship between INVITE and other call setup messages, such as (200 OK), ACK etc. are violated. However, under normal conditions these deviations from ideal behavior are small and exhibit strong positive correlations among call setup messages.

22.2.3 Behavior Under Flash Crowd

Figure 22.3 plots the SIP session establishment behavior under flash crowd. During the initial 10 minutes, the SIP proxy server is loaded with a Poisson call rate (with a mean of

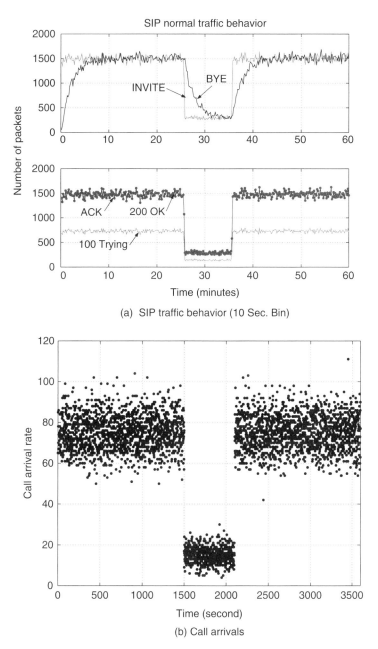

FIGURE 22.2   Normal SIP signaling traffic behavior.

75 CPS). In this period, the call setup messages demonstrate one-to-one relationship with occasional packet drops and retransmissions. In order to simulate a flash crowd event, at the starting of the 10th minute, additional 400 CPS are introduced, bringing the overall call rate to 475 CPS. Even at this high call rate, the SIP proxy server shows remarkable resiliency and tries to behave normally for the next couple of minutes (depending upon the severity of flash crowd). As we know, a transaction stateful server

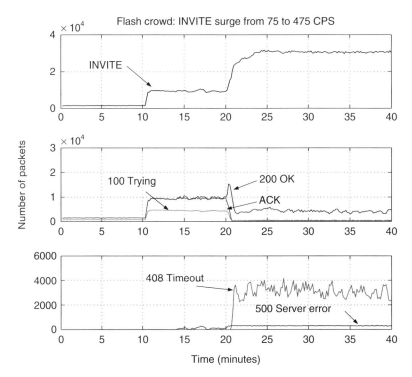

FIGURE 22.3　SIP traffic behavior under flash crowd (10 Sec Bin).

keeps a copy of the received request for some time and a transaction context typically consumes ≈3 kbytes (depending on message type and memory management overhead) [6]. Therefore, after maintaining a certain number of transaction contexts, the SIP proxy server's performance degrades. At this transition point, the call throughput falls quickly and because of the resource exhaustion and processing delays, we observe a sudden jump in INVITE and (200 OK) retransmissions by the SIP UAs. The existing INVITE transactions in the server also start timing out (by sending out [408 Timeout] messages). For example, as shown in Figure 22.3, a certain number (e.g., ≈30 CPS) of INVITE messages are straightforward refused to be entertained by sending (500 Server Error) messages. Only a fraction of INVITE requests are responded with provisional response of (100 Trying) messages (i.e., accepted for further processing). However, amidst the timeouts and retransmissions (due to processing delays), the server's call throughput drops to ≈10 CPS from 475 CPS.

22.2.4 Behavior Under INVITE Flooding

Figure 22.4 plots the SIP session establishment behavior under DDoS attack. At the starting of the 10th minute, the initial Poisson call rate (with the mean of 75 CPS) is mixed with additional calls at the rate of 150 CPS with spoofed source IP addresses. First, the SIP proxy server tries to behave normally by sending (100 Trying) messages for each of the received INVITE requests and by maintaining their transaction state. For each of the accepted spoofed INVITE requests, the SIP proxy server transmits 7 INVITE messages towards the unreachable destination IP address before timing out the transaction

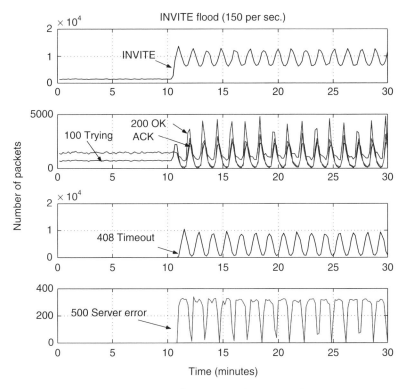

FIGURE 22.4 SIP traffic behavior under `INVITE` flooding attack (10 Sec Bin).

state. Second, as the flooding rate increases, the major part of server resources are held up by the spoofed requests and these are made available for further reuse only when the transactions time out. At the exhaustion of server resources and due to processing delays, the number of (`100 Trying`) declines and a fraction of new `INVITE` requests is refused to be serviced by sending (`500 Server Error`) messages, further, the number of (`200 OK`) messages being forwarded through the SIP proxy server also declines resulting in a fall of call throughput, but at the same time more and more (`200 OK`) retransmissions by the SIP UAs. However, because of the timeouts (i.e., removal of existing transactions) and (`500 Server Error`) (i.e., refusal of accepting new transactions) messages, the proxy server tries to recover and accepts new requests, but due to unabated DDoS traffic, the server's performance degrades again, showing an oscillatory behavior of recovery and degradation.

22.2.5 Difference Between Flash Crowds and DDoS Attacks

During flash crowds and flooding attacks, the SIP proxy server notices a dramatic increase in the number of `SIP INVITE`. As stated, the analysis of a message's content or the number of `INVITE` requests are insufficient to predict the true cause of an `INVITE` surge. Nevertheless, the main difference between these two events are in the nature of `INVITE` requests that results in two different observable protocol behavior as seen in Figures 22.3 and 22.4. Instead of analyzing `INVITE` requests in isolation, our approach studies call setup transactions (e.g., both `INVITE` and `ACK`) involving various message exchanges (to be explained shortly) and revealing many unique characteristics as described in the following:

- In flash crowds, proxy server tries to behave normally and if the request rate remains unabated, then after some time (depending upon the INVITE rate) the server starts sending server error messages and timeouts. Whereas, in flooding attacks, the transmission of timeouts and server error messages happens much quicker because spoofed INVITE messages do not have corresponding (200 OK) and ACK messages and therefore server resources are exhausted much faster.

- Because of the lockup of server resources (e.g., incomplete transactions), INVITE flooding attacks are more harmful even at a smaller rate compared to a flash crowd. For example, as shown in Figure 22.4, the introduction of an extra 150 spoofed INVITES per second degrade the performance of a server, whereas legitimate requests at this rate do not have any observable impact on the server's request processing.

- In a flooding attack, due to locking up of server resources and their release (by denying the admittance of new call requests and the timing out of existing transactions), traffic exhibits oscillatory behavior of recovery and degradation. Contrastingly, in flash crowds, the call throughput decreases to a minimum level and remains relatively invariant with time.

- In a flash crowd, there are considerably higher number of INVITE and (200 OK) retransmissions compared to a DDoS attack because, first, INVITE's destination are real and second, each SIP client behaves in a protocol compliant way by retransmitting INVITEs at exponential backoff timer values.

22.3 Distinguishing Flash Crowds from DDoS Attacks

Our mechanism detects anomalies in collections of packet streams, going through a cyclic behavior consisting of two phases. As shown in Figure 22.5, during the training phase, the training data set consisting of the attribute set is collected over n sampling periods over a normal traffic stream. The duration of each period is Δt. This initial training data set is assumed to be devoid of any attacks and acts as a base for comparison with the next $(n + 1)$th period of the testing data set. Using the soon-to-be-described *Hellinger distance*, we measure the distance between these two data sets. If the measured distance exceeds a threshold, an alarm is raised, and otherwise the testing data set is included in the immediately preceding

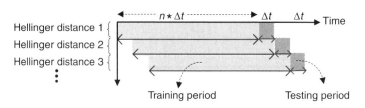

FIGURE 22.5   Relationship between training and testing periods.

$(n − 1)$ sampled traffic data to derive a new training data set. This moving window mechanism helps the training data set to adapt with the dynamics of network traffic. In order for this design to work, the following three parameters are computed online:

1. *The Probabilistic Distribution for Normal Traffic:* These are computed as the ratio of packets that satisfy the feature (i.e., attribute) to the total number of packets received during the training phase.
2. *The Probabilistic Distribution for Abnormal Traffic:* These are computed as averages during the time window immediately following the training period, again as a ratio of packets satisfying the chosen feature to the total number of packets, whereas the deviation of the two probability distributions are computed using the Hellinger distance.
3. *The threshold of deviation to distinguish normal behavior from the abnormal behavior:* to compute a dynamic threshold as the computation progresses through cycles of training and testing phases, using Chebyshev inequality.

22.3.1 Hellinger Distance

Hellinger distance is a metric for measuring the deviation between probability measures that does not make any assumptions about the distributions themselves. It is closely related to the *total variation distance* [7] but with several advantages. Let **P** and **Q** be two probability distributions on a finite sample space Ω, where **P** and **Q** on Ω are N-tuples $(p_1, p_2, \ldots, p_N)$ and $(q_1, q_2, \ldots, q_N)$ respectively, satisfying (in)equalities $p_\alpha \geq 0$, $q_\alpha \geq 0$, $\Sigma_\alpha p_\alpha = 1$ and $\Sigma_\alpha q_\alpha = 1$. The *Hellinger distance* between **P** and **Q** is defined as:

$$d_H^2 (\mathbf{P}, \mathbf{Q}) = \frac{1}{2} \sum_{\alpha=1}^{N} \left(\sqrt{p_\alpha} - \sqrt{q_\alpha} \right)^2 \tag{22.1}$$

The Hellinger distance satisfies the inequality of $0 \leq d_H^2 \leq 1$. The distance is 0 when **P** = **Q**. Disjoint **P** and **Q** shows the maximum distance of 1. Sometimes, the factor $\frac{1}{2}$ is not used in the above equation. A related notion is the *affinity* between probability measures, which is defined as

$$A(\mathbf{P}, \mathbf{Q}) = 1 − d_H^2 (\mathbf{P}, \mathbf{Q}) = \sum_{\alpha=1}^{N} \sqrt{p_\alpha q_\alpha} \tag{22.2}$$

The affinity between two probability measures **P** and **Q** is one (i.e., $A = 1$) if they are equal, and zero if the measures are *totally different*. Further details on *Hellinger distance* can be found in [7,8].

22.3.2 Hellinger Distance Measurement

To detect and distinguish INVITE flooding and flash crowds, we chose only four SIP protocol attributes, namely INVITE, (100 Trying), (200 OK), and ACK. The choice of this attribute set {INVITE, (100 Trying), (200 OK), ACK} was based on the observation that session establishment could be confirmed by the three-way handshake of INVITE/(200

OK)/ACK messages. The selection of (180 Ringing) message did not add any value because its behavior with time was almost the same as (200 OK). In the case of an authenticated call setup, we could select other challenge/response messages as well. In fact, any number of appropriate attributes could have been selected without suffering from the *curse of dimensionality* problem. It is important to note that (100 Trying) was the only message in the attribute set that was generated by the stateful proxy server. Consequently, if a sophisticated attacker tries to circumvent the detection mechanism by sending a mixture of spoofed INVITE messages along with (200 OK) and ACK messages, still, it will not succeed, rather makes detection easier because of the unspoofable (100 Trying) messages.

Here, the probability measure $\mathbf{P}$ is an array of normalized frequencies of p_{INVITE}, $p_{100\ Trying}$, $p_{200\ OK}$, and p_{ACK} (i.e., $p_\alpha = N_\alpha/N_{Total}$ where $\alpha \in \{INVITE, 100\ Trying, 200\ OK, ACK\}$ and $N_{Total} = (N_{INVITE} + N_{100\ Trying} + N_{200\ OK} + N_{ACK})$) over the training data set assuming that we observe N_{INVITE}, $N_{100\ Trying}$, $N_{200\ OK}$, and N_{ACK} packets in $n * \Delta t$ time period. Similarly, during the testing period (i.e., at the $(n + 1)$th sampling duration), $\mathbf{Q}$ is an array of normalized frequencies of q_{INVITE}, $q_{100\ Trying}$, $q_{200\ OK}$, and q_{ACK}. To calculate the HD between $\mathbf{P}$ and $\mathbf{Q}$ at the end of $(n + 1)$th sampling period, we use $d_H^2(\mathbf{P}, \mathbf{Q})$ formula.

Figure 22.6a shows the HD plot for normal behavior of SIP protocol attribute set. During an hour, the maximum observed distance is $4 * 10^{-5}$ and most of the time the HD shows remarkable closeness between the observed and training data sets. The occasional peaks in the plot overlapped the period where the traffic rate is very low (as shown in Figure 22.6) and even a few packet drops and retransmissions result into spikes. At higher rate, the dropping or retransmissions of few packets are masked and do not cause any significant observable deviation. Figure 22.6b is plotted under flash crowd event. The initial 10 minutes show a normal behavior and at the start of the 10th minute, a flash crowd of call requests happens bringing the call rate from 75 CPS to 500 CPS. We observe that toward the end of the 14th minute, the distance starts climbing and reaches a maximum of 0.21 at 15.66 minute. In Figure 22.6c, at the start of the 10th minute, 150 spoofed INVITE requests per second are mixed with the normal request rate of 75 CPS; consequently, in the very next observation period the distance starts climbing and then within a minute reaches to the maximum of 0.086 at the 11th minute.

22.3.3 Detection Threshold Setup

Using HD, we profiled the normal traffic behaviors. Our goal is to seek a threshold of measured distances that could differentiate among different kinds of traffic anomalies and normal behavior. During the testing period, once the measured distance was higher than the threshold distance, we raised an attack alert. The distribution of the measured HDs was used to calculate the mean μ and variance σ^2 of distances. Given these two parameters, we wish to determine the validity of the observed distance d in the *testing period*. To do so, assume that the measured distance X was a random variable with mean $E(X) = \mu$ and variance $\sigma^2 = \mathrm{var}(X)$. Then, the Chebyshev inequality,

$$P(|X - \mu| \geq t) \leq \frac{\sigma^2}{t^2}, \qquad \text{for any } t > 0$$

could be used to compute an upper bound on the probability that some random variable X deviated from its mean by more than any positive value t, given only the mean and variance

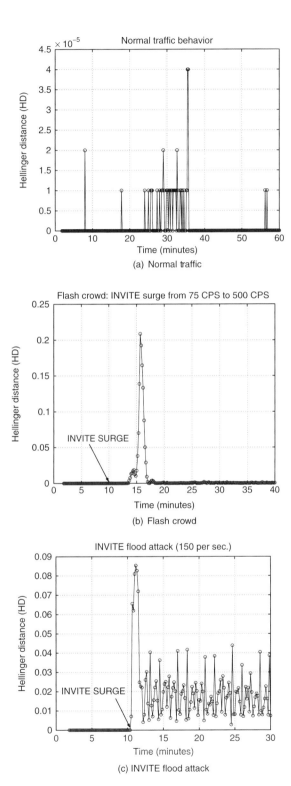

FIGURE 22.6 Hellinger distance calculation.

of X. The Chebyshev's inequality could also be interpreted as the proportion of observed distances in a sample deviating at least t times the standard deviation from the mean would be at most $1/t^2$. Therefore, we could define a confidence band of $\mu \pm 8 * \sigma$ as a *normal region*, where the proportion of observed distances occurring within the region was at least 98.5%. Beyond this normal region, the observed distances were anomalous and assigned a severity level depending upon their distance from the normal region. For example, a *low* severity level could be assigned to those distances which occur within the regions $\mu + 8\sigma < d < \mu + 12\sigma$ or $\mu - 12\sigma < d < \mu - 8\sigma$ and a *high* severity could be assigned for those distances where $d > \mu + 12\sigma$ or $d < \mu - 12\sigma$. The threshold we calculated is based on the theoretical analysis of proportional abundance of the distance distribution. With time, as more empirical observations are available, the distance thresholds and severity levels can be calibrated more precisely.

22.3.4 Detection and Classification

When the arrival rate of `INVITE` messages crosses a predefined maximum safe limit set for a server, an alert flag is raised. The definition of safe limit is governed by many factors, such as server type (i.e., its configuration), peak hour call rate, the number of subscribers, etc. By correlating `INVITE` surge alert with the anomalous protocol behavior alert (raised by HD), we distinguished flash crowds from DDoS attacks. At the moment of an `INVITE` surge alert, if the measured distance of protocol behavior occurred in the *normal* or *low severity* zone, then it was determined to be a flash crowd event. Whereas, in DDoS attacks the distance between protocol attributes would be significant and would occur in the *high severity* zone. For example, with an `INVITE` surge alert, if the protocol behavior distance remains lower than the threshold value of ($\mu + 8 * \sigma = 1.841 * 10^{-6} + 8 * (5.76 * 10^{-6}) \approx 4.8 * 10^{-5}$) then it is a flash crowd event. Whereas, in DDoS attack, the protocol behavior distance crosses the threshold value along with an `INVITE` surge alarm.

22.4 Further Discussion

We now address some of the concerns regarding the proposed detection mechanism.

A1. It could be argued that a sophisticated attacker may circumvent the detection mechanism by sending a mix of attack traffic.

The (`100 Trying`) message in the attribute set is the only per-hop message among the other end-to-end messages. It is generated by a stateful proxy server and makes mixed attack traffic detection easier.

A2. Since our profile creation is adaptive, it is possible that an attacker can circumvent the detection mechanism by gradually training the profile creation.

To overcome this drawback of adaptive profiling, we use a detection scheme that is sensitive enough to detect low-rate attacks too. As shown in Figure 22.7a, we can detect low-rate attacks as low as 10 spoofed `INVITES` per second. In this case, the maximum observed distance is $\approx 4 * 10^{-3}$, still much higher than the threshold value of $4.8 * 10^{-5}$. The detection of low-rate attacks below 10 `INVITE`s per second. is also possible, but may cause many false alarms because of the adverse Internet conditions. For example, the added latencies and packet drops can cause transitory deviation from normal protocol behavior as shown in Figure 22.7b and c. We observe that the worst Internet conditions may cause

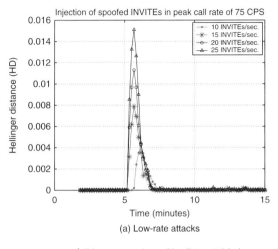

(a) Low-rate attacks

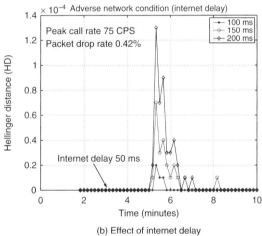

(b) Effect of internet delay

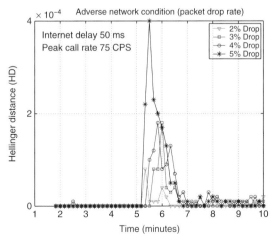

(c) Effect of internet drop rate

FIGURE 22.7 Detection sensitivity.

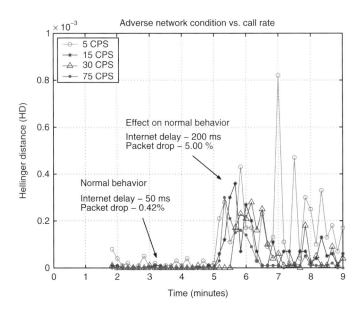

FIGURE 22.8   Call rate vs. adverse network conditions.

the distances to jump up to $\simeq 10^{-4}$. To avoid such false alarms, we restrict ourselves to the detection of abnormal behaviors causing distances at the order of 10^{-3}.

A3. Is it possible to use the detection mechanism in small enterprise networks where the call rate is relatively low compared to voice service provider's network?

In our experiments, we assumed the peak call rate of 75 CPS. However, as shown in Figure 22.8, the proposed detection mechanism is equally applicable to low call rates as typically observed in small enterprise networks. In normal conditions, the low call rate such as 5 CPS generate the distances of $\simeq 10^{-5}$. Under the assumption of worst network conditions (packet drop of 5% and Internet delay of 200 ms), the distance jumps up to $\simeq 10^{-4}$.

The proper setting of threshold value at $1 * 10^{-3}$ achieves three goals: (1) it accommodates the possibility of low call rates, (2) it allows the detection of low-rate attacks, and (3) it reduces the possibility of false alarms due to adverse network conditions.

References

1. H. Jamjoom and K. G. Shin, "Persistent dropping: an efficient control of traffic aggregates," *Proceedings of the ACM SIGCOMM '03 Conference*, 2003.
2. K. Park and H. Lee, "On the effectiveness of route-based packet filtering for distributed DOS attack prevention in power-law internets," *Proceedings of the ACM SIGCOMM '01 Conference*, 2001.
3. IPTEL, "SIP express router," SIP Proxy Server, http://www.iptel.org/ser/.
4. M. Carson and D. Santay, "NIST Net Network Emulation Package," Nist Net Web site: http://snad.ncsl.nist.gov/itg/nistnet/, June 1998.

5. D. G. Andersen, A. C. Snoeren, and H. Balakrishnan, "Best-path vs. multi-path overlay routing," *IMC '03: Proceedings of the 3rd ACM SIGCOMM Conference on Internet Measurement*, 2003.

6. D. Sisalem, J. Kuthan, and S. Ehlert, "Denial of service attacks targeting a SIP VoIP infrastructure: Attack scenarios and prevention mechanisms," *IEEE Network*, vol. 20, September/October 2006.

7. D. Pollard, *Asymptopia*, http://www.stat.yale.edu/pollard/ (book in progress), 1st ed., 2000.

8. M. Fannes and P. Spincemaille, "The mutual affinity of random measures," *eprint arXiv: math-ph/0112034*, December 2001.

23

Don't Let the VoIP Service to Become a Nuisance for Its Subscribers

Hemant Sengar

CONTENTS

Voice over Internet Protocol (VoIP) is emerging as an alternative to regular public telephones. IP telephone service providers are moving fast from low-scale toll bypass deployments to large-scale competitive carrier deployments, thereby giving an opportunity to enterprise networks a choice of supporting less expensive single network solution rather than multiple separate networks. Broadband-based residential customers also switch to IP telephony due to its convenience and cost-effectiveness. Contrary to traditional telephone system (where the end devices are dumb), the VoIP architecture pushes intelligence towards the end devices (i.e., PCs, IP phones etc.) giving an opportunity to create many new services which cannot be envisaged using traditional telephone system. This flexibility, coupled with the growing number of subscribers, VoIP becomes an attractive and potential target to be abused by malicious users, hackers, and criminals alike to harass the legitimate users and to capitalize on its weaknesses.

23.1 Introduction

Several denial-of-service (DoS) and fraud attacks against session initiation protocol SIP-based VoIP systems have been described in [1–3]. A great deal of the discussion of possible attacks revolves around the assumption of lack of proper authentication and second, the independence (i.e., no cross-protocol interaction) of various protocols used in VoIP systems for call control and data delivery. Realizing the potential threats, the VoIP service providers in the market implement security solutions to secure the vulnerable SIP proxy server. Only authenticated subscribers are authorized to access the SIP registrar or SIP proxy server. To a large extent, these server-oriented solutions are successful in their goals of protecting the SIP servers from various attacks. But what if, instead of attacking the registrar or SIP proxy server, an attacker changes his attack target and launches attacks against the subscribers (i.e., call recipients). The common perception that since the call recipients are behind a secured SIP proxy server and hence, they are also secured, is not entirely true. In this chapter, we show many examples of real-world attack scenarios, in which an attacker can launch the following attacks against telephone subscribers and network elements: (1) a *reconnaissance attack* to check the liveliness and current IP address of the victim; (2) a *nuisance attack*, in which an attacker rings victim's phone at odd times; (3) a *flooding attack* to cause DoS against a particular SIP client (i.e., subscriber); (4) a *phishing attack* to steal the subscriber's personal information; (5) CANCEL/BYE DoS attacks; (6) Media stream redirection attack; (7) SIP *fuzzing* DoS attacks; (8) a *voice spam attack*, in which an attacker sends unsolicited messages to a number of subscribers at the same time; and finally, (9) a coordinated *distributed denial-of-service* (DDoS) attack against a particular data or voice network element [4]. The last attack needs a special reference because it is based on the exploitation of INVITE messages and a number of unwitting telephone subscribers.

23.2 The Threat Model

In today's IP telephony world, most of the IP telephone service providers (such as Vonage, AT&T Callvantage, and ViaTalk, etc.) operate in partially closed environment and are

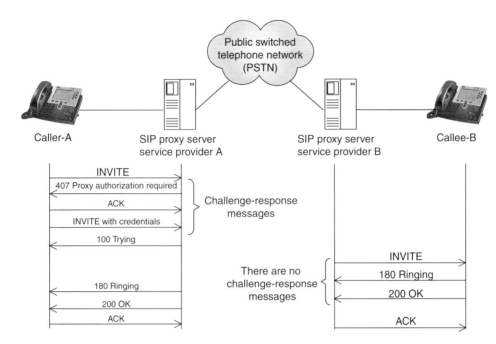

FIGURE 23.1 SIP proxy server authenticates a caller.

connected to each other through the Public Switched Telephone Network (PSTN) as shown in Figure 23.1. For example, let us assume that User A and B belong to two different VoIP service providers A and B, respectively. Though the service providers use IP networks to connect with their users, still the calls between User A and B are expected to traverse the PSTN somewhere in the middle. In an island-based VoIP deployment, the IP traffic is translated into the SS7 traffic [5] (for the transportation over the PSTN) and then back into the IP traffic. It is expected that as VoIP adoption grows the VoIP service providers will interconnect through the peering points.

23.2.1 Lack of Authentication between Callee and Inbound SIP Proxy Server

As shown in Figure 23.1, the subscriber A's call request is received by the outbound SIP proxy server A. The proxy server is an application-layer router that forwards SIP requests toward its destination user agents (UAs). Besides facilitating the communications between UAs, SIP proxies are also useful for enforcing policy. Being an important network element in the creation and management of multimedia communication sessions between clients, the SIP proxy servers are always under threats from malicious users and hackers. Consequently, most of the security efforts are focused on securing it from various attacks. The IP telephone service providers ensure that only authenticated users are authorized to use the SIP proxy server resources. Each of the received INVITE requests are challenged by sending 401 Unauthorized or 407 Proxy Authentication Required messages back to the callers. However, as we observe in Figure 23.1, there are no such authentication mechanism (e.g., challenge-response messages) between the call recipient (i.e., callee-B) and his/her inbound SIP proxy server B. Therefore, malicious users capable of capturing SIP signaling messages between the SIP proxy server and its subscribers can launch various attacks on the call recipients.

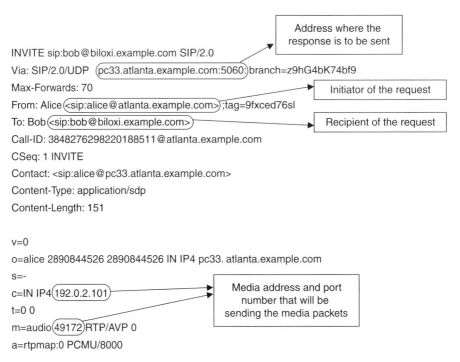

FIGURE 23.2   Structure of an INVITE message.

23.2.2 Exploitation of Call Setup Request (`INVITE`) Message

Before discussing the exploitation of an `INVITE` message, we describe its message structure and the purpose of various header fields. As shown in Figure 23.2, the `Via` header field contains the address where the caller is expecting to receive the response messages of this request, the `From` and `To` header fields contain SIP URIs of the caller and callee, respectively. The `Call-ID` is a globally unique identifier for this call and the `Contact` field contains direct route information to reach the caller. The `INVITE`'s header fields and message body are separated by a blank line. The session description such as media type, codec, and sampling rate etc. are contained in `INVITE`'s message body. The connection information field (i.e., `c=`) contains media connection information such as media's source IP address that will send the media packets. Similarly, the media information field (i.e., `m=`) contains media type and the port number.

The exploitation of `INVITE` message is based on abusing *connection information field* contained in the `INVITE` message body. The SIP proxy server remains at `INVITE` header level and routes this message towards the callee without inspecting message body. The callee's SIP UA parses and interprets the `INVITE` message and records media IP address and port number mentioned in the `c=` and `M=` fields, respectively.* After a SIP session is established, the callee sends audio packets towards this media IP address and port number. By spoofing the connection information field, an attacker can redirect the media stream towards the spoofed (victim) IP address and port number [4].

---

* In some cases where connection information field contains non-routable (private) IP address, the SIP UA relies on received parameter of the first `Via` header field.

23.3 Real-World Attack Scenario

To demonstrate the possible nuisance attacks on subscribers and DDoS attack on network elements, we simulated real-world attack scenarios using IP phones from three different leading VoIP service providers, namely Vonage, AT&T Callvantage, and ViaTalk as shown in Figure 23.3. In the Internet, the SIP signaling messages exchanged between callers and callees are captured at two locations: (1) *location A* is in between callers and their outbound proxies; similarly, (2) *location B* is in between callees and their inbound proxies. At location A, we observe that in order to prevent *replay* attacks, the service providers challenge `INVITE` messages by sending `401 Unauthorized` (in the case of AT&T) or `407 Proxy Authentication Required` messages that include MD5 hash of user's credential and a "nonce" value. This can only be defeated if we have the capability of modifying some header fields (that is not used in MD5 hash computation) and reconstructing the message at real time or by exploiting the implementation of some SIP proxy servers that may accept stale nonce values [6]. However, at location B, there is no such challenge/response messages leaving the subscribers exposed and vulnerable to abuse. In our example of attack scenarios, we exploit the vulnerable and mostly overlooked location B to launch many different types of attacks and some of the attacks are described in the following sections.

23.4 Example Attack Scenarios of Spoofed INVITE Message Body

23.4.1 Case I: Interactive Voice Response (IVR) System

An interactive voice response (IVR) is a phone technology that allows a computer to detect voice and touch tones using a normal phone call. The IVR system can respond with pre-recorded or dynamically generated audio to further direct callers on how to proceed [7]. The IP and traditional (i.e., PSTN) telephone networks are full of IVR systems where

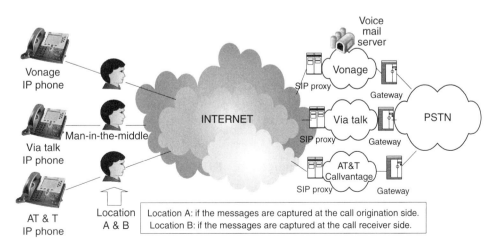

FIGURE 23.3 Real-world attack scenario.

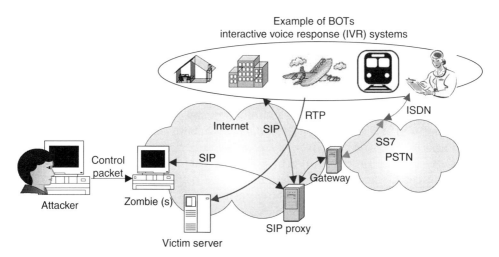

FIGURE 23.4 IVR systems acting as bots.

an user calling a telehone number is briefly interfaced with an automatic call response system. The typical usage of IVR includes call centers, bank and credit card's account information system, air and rail's reservation system, hospital's helpline, University's course registration system, etc.

As shown in Figure 23.4, imagine an attacker knowing this vulnerability sends out few hundred `INVITE` messages (while keeping the same media connection address in the message body) to well-known automatic call response systems and establishes fake call sessions with them. In response, the IVR systems flood the victimized connection address with UDP-based RTP packets. In order to establish a call, an exchange of few call setup SIP messages (i.e., `INVITE/200 OK/ACK`) can result in few hundreds to thousands of RTP packets. Such an attack scenario uses both reflection and amplification effect to make a DDoS attack more potent.

23.4.2 Case II: User's Voicemail System

Sometimes when a callee is busy or he is not available to answer a phone call, the caller is directed to an answering machine or a voicemail system that plays a greeting message and stores incoming voice messages. An attacker may send fake call requests with the same media connection address to hundreds and thousands of individual telephone subscribers distributed over the Internet. The simultaneous playing of individual greeting messages can overwhelm the link's bandwidth connecting to the victim.

23.4.3 Case III: User's Voice Communication—RTP Stream

Even if we assume that the callee is not busy and answers a phone call, the callee's voice stream (e.g., Hello, hello … or some other initial greeting message) can be directed to a target machine. As in the previous examples, an attacker sends fake call requests with the same media connection address to hundreds and thousands of individual telephone subscribers and simultaneous response of subscribers can cause a flooding attack on the victim.

23.4.4 Case IV: SPIT Prevention—The Turing Test

In many aspects a voice spam is similar to an e-mail spam. The technical know-how and execution style of e-mail spam can easily be adapted to launch voice spam attacks too. For example, first, a voice spammer harvests user's SIP URIs or telephone numbers from the telephone directories or by using spam bots crawling over the Internet. In the second step, a compromised host is used as a SIP client that sends out call setup request messages. Finally, in the third step, the established sessions are played with pre-recorded WAV file. However, voice spam is much more obnoxious and harmful than e-mail spam. The ringing of telephone at odd time, answering a spam call, phishing attacks and inability to filter spam messages from voicemail box without listening to each one, are real nuisance and waste of time.

The *Internet Engineering Task Force* (IETF)'s request for comments (RFC) [8] analyzed the problem of voice spam in SIP environment examining various possible solutions that have been discussed for solving the e-mail spam problem and considered their applicability to SIP. One such solution is based on the Turing test that can distinguish computers from human. In the context of IP telephony, the machine-generated automated calls can be blocked by applying an audio Turing test. For example, a call setup request from an unidentified caller is sent to an IVR system where a caller may be asked to answer few questions or to enter some numbers through the keypad. The successful callers are allowed to go through the SIP proxy server and may also be added to a white list.

The VoIP security products such as NEC's VoIP SEAL [9] and Sipera Inc.'s IPCS [10] have implemented audio Turing test as an important component in their anti-spam product to separate machine-generated automated calls from real individuals. However, an attacker may use these devices as reflectors and amplifiers to launch stealthy and more potent DDoS attacks. For example, to determine the legitimacy of a single spoofed `INVITE` message, these devices send few hundreds of RTP-based audio packets (assuming ≈10–20 sec. of audio test) toward the media connection address of an `INVITE` message. A victimized connection address can be flooded with audio packets if an attacker sends one or two spoofed `INVITE` messages (with same media connection address) to several of such devices distributed over the Internet.

23.5 Example Attack Scenarios of an Unauthenticated Last Hop

23.5.1 Case I: Reconnaissance Attack

A subscriber's liveliness (whether the device (i.e., phone or PC) is still connected to the Internet or not) can be checked by sending the captured `INVITE` message and observing the response messages. The `INVITE` message acts more than a *ping* packet as it checks the availability of the SIP machine at a particular host (corresponding to the address obtained from the captured `INVITE` message). However, this approach generates a false ring at the subscriber's telephone. Therefore, as a substitute, the `INVITE` message header fields can also be used to create a `SIP OPTION` message pinging the host at application layer.

23.5.2 Case II: Nuisance Attack

A telephone starts ringing after receiving an `INVITE` message and keeps on ringing until either a callee picks it up or a (`486 Busy Here`) response message is send back to the caller.

To harass a particular subscriber, the captured `INVITE` message (belonging to that particular subscriber) can be replayed again and again ringing his/her telephone at odd times.

23.5.3　Case III: Flooding Attack

The flooding attacks are the most severe threats to VoIP systems, perhaps due to its simplicity and the abundance of tool support. A captured `INVITE` message can be played in quick session to flood a victim with `INVITE` requests. This simple attack is potent enough to bring down the telephone adapter and cause a DoS attack on a particular subscriber.

23.5.4　Case IV: Phishing Attack

In the previous attacks, the integrity of `INVITE` messages remained intact. However, by changing the `INVITE` parameters, it is possible to launch a phishing attack too. For example, in phishing attack, an attacker modifies the `INVITE` header fields to represent himself as a financial institution and tries to steal the subscriber's personal information. An unwitting subscriber believing on spoofed caller-ID, easily provides account number and PIN.

23.5.5　Case V: `CANCEL/BYE` DoS Attacks

The `CANCEL` method is used to terminate pending searches or call attempts. Specifically, it asks the UAS to cease the request processing and generate an error response to the request. In general, a `CANCEL` is for an outstanding `INVITE` request that can be generated for an UA or the proxy servers that have received 1xx response, but still waiting for the final 2xx response. When a caller has received the response message `200 OK`, the session is considered to be established. An established media session is terminated upon receiving an end-to-end `BYE` message. A malicious user can send a fabricated `CANCEL/BYE` message to cause a DoS attack on a specific telephone subscriber.

23.5.6　Case VI: Media Stream Redirection Attack

An attacker may send a fabricated `re-INVITE` message (with spoofed media IP address) to a callee redirecting the RTP flow to another location. As a result, the two-way communication between the caller and a callee is broken and possibly the attacker may also barge in and establish a communication channel with the callee.

23.5.7　Case VII: SIP Fuzzing DoS Attack

In fuzzing attack, the SIP protocol machine at the callee side is fed with crafted SIP messages carrying unexpected and faulty parameter values. The vendor's implementation of the SIP protocol machine is targeted for security vulnerabilities such as buffer overflow, unhandled exceptions etc. causing malfunctioning or crashing of the protocol machine.

23.5.8　Case VIII: Spam Attack

Contrary to popular belief that the launching of spam attack requires specialized tool, we show that by replaying SIP signaling messages and prerecorded voice messages, spam voice messages can be send to telephone subscribers. For example, as shown in Figure 23.5, the replay of `INVITE` and `ACK` messages (maintaining the same relative order with respect

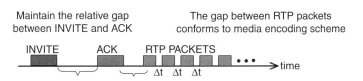

FIGURE 23.5 Spam attack on a subscriber.

to time) establishes a spam call session. Once a call session is established, a RTP stream is pushed toward the subscriber.

Further, it should be noted that the above attack list is not exhaustive. An attacker may monitor the signaling traffic flow to gain information such as the nature of traffic and load besides behavior and identity of subscribers.

23.6 Solution: Don't Ignore the Call Recipients

The use of IPSec or TLS at the first and last hop of the call resolves the subscriber authentication problem and hence, the MITM attacks also. However, these solutions are not widely deployed; instead, to avoid the overhead and complexity of these solutions, the VoIP service providers prefer UDP for the transportation of SIP messages and the digest authentication mechanism for SIP security. Furthermore, the digest authentication mechanism is used extensively only at the first hop of the call (i.e., at the caller side) to authenticate the received INVITE messages at the SIP proxy server. However, the attacks mentioned in this chapter can still be prevented, if the call recipient UAs also have the same capability of authenticating received INVITE messages. For example, before an INVITE message rings the phone, the SIP UA can authenticate the INVITE message by sending a challenge message back to its own SIP proxy server. The SIP proxy server acknowledges the receipt of the challenge and constructs an INVITE message with proper credentials. Therefore, only the authenticated INVITE message with valid credentials are allowed to initialize the call setup phase at the recipient SIP UA. The exchange of challenge/response messages at the last hop of the call (i.e., at the call recipient side) prevents many of the MITM attacks on subscribers.

References

1. O. Arkin, "E.T. can't phone home—VoIP security," Presentation, Black Hat Briefings, Las Vegas, NV, USA, 2002.
2. O. Arkin, "Why E.T. can't phone home?—Security risk factors with IP telephony," Presentation, AusCERT Australia, 2004.
3. CISCO, "Security in SIP-based networks," White Paper, 2002, http://www.cisco.com/warp/public/cc/techno/tyvdve/sip/prodlit.
4. H. Sengar, "Beware of a new and readymade army of legal BOTs," *USENIX; login: Magazine*, vol. 32, no. 5, October 2007, pp. 35–42.

5. H. Sengar, R. Dantu, D. Wijesekera, and S. Jajodia, "SS7 over IP: Signaling interworking vulnerabilities," *IEEE Network Magazine,* vol. 20, no. 6, November, 2006, pp. 32–41.
6. SIPERA Systems, "Some implementations of SIP proxy may honor replayed authentication credentials," May 2007, http://www.sipera.com/index.php?action=resources,threat_advisory& tid=183&.
7. Wikipedia Encyclopedia, "Interactive voice response," April 2007, http://en.wikipedia.org/ wiki/Interactive_voice_response.
8. J. Rosenberg and C. Jennings, "RFC 5039: The session initiation protocol (SIP) and spam," IETF Network Working Group, 2008.
9. NEC Corporation, "NEC develops world-leading technology to prevent IP phone SPAM," Product News, 2007, http://www.nec.co.jp/press/en/0701/2602.html.
10. SIPERA Systems, "Products to address VoIP vulnerabilities," April 2007, http://www.sipera. com/index.php?action=products,default.

Index